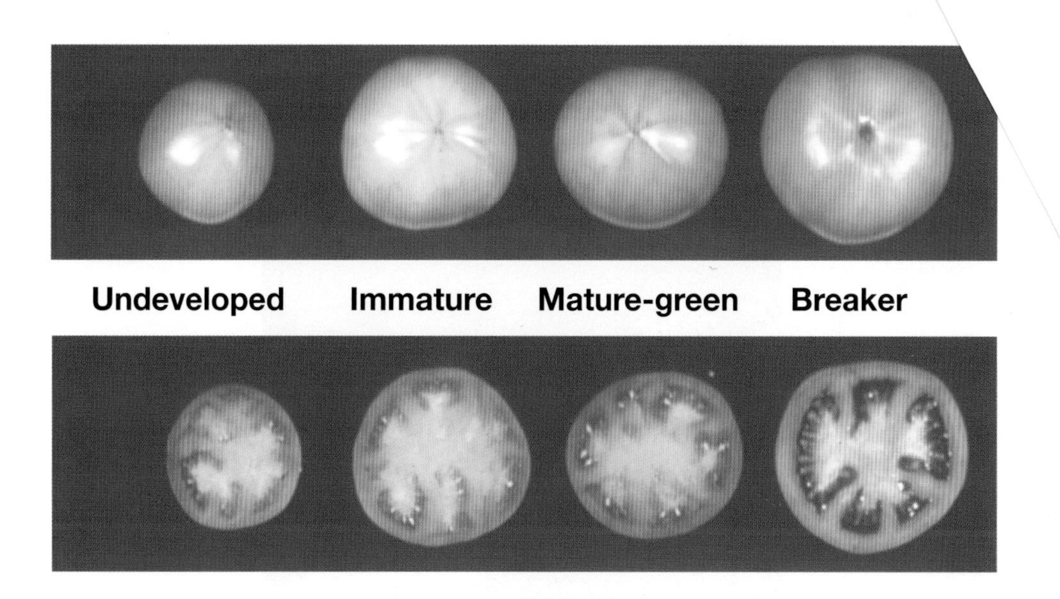

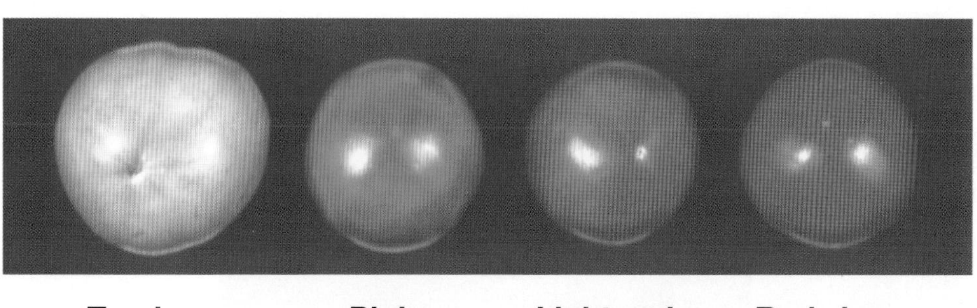

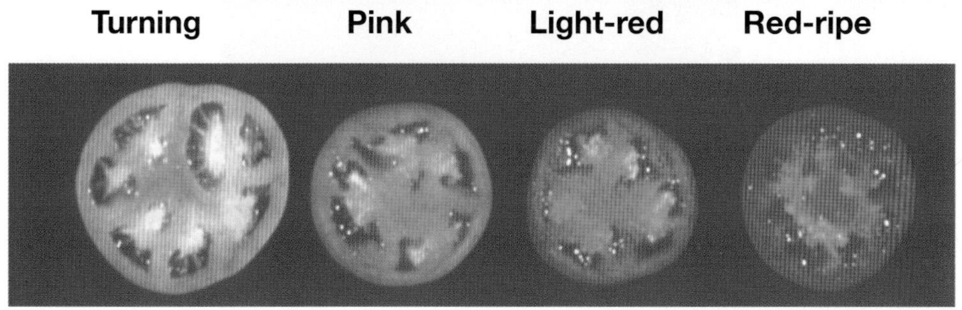

Plate 1. External and internal colour and tissue changes in ripening tomato fruit. (See Chapter 5.)

Plate 2. Larva of *Spodoptera eridania* (a) and feeding damage (b) on tomato fruit. (See Chapter 7.)

Tomatoes

Crop Production Science in Horticulture Series

Series Editors: Jeff Atherton, Professor of Tropical Horticulture, University of the West Indies, Barbados, and Alun Rees, Horticultural Consultant and Editor, *Journal of Horticultural Science.*

This series examines economically important horticultural crops selected from the major production systems in temperate, subtropical and tropical climatic areas. Systems represented range from open field and plantation sites to protected plastic and glass houses, growing rooms and laboratories. Emphasis is placed on the scientific principles underlying crop production practices rather than on providing empirical recipes for uncritical acceptance. Scientific understanding provides the key to both reasoned choice of practice and the solution of future problems.

Students and staff at universities and colleges throughout the world involved in courses in horticulture, as well as in agriculture, plant science, food science and applied biology at degree, diploma or certificate level will welcome this series as a succinct and readable source of information. The books will also be invaluable to progressive growers, advisers and end-product users requiring an authoritative, but brief, scientific introduction to particular crops or systems. Keen gardeners wishing to understand the scientific basis of recommended practices will also find the series very useful.

The authors are all internationally renowned experts with extensive experience of their subjects. Each volume follows a common format covering all aspects of production, from background physiology and breeding, to propagation and planting, through husbandry and crop protection, to harvesting, handling and storage. Selective references are included to direct the reader to further information on specific topics.

Titles available:

1. **Ornamental Bulbs, Corms and Tubers** A.R. Rees
2. **Citrus** F.S. Davies and L.G. Albrigo
3. **Onions and Other Vegetable Alliums** J.L. Brewster
4. **Ornamental Bedding Plants** A.M. Armitage
5. **Bananas and Plantains** J.C. Robinson
6. **Cucurbits** R.W. Robinson and D.S. Decker-Walters
7. **Tropical Fruits** H.Y. Nakasone and R.E. Paull
8. **Coffee, Cocoa and Tea** K.C. Willson
9. **Lettuce, Endive and Chicory** E.J. Ryder
10. **Carrots and Related Vegetable *Umbelliferae*** V.E. Rubatzky, C.F. Quiros and P.W. Simon
11. **Strawberries** J.F. Hancock
12. **Peppers: Vegetable and Spice Capsicums** P.W. Bosland and E.J. Votava
13. **Tomatoes** E. Heuvelink

Tomatoes

Edited by
Ep Heuvelink
Wageningen University, The Netherlands

CABI Publishing

CABI Publishing is a division of CAB International

CABI Publishing
CAB International
Wallingford
Oxfordshire OX10 8DE
UK

Tel: +44 (0)1491 832111
Fax: +44 (0)1491 833508
E-mail: cabi@cabi.org
Website: www.cabi-publishing.org

CABI Publishing
875 Massachusetts Avenue
7th Floor
Cambridge, MA 02139
USA

Tel: +1 617 395 4056
Fax: +1 617 354 6875
E-mail: cabi-nao@cabi.org

A catalogue record for this book is available from the British Library, London, UK.

Library of Congress Cataloging-in-Publication Data

Tomatoes / edited by Ep Heuvelink.
p. cm. -- (Crop production science in horticulture ; 13)
ISBN 0-85199-396-6 (alk. paper)
I. Heuvelink, Ep. II. Title. III. Series.
SB349.T678 2005
635′.642--dc22

2004022555

ISBN 0 85199 396 6

Typeset by Columns Design Ltd, Reading, UK.
Printed and bound in the UK by Cromwell Press, Trowbridge.

Contents

† Deceased before publication.

Contributors

J.M. Costa, Instituto Superior da Agronomia, Dep. Botânica e Engenharia Biológica, Universidade Técnica de Lisboa, Tapada da Ajuda, 1349-017 Lisbon, Portugal

A.A. Csizinszky, University of Florida, IFAS, Gulf Coast Research and Education Center, 5007-60th Street East, Bradenton, FL 34203, USA

M.A. Dorais, Horticulture Research Centre, Envirotron Building, Room 2120, Laval University, Quebec G1K 7P4, Canada

J.B. Jones, Department of Plant Pathology, University of Florida, 2553 Fifield Hall, PO Box 110680, Gainesville, FL 32611-0680, USA

E. Heuvelink, Department of Plant Sciences, Horticultural Production Chains Group, Wageningen University, Marijkeweg 22, 6709 PG Wageningen, The Netherlands

P. Lindhout, Laboratory of Plant Breeding, Wageningen Agricultural University, PO Box 386, 6700 AJ Wageningen, The Netherlands

M.M. Peet, Department of Horticultural Science, North Carolina State University, Box 7609, Raleigh, NC 27695-7609, USA

M.E. Saltveit, Department of Vegetable Crops, Mann Laboratory, University of California, Davis, CA 95616-8631, USA

D.J. Schuster, University of Florida, IFAS, Gulf Coast Research and Education Center, 5007-60th Street East, Bradenton, FL 34203, USA

J.C. van Lenteren, Laboratory of Entomology, Wageningen University, PO Box 8031, 6700 EH Wageningen, The Netherlands

G.W.H. Welles, Applied Plant Research, PO Box 8, 2670AA Naaldwÿk, The Netherlands

Preface

Tomatoes are one of the most widely produced and consumed 'vegetables' in the world, both for the fresh fruit market and the processed food industries. Furthermore, tomato fruits or plants are occasionally used for decoration (ornamental value), but perhaps the most exceptional use we find in Spain. Every year, the tiny village of Buñol in Valencia hosts the largest tomato war in the world: 'La Tomatina'. In this festival, at the peak of the tomato season, for 2 hours, participants pelt each other with ripe, red fruit and the streets turn into rivers of tomato juice.

There is a vast amount of literature on tomato resulting from its economic importance, but also because it is seen as a model crop in plant genetic, physiological and pathological studies. In 1986, Jeff Atherton and Jehoshua Rudich did an excellent job as editors bringing, for the first time, fundamental and practical knowledge on field-grown and greenhouse-grown tomatoes together in a 661-page 'tomato bible' (Atherton and Rudich, 1986). Now, almost 20 years later, our knowledge on tomato has greatly extended, and completely new areas have developed, like genetic modification and biological pest control. Hence, the present book attempts to give an update – however, more compact – so as to fit in the CABI Publishing Series *Crop Production Science in Horticulture*. Most of the books in this series are monographs or have two or three authors. This volume on tomato has one editor, and ten additional authors contributing to one or more chapters. Maybe this reflects the vast knowledge present in all fields for this crop, such that no single author could cover the subject with authority. Contributing authors are experts in their particular subject areas, and come from different regions of the world, hence giving the book a more international nature than would have been possible with just one or two authors.

I thank the authors of the different chapters sincerely for their time spent on this book, as well as for their patience. I am grateful to CABI Publishing for giving me the opportunity to be the editor of the tomato volume in their series. Tim Hardwick should be mentioned particularly: without his patience and stimulating words and e-mails, this book would never have been finished.

I hope readers will find in this book the information they were looking for. I hope they enjoy reading this new book on tomato and I look forward to comments and suggestions for improvement.

I wish to dedicate this book to the late Prof. Hugo Challa, who has been my teacher in science.

Ep Heuvelink
Wageningen, September 2004

REFERENCE:

Atherton, J.G. and Rudich, J. (eds) (1986) *The Tomato Crop. A Scientific Basis for Improvement*. Chapman & Hall, London, 661 pp.

1

INTRODUCTION: THE TOMATO CROP AND INDUSTRY

J.M. Costa and E. Heuvelink

THE CROP

Classification and taxonomy

The tomato belongs to the family *Solanaceae* (also known as the nightshade family), genus *Lycopersicon*, subfamily *Solanoideae* and tribe *Solaneae* (Taylor, 1986). The *Solanaceae* family includes other important vegetable crops such as chilli and bell peppers (*Capsicum* spp.), potato (*Solanum tuberosum*), aubergine (*Solanum melongena*), tomatillo (*Physalis ixocarpa*) and tobacco (*Nicotiana tabacum*). The taxonomic classification of the tomato is still debated. In 1753 the Swedish botanist Linnaeus named it *Solanum lycopersicon*, but 15 years later Philip Miller replaced the Linnaean name with *Lycopersicon esculentum* (Taylor, 1986). Although taxonomists have recently reintroduced its original name, *Solanum lycopersicon* (Heiser and Anderson, 1999), the commonly accepted and still valid name of *Lycopersicon esculentum* will be used in this book.

Genetic variation within the *Lycopersicon* genus

The *Lycopersicon* genus includes a relatively small collection of species: the cultivated tomato *L. esculentum* Mill. and several closely related wild *Lycopersicon* species, namely *L. esculentum* var. *cerasiforme*, *L. pimpinellifolium* (Jusl.), *L. cheesmannii*, *L. parviflorum*, *L. chmielewski*, *L. hirsutum* Humb., *L. chilense* Dun. and *L. peruvianum* (L.) Mill. (Taylor, 1986).

The genus was initially divided into two major subgenera regarding the colour of the fruits: *Eulycopersicon* (for colour-fruited species) and *Eriopersicon* (for green-fruited species) (Müller, 1940). A more comprehensive and objective division was proposed by Rick (1976), which included two major complexes: (i) the esculentum complex, for the species that cross easily with the commercial tomato; and (ii) the peruvianum complex, for the species that could not be easily crossed with the commercial tomato.

The cultivated tomato reached its present status after a long period of domestication. The improvement of the crop started with the selection of preferred genotypes. Because tomato is a predominantly inbreeding species the genetic variation tends to decrease, even without selection. Thus cultivated tomato species have an extremely limited genetic variation. This lack of variation at the DNA level contrasts with the large morphological variation observed among tomato accessions in genebanks. Specific traits from the wild *Lycopersicon* species have been extensively studied and used to improve tomato, e.g. fruit size, disease resistance, taste and colour (Stevens and Rick, 1986).

Origin, history and uses

All related wild species of tomato are native to the Andean region that includes parts of Chile, Colombia, Ecuador, Bolivia and Peru (Sims, 1980). The most likely ancestor is the wild *L. esculentum* var. *cerasiforme* (cherry tomato), which is indigenous throughout tropical and subtropical America (Siemonsma and Piluek, 1993). Although the ancestral forms of tomato grew in the Peru–Ecuador area, the first extensive domestication seems to have occurred in Mexico (Sims, 1980; Harvey *et al.*, 2002).

The Spanish introduced the tomato into Europe in the early 16th century (Harvey *et al.*, 2002). European acceptance of the tomato as a cultivated crop and its inclusion in the cuisine were relatively slow. Tomatoes were initially grown only as ornamental plants: the fruits were considered to be poisonous, because of the closely related deadly nightshade (*Solanum dulcamara*). Since the mid-16th century tomatoes have been cultivated and consumed in southern Europe, though they only became widespread in north-western Europe by the end of the 18th century (Harvey *et al.*, 2002). In the 17th century, Europeans took the tomato to China, South and South-east Asia, and in the 18th century to Japan and the USA (Siemonsma and Piluek, 1993). The production and consumption of tomatoes expanded rapidly in the USA in the 19th century, and by the end of that century processed products such as soups, sauces and ketchup were regularly consumed (Harvey *et al.*, 2002).

Tomatoes are one of the most widely eaten vegetables in the world. Their popularity stems from the fact that they can be eaten fresh or in a multiple of processed forms. Three major processed products are: (i) tomato preserves (e.g. whole peeled tomatoes, tomato juice, tomato pulp, tomato purée, tomato paste, pickled tomatoes); (ii) dried tomatoes (tomato powder, tomato flakes, dried tomato fruits); and (iii) tomato-based foods (e.g. tomato soup, tomato sauces, chilli sauce, ketchup).

Tomatoes are commonly used as a 'model crop' for diverse physiological, cellular, biochemical, molecular and genetic studies because they are easily grown, have a short life cycle and are easy to manipulate (e.g. by grafting, cuttings) (Kinet and Peet, 1997). They are used as model plant species to study the physiology and biochemistry of seed development, germination and dormancy (Suhartanto, 2002). Therefore, the tomato is an excellent tool to improve knowledge on horticultural crops (Taylor, 1986; Kinet and Peet, 1997).

General characteristics of fresh and processed tomato

The cultivated tomato is a perennial diploid dicotyledon ($2n = 24$). It is grown as an annual in temperate climates, as plants and fruit suffer physiological injury when exposed to non-freezing temperatures below 12°C. Varieties of processing and fresh-market tomatoes have different growth habits. The major traits of processing tomatoes are determinate growth, dwarf habit, concentrated and uniform fruit set and ripening, tough skins, and a high soluble solids content (George, 1999). Processing tomatoes are grown in open-field systems and are usually direct drilled, but transplants are commonly used in more advanced production systems. They do not require trellising or staking and are harvested mechanically.

Fresh-market varieties grown in greenhouses are generally indeterminate and require trellising, but varieties grown in the field are determinate. In both, plants are started as transplants. Staking is beneficial, to keep fruit away from the soil, and pruning (removal of several or all side shoots) is used to increase fruit size. Harvesting is done by hand. There is a large number of varieties and cultivars for fresh production. They range from small, sweet-tasting cherry tomatoes to large beefsteak tomatoes with different colours, shelf-life and flavours (Dorais *et al.*, 2001). Five major types can be distinguished (http://www.britishtomatoes.co.uk/facts/varieties.shtml):

- **Classic round tomatoes** – The most popular varieties, with a round shape, containing two to three locules and with an average fruit weight of 70–100 g and diameter 4.7–6.7 cm. Used in salads, or for grilling, baking or frying and as ingredient for soups or sauces.
- **Cherry and cocktail tomatoes** – Smaller than the classic tomatoes, with weight varying between 10 and 20 g and diameter 1.6–2.5 cm. Cherry tomatoes are smaller than cocktail tomatoes, but both are very sweet. Cherry tomatoes are generally red, but golden, orange and yellow varieties exist. Eaten whole and raw or cooked. Cocktail tomatoes can be halved for salads, or skewered whole for grilling. Currently almost all cocktail varieties are sold attached to the stem ('on the vine'; not in USA).
- **Plum and baby plum tomatoes** – Typical oval shape but the baby plum tomatoes are smaller. The flesh is firm and less juicy in the centre. Used for barbecues, and processed for pizzas and pasta dishes.
- **Beefsteak tomatoes** – Larger than the traditional round tomato, weighing 180–250 g and containing five or more locules. Used for

stuffing and baking whole, and sliced for salads and sandwiches. Wide range available, varying in shape, colour (red, pink), texture and flavour.

- **Vine or truss tomatoes** – These may be any of the types mentioned above, but are marketed still attached to the fruiting stem, as it is the stem that gives the distinct tomato aroma, rather than the fruits themselves.

WORLDWIDE OVERVIEW OF TODAY'S TOMATO INDUSTRY

The global production of tomatoes (fresh and processed) has increased by about 300% in the last four decades. The annual worldwide production of tomatoes in 2003 has been estimated at 110 million t with a total production area of about 4.2 million ha (Table 1.1). It is possible that these figures underestimate the real production area and tonnage, considering that tomatoes are also grown on very small plots and gardens throughout the tropics and subtropics with a large total production that is consumed locally and is an important food source for the population. The global trade of tomatoes and tomato products reached US$4.2 billion, which represents a 33% increase relative to the beginning of the 1990s (FAS/USDA, 2003). The three leading producers are China, the USA and Turkey. China accounts for about 25% of the world's total production and cultivated area (Table 1.1) and it is also the largest consumer (Wijnands, 2003). The USA, Italy, Spain and Turkey dominate the world processing tomato industry. The Netherlands is the world leader in intensification of fresh tomato production in glasshouses, with annual yields above 55 kg/m^2.

Table 1.1. The top ten tomato-producing countries, with estimated percentages dedicated to processing in 2003 (source: FAO).

Country	Production (million t)	Percentage for processing	Area harvested (ha)
China	25.9	10	1,105,153
USA	12.3	95[a]	177,000
Turkey	9.0	25	225,000
India	7.4	7–10[b]	520,000
Italy	6.9	70[a]	130,932
Egypt	6.4	–	181,000
Spain	3.8	40[a]	64,100
Brazil	3.6	30[c]	59,766
Iran	3.0	–	112,000
Russian Federation	2.2	–	160,000
World	110.5	–	4,161,295

[a] United States Department of Agriculture (USDA).
[b] Dhaliwal *et al.*, 2003.
[c] Mediterranean International Association of the Processing Tomato (AMITOM).

The tomato industry is one of the most advanced and globalized horticultural industries. Most production is located in temperate zones (http://www.tomatonews.com) that have long summers and mainly winter precipitation. However, cultivation practices, the ratio between production for processing or fresh consumption (Table 1.1) and the organization and structure of the industry and markets differ widely among countries.

Europe and Turkey

European tomato production can be divided into two major production systems. The Northern system is capital intensive (Fig. 1.1), using modern technology (greenhouse structures, climate control, crop protection). It is highly productive and focused on fresh tomato production. The Southern (Mediterranean) system produces fruit in the open field for processing, and in plastic-covered structures for fresh consumption (Harvey *et al.*, 2002).

Fig. 1.1. Training and side-shoot removal in a greenhouse crop grown according to the high-wire system in The Netherlands (photo: PPO, Naaldwijk).

Italy

Italy is the leading tomato producer in Europe, with a production of 6.9 million t, of which 70% is for processing (Table 1.1). The main producing areas for processing tomatoes are located in the south, namely in Puglia (> 50% of the total production) and Campagnia (8% of total production). In the north, Emilia Romagna is the main producing area with 30% of the total Italian production for processing tomatoes (AMITOM, 2003; Santella, 2003). Cultivation in the north is highly mechanized and uses hybrid seeds and transplants, resulting in higher yields (75–100 t/ha) than in the south (about 65 t/ha) (AMITOM, 2003). Planting starts in early May, harvest commences in mid July and the season ends by the middle or end of September (Santella, 2003). Regarding fresh-market tomatoes, the total production in greenhouses was 547,000 t in 2000, which represented one-third of Italian vegetable production in greenhouses (La Malfa, 2003). Also relevant is the fast expansion of cherry tomatoes in Sicily (La Malfa, 2003).

Spain

Spain is the second largest producer of tomatoes in Europe, with 3.8 million t in 2002 (Table 1.1). In contrast to Italy, a large part of Spanish production is for the fresh market (Pazos, 2003). Extremadura, in western Spain, produces 80% of the country's total crop of processing tomatoes (Pazos, 2003), and the Ebro river valley is another important production area. Mechanization and new varieties contribute to yields of about 60 t/ha (Pazos, 2003).

The south-eastern part of Spain, in particular the provinces of Almería and Murcia, have emerged as important and competitive areas for production and export of fresh greenhouse tomatoes. In 2001/02, Almería exported 229,000 t of greenhouse tomatoes, of which 70% went to France, Germany, The Netherlands and the UK (Cajamar, 2002; Eurosur Consultores, 2002). In Almería, the area dedicated to protected tomato cultivation in 2000/01 was slightly larger than 8000 ha (Eurosur Consultores, 2002). In general, growers use simple plastic structures called *parral* (Fig. 1.2) but fast changes are occurring and the area of modern plastic multi-tunnels is increasing. The area of glasshouses remains minor (Costa and Heuvelink, 2000). The area of integrated crop protection has also been increasing in recent years. Cultivation in greenhouses is mostly in *enarenado*: a 30 cm layer of soil put on top of the natural soil, then 2–3 cm of organic compost and a top layer of around 10 cm of sand. The production peak is achieved in December/January. Cooperatives have a strong position in the production and marketing of both fresh and processing tomatoes (Costa and Heuvelink, 2000; Pazos, 2003). Fresh tomatoes for export are also cultivated on about 3100 ha on the Canary Islands (more than half on Gran Canaria island). The production is 13 kg/m^2 and production methods are similar to those used in southern Spain (Rodríguez, 2002).

Fig. 1.2. Tomato production in a plastic greenhouse (*parral*) in Almería (Spain).

Greece and Portugal

Greece and Portugal are also important tomato producers in Europe. In 2002, Greece produced a total of 1.5 million t, of which 860,000 t were processed (Sekliziotis, 2003). In the same year Portugal produced 830,000 t, and the anticipated production for 2003/04 is 1.2 million t (Monte Gomes, 2003). Ribatejo and Baixo Alentejo (southern Portugal) are the main production regions for processing tomato, whereas Algarve is the most important region for fresh tomato production in greenhouses (mostly under plastic). In 2001, the production of fresh tomato was 79,000 t and occupied an area of 1450 ha (Anuário Vegetal, 2002), of which about 40–50% was produced in greenhouses (A. Monteiro, personal communication).

The Netherlands and Belgium

Regarding the Northern tomato regime in Europe, The Netherlands leads in production in greenhouses. In 2002, about 1200 ha were dedicated to tomato production, with a total production of 550,000 t of fresh tomato (CBS, 2003).

This makes The Netherlands the world leader in intensification, with an annual average of 50 kg/m^2, and in many cases production is > 60 kg/m^2. Crops are started in December; harvest starts in March and continues up to November. The use of modern technology (100% glasshouse, computerized climate control, CO_2 enrichment, integrated crop protection) characterizes this Dutch production system. Some growers use assimilation light to make year-round production possible. Growers are well-educated and organized and innovation is promoted by research that is often initiated by the companies themselves. The main export market is Germany, even though Spanish tomatoes have gained a market share from the Dutch (Wijnands, 2003). Growers in The Netherlands focus on high quality segments and selling their produce more by contract than by auction, in order to attain the quality requirements of supermarket chains. The Netherlands also re-exports fresh tomatoes.

Tomato production in Belgium expanded very quickly from the mid-1980s and tomatoes became the most important vegetable produced in Belgium (Taragola and Van Lierde, 2000). Like The Netherlands, Belgian growers focus on the production of fresh-market tomatoes in greenhouses and Belgium has about the same level of technology as its northern neighbour. At present the total greenhouse area is estimated at about 690 ha (Cantliffe and Vansickle, 2003a). Like the Dutch, the Belgian tomato industry is export driven (Harvey *et al.*, 2002), but competition with Spanish tomato growers became very fierce and resulted in a serious crisis in 1995 (Taragola and Van Lierde, 2000).

France

In 2003, total tomato production in France was close to 900,000 t, of which about 70% was for the fresh market and the remaining 30% was for the processing industry (Hénard, 2003). France, with its 300,000 t of processing tomatoes, is a small producer compared with neighbouring Mediterranean countries. However, it has 2057 ha of greenhouses for fresh tomato production which produced about 560,000 t in 2003 (Hénard, 2003). According to the same source, field production for fresh consumption was about 40,000 t. France is a net importer of canned tomatoes, tomato paste and fresh tomatoes (Hénard, 2003), coming mostly from Spain and Morocco.

Germany and the UK

Both of these countries have experienced a decrease in production area in recent decades. Germany produces tomatoes in about 280 ha of greenhouses (mostly glasshouses) and the greenhouse area in the UK is even smaller (only 200 ha). Due to the population size of these two countries (together 140 million inhabitants), both are large-scale importers of European fresh tomatoes, mainly from Spain and The Netherlands. There are differences between the British and German markets: Germany is strongly price oriented, whereas the British market is quality oriented (Harvey *et al.*, 2002; Wijnands, 2003).

Turkey

Turkey is the third-largest tomato producer worldwide (Table 1.1). One-fourth of the 9 million t produced annually are for processing (Sirtioglu, 2003). The main regions for processing tomatoes are in Marmara (the townships of Balikesir and Canakkale), producing 2 million t (Yoltas *et al.*, 2003) and in Aegean (townships of Bergama and Turgutlu) and Central Anatolia (Tokat). Small family farms, using a lot of hand labour (e.g. for planting and harvesting), characterize the production system but there are also large growers producing under contract (Sirtioglu, 2003). Seedlings are normally started in mid-March and are transplanted after mid- to late April when there is no risk of frosts. Harvest is between late July and early October (Sirtioglu, 2003). Paste is the most exported item and Japan and Russia are the main markets. Production of tomatoes for fresh consumption is expanding in southern Turkey. Tomatoes are the main greenhouse vegetable, with a production of about 1.2 million t, and 80% of the greenhouse producing area is in the province of Antalya, close to the Mediterranean Sea (Öztürk *et al.*, 2001). Greenhouse cultivation takes place during winter and early spring (Sirtioglu, 2003).

South Asia

China

Worldwide, China is the largest producer of tomatoes with its 25.8 million t (Table 1.1). In 2002, China became also the third largest producer of processed tomato products, just behind the USA and Italy, as a result of the establishment of a huge processing operation in the province of Xinjiang which processes the crop from about 46,700 ha (Deloitte & Touche, 2003). The production of tomatoes for processing represented 10% of total production in 2002 and was expected to reach 3.0 million t in 2003 (http://www.tomatonews.com).

Xinjiang is the most important province for the production of tomato paste (about 90% of China's total) but tomatoes for processing are also grown in Inner Mongolia and Gansu province. Xinjiang's cultivation calendar is very similar to that of California: in March, fields are prepared for planting, the growth period is April to July, and harvest and processing occur between August and October. Several multinational companies are already present in Xinjiang and processing technology has been imported from the USA and Italy. The high competitiveness of China regarding the processing tomato industry resides in low production costs (mostly low labour costs). This is causing concern among growers from other countries and it has already affected world prices.

Tomatoes for fresh consumption are mainly produced in the provinces of Shandong (Fig. 1.3), Henan and Sichuan and also the peri-urban areas of

Shanghai and Beijing (Ma *et al.*, 2000). Shandong (north-east China) is the most important province for tomato production in greenhouses. Production is based on small farms (600–2000 m^2) and growers use mainly plastic tunnels or the so-called solar greenhouse, which uses the sun as the unique source of energy and possesses three non-transparent walls (the back and two side walls built of clay soil or brick in order to retain heat). The roof structure is of wood or metal and the covering material is polyethylene (PE) or polyvinylchloride (PVC) (Costa *et al.*, 2004). F_1 hybrids are already in use (Ma *et al.*, 2000) but better greenhouse cultivars (e.g. better adapted to low light and temperature conditions, disease-resistant) are needed to improve yields (Costa *et al.*, 2004).

India

India is the fourth largest tomato producer worldwide (Table 1.1). Tomatoes are the second largest vegetable crop in India. The most important production states are Bihar, Karnataka, Uttar Pradesh, Orissa, Andhra Pradesh, Maharashtra, Madhya Pradesh and Assam. The expansion of tomato production in India was to a great extent attributed to the private seed sector,

Fig. 1.3. Tomato production in a solar greenhouse in Shandong (China) using organic substrate (eco-soilless system).

which developed high-yielding cultivars (Nagaraju *et al.*, 2002). Tomatoes are normally grown in rotation with other vegetables. The use of hybrid seeds is increasing and enables yields of 24.5 t/ha, which is double the yield of non-hybrid varieties (Nagaraju *et al.*, 2002). Staking is a common procedure for growing hybrids. The processing industry is not well developed and only a minor proportion (6–10%) is used for processing into products such as purée, paste, ketchup, sauce and pickles. The lack of suitable varieties for processing may be one of the reasons for this situation (Dhaliwal *et al.*, 2003). The processing technology is also outdated.

Japan and South Korea

The Japanese tomato processing industry is small in world terms, with 60,000 t processed in 2002 (AMITOM, 2003; Ito, 2003). However, production of fresh tomatoes is very important. The production area in 2002 was 13,300 ha and the total output was 780,000 t (Ito, 2003). Cherry tomato production reached more than 85,000 t (Ito, 2003). Japan also imports tomato paste from China (Ito, 2003) and South Korea is the largest supplier of fresh tomatoes to Japan (3200 t in 2002).

South Korea focuses on greenhouse production of fresh tomatoes. In 2000, according to official statistics, the total planted area of tomatoes was 4916 ha, of which greenhouse cultivation occupied 4700 ha with a total production of 269,500 t (Ho-Shin, 2001; NRHI, 2004).

North America

USA

The USA, besides being the world's second largest tomato producer (Table 1.1), is the world's leader in the production and export of processed tomatoes. Almost the totality of its production is for processing. The state of California produces 95% of the processing tomatoes and 30% of the fresh tomatoes grown in the USA (AMITOM, 2003). California combines a good climate (long, warm and dry growing seasons) with the use of hybrids, transplants and modern technology (e.g. laser levelling of fields, precision planting, advanced irrigation, fertilization, crop protection systems and modern fruit sorting systems) (Mitchell *et al.*, 2001; Murray *et al.*, 2001). Moreover, research and extension institutions have close ties with the industry, which favours innovations (Murray *et al.*, 2001; AMITOM, 2003), and the breeding industry is very competitive (Murray *et al.*, 2001). In California, processing tomatoes are grown in rotation with other crops such as cotton, garlic, onions, melons and wheat in San Joaquin valley, or with wheat and edible dry beans in the Sacramento valley (Mitchell *et al.*, 2001).

Fresh tomatoes are mainly produced in California and Florida and to a lesser extent in Michigan, Pennsylvania, Virginia and Illinois. Fresh tomatoes were initially grown in open-field systems and close to the markets but imported greenhouse tomatoes from The Netherlands, Israel, Canada and Spain are also available on the market (Cantliffe and Vansickle, 2003b). Mexico is a major supplier of field-grown fresh tomatoes to the USA (83% of total US fresh tomato imports in 1999). Imports compete directly with field-grown tomatoes from Florida and the greenhouse product (Cantliffe and Vansickle, 2003b). The production of greenhouse tomatoes was 159,000 t in 2001, produced on a greenhouse area of 340 ha located mainly in California, Florida, Colorado, Arizona, Texas and Pennsylvania (USDA, 2002; Cantliffe and Vansickle, 2003a).

Canada

Canada is a relevant producer of tomato for processing and for fresh consumption. The glasshouse production of fresh tomatoes is located in Ontario (about 70% of the area), British Columbia and Quebec (USDA, 2002), because these provinces are close to the US market and have good climate (Agrifood-Canada, 2001). In 2001, the greenhouse area for tomato production was 468 ha and total production in 2002 was 217,747 t (Statistics Canada, 2002). The greenhouse production system is modern and comparable to that of the northern European countries, with intensive use of capital and technology (modern greenhouse structures, computer-controlled systems, hydroponics, supplementary lighting, integrated pest management). In 2001, Canada produced about 450,000–500,000 t of tomatoes for processing. The most important areas are the counties of Kent and Essex in Ontario. Transplants are used and the harvest is fully mechanized (AMITOM, 2003). Only 5% of the cultivated area is irrigated, because rainfall in spring and summer is adequate. Growers sell their produce by contract with processors.

Central and South America

Chile and Mexico

Chile and Mexico are important tomato producers in Central America. Tomatoes are the most important vegetable crop in Chile and occupy a total area of 20,000 ha (Krarup, 2002). About 12,000 ha are for processing and the remaining 8000 are for fresh consumption. Fresh tomatoes are produced in open-field systems (6000 ha) or in plastic greenhouses (2000 ha) (Krarup, 2002). The total yield of processing tomatoes in 2002 was 600,000 t and is mainly exported as tomato paste.

According to the Food and Agriculture Organization of the United Nations (FAO), the area of tomato cultivation in Mexico was 70,000 ha in 2003, with a total production of 2.14 million t. Red, green and cherry tomatoes are produced. Mexico is at present the most important supplier of field-grown fresh tomatoes to the USA and the greenhouse area is expanding in order to supply the North American market (Agrifood-Canada, 2001). At present the greenhouse area is estimated at 720 ha (Cantliffe and Vansickle, 2003a). The production of processing tomatoes is estimated at 100,000 t (AMITOM, 2003).

Brazil

Brazil produced about 1.1 million t of processing tomatoes in 2001, which represented about one-third of total production. The cultivated area in 2001 was 15,130 ha (AMITOM, 2003). The central part of Brazil, due to its mild climate, flat topography and good soils, is the major producing area for processing tomatoes (Washington *et al.*, 2001). The main producing areas are the north-eastern states of Pernambuco and Bahía in Cerrado and the state of São Paulo (AMITOM, 2003). Cerrado represents > 80% of total production and the tendency is for this to continue to increase (AMITOM, 2003). Mechanization (of transplanting and harvesting), the use of new hybrid varieties and the use of drip irrigation enables yields to reach 100 t/ha in some cases (Washington *et al.*, 2001). The tomatoes are sold by contract with the processors and the majority are processed into paste. Regarding the production of fresh tomatoes, long shelf-life tomatoes and cherry tomatoes are increasing in importance. The marketing of fresh tomatoes occurs via wholesale markets, so-called *entrepostos* or *Centrais de Abastecimento* (CEASAs).

Oceania

Australia

In Australia, tomatoes are the second most important vegetable crop after potatoes. In 2001/02 the total area was 8500 ha with a production of 425,000 t (Australian Bureau of Statistics, 2003). The three main tomato-growing states are Victoria (about 50% of the cultivated area), Queensland and New South Wales. The Australian processing industry is relatively small but very competitive. In 2002, the planted area was about 4500 ha (AMITOM, 2003) and the overall production was 260,000 t (http://www.entapack.com.au). The main production areas are located in northern Victoria and southern New South Wales (in the south-east part of the country). The production system is similar to that of California and US varieties are used (AMITOM, 2003). Drip irrigation is used on half of the production area, which results in higher yields (about 80–90 t/ha) and saves water and labour (Ashcroft *et al.*, 2003). Transplants are used on about 40% of the total cropped area and harvest (from January to April) is done by machine (AMITOM, 2003). In 2002, 64% of the crop was processed into tomato paste, 21% into whole peeled tomato and 15% into sauce, juice, soup, dried and other uses.

New Zealand

In New Zealand, fresh tomatoes are predominantly grown in greenhouses but a small volume is grown outdoors. Tomatoes are the most important vegetable produced in greenhouses and they occupy an area of about 150 ha (production 40,000 t). Production is mostly for the internal market but about 5% is exported to Australia and the Pacific Islands (Nederhoff, 2003).

Africa and the Middle East

The major tomato producers (fresh or for processing) in the Middle East and North Africa are Egypt, Tunisia, Morocco and Jordan. Egypt is the largest producer of tomatoes on the African continent with a total production of slightly over 6 million t (Table 1.1). The tomato production is mainly located in the bordering areas of the Nile Delta. Another important tomato producer in North Africa is Tunisia, with a total production of 810,000 t in 2002 (FAO, 2003), of which about 75% is for processing (AMITOM, 2003). The main producing area for processing tomato is Cap-Bon in the north-east part of the country (AMITOM, 2003).

In the Middle East, Israel is a relevant producer of processing tomato (about 140,000 t in 2001) (AMITOM, 2003). Tomatoes are grown on collective farms and the production system is mechanized and uses drip irrigation; yields can reach up to 100 t/ha (AMITOM, 2003). The production area for fresh tomatoes in greenhouses (plastic) is about 300 ha (http://www.e-campo.com/media/news/nl/althorticultura23.htm).

In Morocco, fresh tomatoes are the most important early-season vegetable produced in greenhouses, with an estimated area of about 5000 ha located along the Atlantic coast and in the Souss-Massa Valley (east of Agadir) (Anonymous, 2003). Total production in 2001 was estimated at 765,000 t, of which about 150,000 t was processed tomatoes (http://www.interex.be). Tomatoes for processing are mostly produced in the regions of Gharb (4700 ha) and Tangiers/Tetouan (800 ha) (AMITOM, 2003).

South Africa has a relevant production of processing tomato with 200,000 t (AMITOM, 2003). About 70% of the area is still harvested by hand, because the cultivars used are not adapted to mechanized harvest. The average yield varies between 65 and 87 t/ha. Sub-Saharan Africa hardly produces tomatoes but the area is a major importer of tomato paste (AMITOM, 2003).

FUTURE PERSPECTIVES FOR THE TOMATO INDUSTRY

Global production is expected to increase for both fresh-market and processing tomatoes. Based on investments made in the processing sector and improvements in production systems and cultivars, China may be the main source of such increases. Expansions in relatively inexpensive production areas, together with increasing production costs in more industrialized countries, are a concern to many growers, especially those producing processing tomatoes.

Spain and The Netherlands will probably continue to be the leading suppliers of fresh tomatoes to the European markets, although more competition is expected from outside the European Union. Fresh tomato production in Turkey and in African countries on the Mediterranean may increase, and will compete with Dutch and, in particular, Spanish production. For example, since early October 2003, Morocco has been able to export more fresh tomatoes (up to 175,000 t from October to May) to the European Union as a result of a bilateral agreement that granted greater European market access for Moroccan tomatoes and in turn the liberalization of the Moroccan market for EU grains (Hénard, 2003). Nevertheless, Spain (especially Almería and Murcia) will keep actively developing its greenhouse technology, quality control and marketing (Goméz-Escolar, 2002; Wijnands, 2003). In turn, The Netherlands will continue to play a leading role in technological innovation and product quality (e.g. closed greenhouse systems, efficient logistics, and trace and tracking systems), aiming for more sustainable production with regard to energy and pesticide inputs. Increased sustainable production (regarding the use of water, fertilizers and pesticides), an increase in the area of organic production, product quality (safety) and more efficient supply chains should be among the most important items to be considered in European tomato production in the future.

In the Americas, expansion of production in Mexico to supply North American fresh markets may be an important future development. American and Canadian investors (often large tomato producers in their own countries) are expanding Mexico's greenhouse capacity to complement their own production schedules, and to take advantage of Mexico's lower production costs. Expansion of free-trade zones in the Americas promises to provide lower-cost tomato products and fruit to consumers while increasing competition among producers, processors and marketers. However, food safety concerns among consumers may limit trade with markets perceived to be lax in implementing proper controls. As an example, Canada is researching food safety, fruit quality (flavour and nutritional value) and environmentally responsible production (sustainable production systems) to increase consumer confidence in Canadian products and to increase the industry's ability to meet or exceed market requirements. More emphasis will be put on organic farming (M. Dorais, personal communication). The same trend can be observed in the USA. In fact, the higher demand for quality may imply that in future the major competitor to the North American tomato is not the Mexican but rather the Spanish tomato, as suggested by Cantliffe and Vansickle (2003a).

The European market may remain closed to genetically modified (GM) tomatoes for some time, but GM tomatoes will become increasingly important in the Americas, China and other countries. Even if GM tomatoes are not actually produced commercially, information arising from genetic modification will be translated into the breeding of better tomato varieties through conventional means.

Protectionism is expected to continue in the European and American tomato markets, even though markets can become more open based on multilateral agreements. Until now, Latin American exports to the EU have faced tariffs that are five times higher for tomato sauce than those levied on fresh tomatoes (White *et al.*, 2002). Simultaneously, fresh tomatoes may face prohibitive tariffs in the EU during several months of the year, to protect mainly Italian and Spanish producers from Latin American and less so from African producers (e.g. Morocco), who benefit from the EU's ACP (African, Caribbean and Pacific Group of States) agreement (Watkins and Von Braun, 2003). However, 'new ways' of protectionism may arise from the product quality requirements imposed by different countries.

REFERENCES

Agrifood-Canada (2001) *Profile of the Canadian Greenhouse Tomato Industry.* Available at: http://www.agr.gc.ca/misb/hort/

AMITOM (Mediterranean International Association of the Processing Tomato) (2003) Available at: http://www.amitom.com/

Anonymous (2003) HighTech Ag in Morocco. *NewAG International* September 2003, 42–60.

Anuário Vegetal (2002) *Crop Production Year Book.* GPPAA, Ministério da Agricultura, do Desenvolvimento Rural e das Pescas, Lisbon, Portugal, 253 pp.

Ashcroft, W.J., Fisher, P.D., Flett, S.P. and Aumann, C.D. (2003) The effects of grower practice and agronomic factors on the performance of successive tomato crops. *Acta Horticulturae* 613, 123–129.

Australian Bureau of Statistics (2003) Agricultural Commodities, Australia. Available at: http://www.abs.gov.au/Ausstats/

Cajamar (2002) *Análisis de la Campana Hortícola de Almería 2001/2002. Informes y Monografias.* Instituto de Estudios Cajamar, Almeria, Spain, 38 pp.

Cantliffe, D. and Vansickle, J. (2003a) *Mexican Competition: Now from the Greenhouse.* Tomato Institute, Florida, pp. 2–3. Available at: http://gcrec.ifas.ufl.edu/TOMATO%202003.pdf

Cantliffe, D. and Vansickle, J. (2003b) *Competitiveness of the Spanish and Dutch Greenhouse Industries with the Florida Fresh Vegetable Industry.* Available at: http://edis.ifas.ufl.edu/CV284

CBS (2003) *Statistiek Jaarboek 2003.* Central Bureau voor de Statistiek, The Hague, The Netherlands, 559 pp.

Costa, J.M. and Heuvelink, E. (2000) *Greenhouse Horticulture in Almería (Spain: Report on a Study Tour 24–29 January 2000).* Horticultural Production Chains Group, Wageningen University, Wageningen, The Netherlands, 119 pp.

Costa, J.M., Heuvelink, E. and Botden, N. (2004) *Greenhouse Horticulture in China – Situation and Prospects (Report on a Study Tour 11 October–2 November 2003).* Horticultural Production Chains Group, Wageningen University, Wageningen, The Netherlands, 168 pp.

Deloitte & Touche (2003) *Express China News.* Available at: http://www.deloitte.com/dtt/cda/doc/content/Issue_03–3(5).pdf

Dhaliwal, M.S., Kaur, P. and Singh, S. (2003) Genetic analysis of biochemical constituents in tomato. *Advances in Horticulture Science* 17(1), 37–41.

Dorais, M., Papadopoulos, A.P. and Gosselin, A. (2001) Greenhouse tomato fruit quality. *Horticultural Reviews* 26, 239–319.

Eurosur Consultores (2002) Estadísticas y Gráficos. In: Martinez, J.L. (ed.) *Anuario de la Agricultura Almeriense 2002.* La Voz de Almeria, Almeria, Spain, pp. 109–162.

FAO (2003) Homepage. Available at: http://apps.fao.org/

FAS/USDA (2003) *Processed Tomato Products Outlook and Situation in Selected Countries.* Available at: http://www.fas.usda.gov/htp

George, R.A.T. (1999) *Vegetable Seed Production,* 2nd edn. CAB International, Wallingford, UK, 328 pp.

Goméz-Escolar, J.M.P. (2002) Los retos de futuro del sector hortofrutícola español. In: Álvarez-Coque, J.M. (ed.) *La Agricultura Mediterránea en el sieglo XXI.* Mediterraneo Economico, Instituto de Estudios Cajamar, Almeria, Spain, pp. 140–158.

Harvey, M., Quilley, S. and Beynon, H. (2002) *Exploring the Tomato. Transformations of Nature, Society and Economy.* Edgar Publishing, Cheltenham, UK, 304 pp.

Heiser, C. and Anderson, G. (1999) 'New' solanums. In: Janick, J. (ed.) *Perspectives on New Crops and New Uses.* ASHS Press, Alexandria, Virginia, pp. 379–384.

Hénard, M.C. (2003) *France: Tomatoes and Products.* Semi-Annual 2003. USDA Foreign Agricultural Service, Gain Report, December 2003. Available at: http://www.fas.usda.gov/gainfiles/200312/146085520.doc

Ho-Shin, H. (2001) Vegetable seed industry in the Republic of Korea. *Chronica Horticulturae* 41(3), 13–15.

Ito, K. (2003) *Japan: Tomatoes and Products.* Annual 2003. USDA Foreign Agricultural Service, Gain Report, July 2003. Available at: http://www.fas.usda.gov/gainfiles/200307/145985581.pdf

Kinet, J.M. and Peet, M.M. (1997) Tomato. In: Wien, H.C. (ed.) *The Physiology of Vegetable Crops.* CAB International, Wallingford, UK, pp. 207–258.

Krarup, C. (2002) Chile: counter season to Mediterranean horticulture. *Chronica Horticulturae* 42(1), 31–35.

La Malfa, G. (2003) Origine, fisionomia e sviluppo dell'orticoltura protetta. *Italus Hortus* 10(3), 14–26.

Ma, S., Huang, J., Wang, D. and Qu, D. (2000) China. In: Ali, M. (ed.) *Dynamics of Vegetable Production, Distribution, Consumption in Asia.* Asian Vegetable Research Center, Shanhua, Taiwan, pp. 69–97.

Mitchell, J.P., Lanini, W.T., Miyao, E.M., Brostrom, P.N., Herrero, E.V., Jakso, J., Roncoroni, E. and Temple, S.R. (2001) Growing processing tomatoes with less tillage in California. *Acta Horticulturae* 542, 347–351.

Monte Gomes, M. (2003) *Portugal: Tomatoes and Products.* Annual. USDA Foreign Agricultural Service, Gain Report, May 2003. Available at: http://www.fas.usda.gov/gainfiles/200305/145885763.pdf

Müller, C.H. (1940) *A Revision of the Genus* Lycopersicum. USDA Miscellaneous Publication No. 382, USDA, Washington, DC.

Murray, M., Cahn, M.D., Miyao, G., Brittan, K. and Mullen, R.J. (2001) The University of California processing tomato workgroup: an example of ongoing effective collaborations. *Acta Horticulturae* 542, 253–259.

Nagaraju, N., Venkatesh, H.M., Warburton, H., Muniyappa, V., Chancellor, T.C.B. and Colvin, J. (2002) Farmer's perceptions and practices for managing tomato leaf curl virus disease in Southern India. *International Journal of Pest Management* 48(4), 333–338.

Nederhoff, E. (2003) Scaling-up greenhouses in New Zealand. *Fruit&VegTech* 3(5), 18–19, 21.

NRHI (2004) *Agricultural Overview in Korea.* National Research Horticultural Institute. Available at: http://www.nhri.go.kr/english/main.htm

Öztürk, A., Deviren, A. and Özçelik, A. (2001) Methyl bromide in Turkey. Available at: http://www.un.org.tr/unido/documents/METHYL%20BROMIDE%20IN%20TURKEY1.doc

Pazos, D. (2003) *Spain: Tomatoes and Products.* Annual (2003). USDA Foreign Agricultural Service, Gain Report, June 2003. Available at: http://www.fas.usda.gov/gainfiles/200306/145885865.pdf

Rick, C.M. (1976) Tomato (family Solanaceae). In: Simmonds, N.W. (ed.) *Evolution of Crop Plants.* Longman Publications, New York, pp. 268–273.

Rodríguez, R.R. (2002) Tomato growing in the Canary Islands. In: Ferre, F.C. (ed.) *Espana: Orchard of Europe.* Ministerio de Agricultura, Pesca y Alimentación of Spain/SEEI, Madrid, Spain, 229 pp.

Santella, R. (2003) *Italy: Tomatoes and Products.* Annual 2003. USDA Foreign Agricultural Service, Gain Report, June 2003. Available at: http://www.fas.usda.gov/gainfiles/200306/145985168.pdf

Sekliziotis, S. (2003) *Greece: Tomatoes and Products.* Annual 2003. USDA Foreign Agricultural Service, Gain Report, June 2003. Available at: http://www.fas.usda.gov/gainfiles/200306/145885860.pdf

Siemonsma, J.S. and Piluek, K. (1993) *Prosea: Plant Resources of South-East Asia, No. 8: Vegetables.* Pudoc Scientific Publishers, Wageningen, The Netherlands, pp. 1199–1205.

Sims, W.L. (1980) History of tomato production for industry around the world. *Acta Horticulturae* 100, 25.

Sirtioglu, I. (2003) *Turkey: Tomatoes and Products.* Annual 2003. USDA Foreign Agricultural Service, Gain Report, May 2003. Available at: http://www.fas.usda.gov/gainfiles/200305/145885672.pdf

Statistics Canada (2002) *Greenhouse, Sod and Nursery Industries.* Catalogue no. 22–202-XIB, 26 pp.

Stevens, M.A. and Rick, C.M. (1986) Genetics and breeding. In: Atherton, J. and Rudich, G. (eds) *The Tomato Crop. A Scientific Basis for Improvement.* Chapman & Hall, New York, pp. 35–109.

Suhartanto, M.R. (2002) Chlorophyll in tomato seeds: marker for seed performance? PhD Thesis, Wageningen University, The Netherlands, 150 pp.

Taragola, N. and Van Lierde, D. (2000) Competitive strategies in the sector of greenhouse tomato production in Belgium. *Acta Horticulturae* 524, 149–155.

Taylor, I.B. (1986) Biosystematics of the tomato. In: Atherton, J. and Rudich, G. (eds)

The Tomato Crop. A Scientific Basis for Improvement. Chapman & Hall, New York, pp. 1–34.

USDA (2002) *International Agricultural Trade Report*. January 2002. Available at: http://www.fas.usda.gov/htp/horticulture/fresh vegetables/grntomIATR.PDF

Washington, L.C.S., Giordano, L.B., Silva, J.B.C.S. and Marouelli, W.A. (2001) The use of drip irrigation for processing tomatoes in Brazil. *Acta Horticulturae* 542, 103–105.

Watkins, K. and Von Braun, J. (2003) *Time to Stop Dumping on the World's Poor*. 2002–2003 IFPRI Annual Report Essay. Available at: http://www.ifpri.org/pubs/books/ar2002/ar02e1.pdf

White, G.B., Nelson, L.B. and Schluep, I. (2002) *Impacts of Trade Liberalization on the New York Horticultural Sector*. Staff Paper, Department of Applied Economics and Management, Cornell University, Ithaca, New York, 25 pp. Available at: http://aem.cornell.edu/research/researchpdf/sp0204.pdf

Wijnands, J. (2003) The international competitiveness of fresh tomatoes, peppers and cucumbers. *Acta Horticulturae* 611, 79–90.

Yoltas, T., Aydin, M., Seferoglu, S., Seker, G. and Topcu, N. (2003) The nutritional status of processing tomato production areas in Marmara Region. *Acta Horticulturae* 613, 173–176.

Genetics and Breeding

P. Lindhout

INTRODUCTION

Tomato breeding is a success story, as illustrated by the many cultivars now available that are grown all over the world in a variety of conditions and that harbour a wide range of traits such as vigorous growth, resistances to pests and diseases, and high fruit quality. Tomato breeding follows the classical approach of identifying superior genotypes, crossing and progeny selection. Nowadays molecular marker techniques facilitate breeding and genetic transformation is possible, though rarely used for commercial breeding as yet. The new developments are in the area of genomics.

GENETIC VARIATION WITHIN THE GENUS *LYCOPERSICON*

The cultivated tomato (*L. esculentum*) reached its present form and place in the human diet after a long period of domestication. Initial development was probably, in part, from selection for preferred genotypes in the existing germplasm. In a predominantly inbreeding species such as tomato, genetic variation tends to decrease, even without selection. The species also suffered from severe genetic bottlenecks in the past 500 years as the crop was carried from the New World to Europe and then back to North America. As a consequence, genetic variation in the cultivated tomato is extremely limited. It is likely that there was no deliberate crossing with wild related species until the 20th century.

The lack of diversity can be visualized using DNA technologies: very few polymorphisms are identified even with the most sensitive DNA techniques. The lack of genetic variation detectable at the DNA level is in contrast to the wealth of morphological variation observed among tomato accessions in genebanks. This can easily be explained: tomato has been a major vegetable crop in the 20th century and, until recently, was maintained as self-pollinated varieties. Spontaneous or induced mutants creating unusual phenotypes are readily detected in self-pollinated species. Such mutants in tomatoes have been carefully documented, preserved and propagated. Most of these will have been caused by recessive point mutations, which are often associated with obvious deviating phenotypes. A large population of deviating phenotypes as observed in a genebank or in the genetic stock of a breeder gives the impression of abundant variation (Rick, 1991). However, at the DNA level, only point mutations are present, which may easily remain unnoticed among the nearly one billion base pairs in the total tomato genome. A number of these mutants have been used for crop improvement. Linkage relationships of many of the mutants were studied from the 1950s to the 1980s, resulting in the creation of the tomato morphological map. This, and the more recent molecular maps discussed below, can be accessed on the Solanaceae Genomics Network database (http://www.sgn.cornell.edu/).

DIVERSITY AMONG AND WITHIN WILD SPECIES OF THE GENUS *LYCOPERSICON*

The lack of diversity within tomatoes is not a barrier to breeding progress, due to the variation readily available from the species' wild relatives. The variation among species within the genus *Lycopersicon* is tremendous, especially within the self-incompatible species (Fig. 2.1). Numerous investigators have studied specific traits present in wild *Lycopersicon* spp. and are exploiting this variation for tomato improvement. Nearly all of the wild species have contributed valuable traits for crop improvement, from the closely related red-fruited self-pollinated *L. pimpinellifolium* to the more distantly related green-fruited self-incompatible species *L. pennellii* and *L. peruvianum*. Crosses between *L. esculentum* and its closer relatives, such as *L. pimpinellifolium* and *L. parviflorum* (Fig. 2.2), are readily successful. Crosses between *L. esculentum* and *L. chilense* or *L. peruvianum* are more challenging and often require embryo rescue to obtain the F_1 progeny. In later generations, additional barriers that are caused by genetic interactions may further impede progress.

As a consequence of their self-incompatibility and outcrossing mode of reproduction, accessions of the outcrossing species may harbour a higher level of genetic variation than the largely self-pollinating species. Miller and Tanksley (1990) assessed this variation using restriction fragment length polymorphisms (RFLPs) as molecular markers (Fig. 2.3). It is striking that more genetic variation was observed within a single accession of the self-incompatible species than in all accessions of tomato.

In addition to the species within the genus *Lycopersicon*, introgressions have been made with *Solanum lycopersicoides* and *S. sitiens* (Ji *et al.*, 2004). The set of introgression lines from this wild relative of tomato is described, and can be requested from the Genetic Stock Center in Davis, California (http://tgrc.ucdavis.edu; e-mail: tgrc@vegmail.ucdavis.edu).

Fig. 2.1. Genetic variation in fruit colour, shape and size in the genus *Lycopersicon*. The top three rows are *L. esculentum* cultivars. The bottom row represents some wild species.

Fig. 2.2. Some wild *Lycopersicon* species: (a) *L. pimpinellifolium*; (b) *L. parviflorum*.

In conclusion, there is limited genetic variation in tomato itself but considerable genetic variation among and within the wild *Lycopersicon* spp. that can be exploited, with varying degrees of difficulty, for crop improvement. Breeders would naturally use the sources closest to tomato, for ease, efficiency and rapidity of the transfer. However, none of the species is out of reach for breeding when sufficient time, labour and space are available.

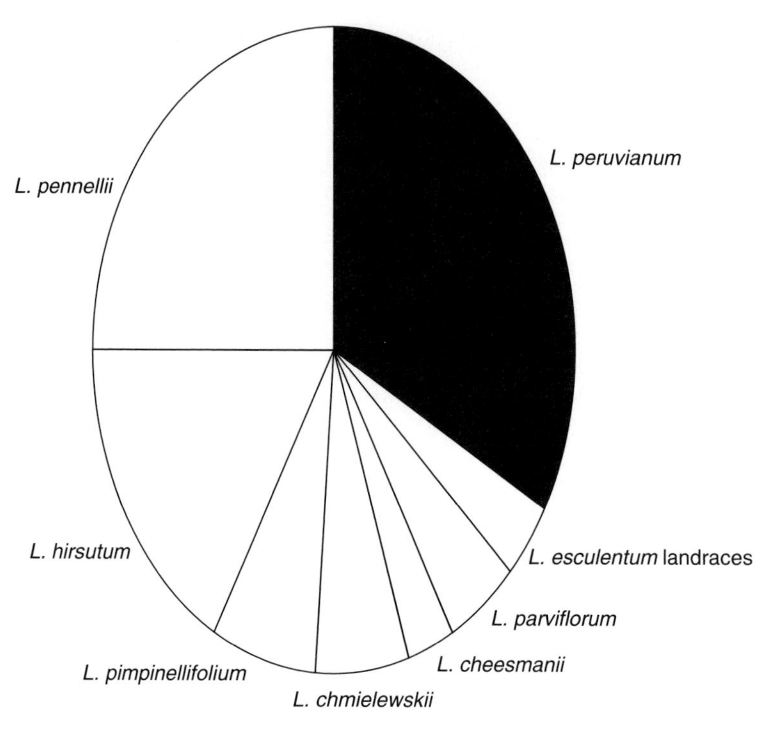

Fig. 2.3. Relative genetic variation in the genus *Lycopersicon* (revised from Miller and Tanksley, 1990).

COLLECTION AND PRESERVATION OF GENETIC RESOURCES

The collection, description, propagation and distribution of genetic material are of the utmost importance if the variation it contains is to be used successfully in tomato breeding. In the latter half of the 20th century, numerous expeditions were made to the Andes region to collect accessions of various wild *Lycopersicon* spp. It is noteworthy that Dr Charlie Rick, from the University of California, understood the importance of germplasm collection and he collected and described numerous accessions of the wild *Lycopersicon* spp., which are still maintained at the Genetic Stock Center in Davis, California (Rick, 1991). These materials are freely available upon request for the scientific community and for the commercial breeding industry. Tomato germplasm is maintained in several other world genebanks and is available on request.

Within a genebank, the accessions should be fully and accurately described. In addition to information on location site, full passport data, and taxonomic classification, this includes data about the native habitat and geographical and climatic characteristics. In addition, evaluations about the morphology (e.g. leaf shape, plant growth habit, fruit size and shape) should be described (Table 2.1). Curators must be concerned with maintaining variability within accessions. More plants and a means of genetic isolation are needed for propagation and maintenance of outcrossing self-incompatible accessions for comparison with self-compatible plants. Data on the reactions of accessions to abiotic and biotic stressors such as fungal pathogens or insect pests are important. However, these evaluations are labour, time and space consuming, and genebanks are generally underfunded for performing their primary tasks of germplasm maintenance and distribution. Therefore these data are lacking, or must be gathered after the accessions have been obtained from the germplasm bank by researchers.

Table 2.1. A descriptors list for some fruit traits evaluated at the Centre for Genetic Resources, Wageningen, The Netherlands.

Trait	Description
Immature fruit colour	(+ = present, 0 = pale, 1–9 = intensity of green collar/greenback: 1 = very light, 9 = very dark)
Mature fruit colour	(A = red, B = orange, C = pink, D = yellow, E = yellow green, F = red with yellow spots, G = light green)
Fruit setting ability	(0 = no flowers set, 5 = 50% of flowers set, 9 = all flowers set)
Fruit ribbing	(0 = absent, 1 = very little, ..., 9 = very high)
Fruit top hardness	No description available
Fruit cracking tendency	(0 = no expression, 1 = very low, ..., 9 = very high expression)
Fruit firmness	No description available
Fruit shape	Sonatine = 5 (1 = very flat, ..., 9 = very long)
Fruit section shape	(1 = very round, 3 = round, 5 = angular, 7 = irregular, 9 = very irregular)
Fruit green edge	(+ = present, 0 = absent, 1–9 = intensity of green edge: 1 = very light, 9 = very dark)
Fruit locule number	Count and estimate the average number, using a few plants
Fruit yellow top	(0 = absent, 1 = very few, 3 = few, 5 = half of the fruit, 7 = most fruits, 9 = all fruits)
Fruit fleshiness	Sonatine = 3 (1 = very low, ..., 9 = very high)
Fruit outerwall thickness	Measurement (9 = 9 mm or more)
Pericarp thickness	Sonatine = 3 (1 = very little, 3 = little, 5 = intermediate, 7 = flesh, 9 = very fleshy)
Fruit puffiness	(0 = no expression, 1 = very low, ..., 9 = very high expression)
Pedicel jointlessness	(0 = absent, + = present)
Blotchy ripening	(0 = absent, + = present)
Harvest time begins	(2,3,4,5 = 4,3,2,1 weeks before Sonatine, 6 = Sonatine, 7,8,9 = 1,2,3 weeks after Sonatine)
Blossom-end rot	(0 = absent, + = present, ++ = severe)

In considering the use of wild germplasm for crop improvement, it should be noted that useful genes of the wild *Lycopersicon* spp. might be masked in accessions of the species. For example, the small-fruited wild species *L. pimpinellifolium* has genes that increase the fruit size when these genes are introgressed into a cultivated tomato background. A series of papers from Steven Tanksley describes the method of using advanced backcross lines to detect and transfer such variation (Monforte and Tanksley, 2000, and references therein).

CYTOGENETICS

Investigations into tomato cytogenetics were initiated at the beginning of the 20th century, when the genetic constitution of the tomato was determined: it has a diploid genome with 12 chromosome pairs and a nuclear genome DNA content of about 900 Mb. At the pachytene stage, tomato chromosomes are very distinct, and the length and physical features of the chromosomes are sufficiently distinctive that the different chromosomes can be identified. The total length of the tomato chromosomes at the metaphase stage is about 50 μm. Haploid plants can be obtained at low frequencies in several ways, such as by X-ray treatment of pollen grains before fertilization and from polyembryonic seeds, although the latter occurs at an extremely low rate. Haploids are sterile, but homozygous diploids derived from haploids are highly fertile. Haploids are useful for fast screening of recessive characters, since at the diploid level several generations of selfing are required to uncover the recessive traits. However, the success of haploids in tomato is very low and hence haploidization is rarely used.

Crossing a diploid with a tetraploid genotype can produce triploids. Although 12 trivalents are expected at diakinesis, bivalents and univalents are frequently observed. Trisomics and tetrasomics, as well as other aneuploids, are produced from the progeny of triploids. Tetraploids are generally obtained after colchicine treatment. Alternatively, a high frequency of tetraploids can be obtained from tissue or cell culture regenerants. Tetraploids are characterized by thicker and darker leaves, reduced internode length, increased stem thickness, reduced fertility and reduced fruit set and fruit quality. Consequently tetraploid tomatoes are not commercially useful. This is also true for all other non-diploid tomatoes.

Large collections of aneuploids were generated in the 1960s, some of which are maintained in the Genetic Stock Center. Triploids are the best source for trisomics, since almost all trisomics can be expected from the triploid progeny. Trisomics have been used to assign linkage groups to chromosomes. The addition of extra chromosomes to the diploid set has already a huge effect on the tomato phenotype and meiosis. It is accepted that three extra chromosomes is the limit that can be added to the diploid set. A very limited number of monosomics have been obtained. Generally, monosomics are very unstable, again confirming the strict diploid nature of tomato. Tomato monosomics have been obtained by fusion of tomato and potato somatic cells, followed by a series of backcrosses to the potato genome (48 chromosomes) as recurrent parent. By using DNA markers and genomic *in situ* hybridization (GISH), a complete set of 12 monosomic lines was obtained, each carrying a different tomato chromosome in a potato background (Fig. 2.4) (Ali *et al.*, 2001).

Nowadays the location of unique DNA sequences can be visualized by fluorescent *in situ* hybridization (FISH). If the positions of these sequences on the genetic map are known, the physical distance between genetic markers can be measured. In this way the relation between physical and genetic distances over the tomato genome can be determined (Budiman *et al.*, 2004).

MOLECULAR LINKAGE MAPS

Originally, the tomato morphological map was generated using visually distinguishable morphological traits. It was therefore constructed from data

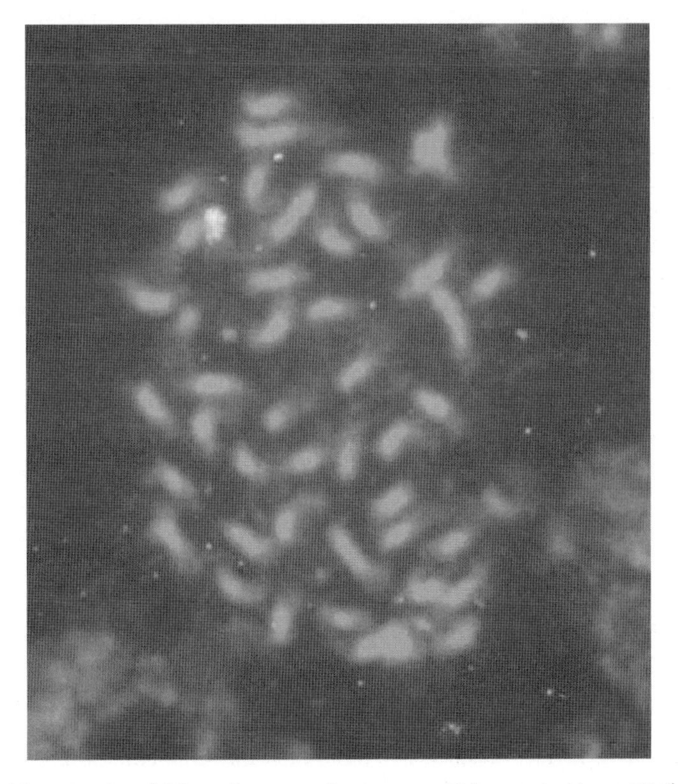

Fig. 2.4. Monosomic addition of tomato chromosome 6 in potato ($4n = 48$) (from Hans de Jong, Wageningen University, The Netherlands).

from numerous linkage analyses, most of which are based on segregation of three to six markers. Aneuploids were used to assign the linkage groups constructed to specific chromosomes. Later, isozymes were identified, predominantly in wild taxa, and added to the morphological map (Tanksley and Mutschler, 1990). These isozymes are considered to be the first generation of molecular markers. The advantages of these molecular markers are codominance, unambiguous phenotypic expression, lack of epistasis, lack of effects on plant vigour and morphology and hence the possibility of accumulating markers in a single line. The limitation of isozymes as molecular markers is that there is a limited number of them available – certainly not enough for a dense map.

With the advent of DNA-based molecular markers, complete genetic maps could be generated using data from only one segregating population. The first high-density map was established in 1991 by Tanksley *et al.* (1992). It comprised mainly RFLP markers and some morphological markers to anchor the genetic linkage groups to chromosomes. RFLPs are polymorphisms based on DNA differences, which are visualized after digestion with specific restriction enzymes, followed by hybridization to cDNA fragments. The map of Tanksley *et al.* (1992) contained 1030 RFLP markers. The markers on this map are applicable in any population within the *Lycopersicon* spp. This is due to the locus specificity of RFLP markers. The order of these markers is conserved over all *Lycopersicon* spp. Moreover, the genetic relationship of tomato with other related species such as potato, pepper and aubergine is so large that the same RFLP markers can be used in all these species and hence the genetic linkage maps compared. The tomato map is very similar to the potato map, but is different from the pepper map. Still, about 18 linkage groups can be identified with the same marker order within these groups in tomato and in pepper. However, the gross position of these linkage groups is scrambled when both species are compared.

These similarities between genomes of related *Solanaceae* also allows a comparison of the map position of similar genes. Indeed, the map positions of genes involved in the basic metabolism or in other similar processes appear to be conserved as well. Taken together, these results imply that the tomato genome shows a great synteny with related species. In other words, the gene repertoire and gene order show a high level of similarity. All these data point to an ancient common ancestor of these related *Solanaceae* species. This is now an area of rapid advance and within the next few years considerably more detailed synteny maps comparing the *Solanaceae* species will be completed. Remarkably, the tomato also shows microsynteny with *Arabidopsis*, which allows researchers to use information from *Arabidopsis* in tomato research.

A disadvantage of RFLP is the laborious and time-consuming technique. Meanwhile, more efficient PCR-based technologies have been developed with the advantage that only minor quantities of DNA are needed and the laboratory techniques are simple and rapid. This has facilitated the generation of a high-density map in any region of interest. Haanstra *et al.* (1999) established an integrated map comprising 1078 AFLP and 67 RFLP markers by using two interspecific F_2 populations of *L. esculentum* × *L. pennellii.*

The tomato information explosion continues. Large numbers of sequences of expressed genes (expressed sequence tags, ESTs) have been determined (Van der Hoeven *et al.*, 2002). Locus-specific single nucleotide polymorphisms (SNPs) are now being made available on DNA chips. International corporations are focusing on identifying protein and metabolic fingerprints of tomato (see also Fig. 2.9). Together with the DNA sequence information, this research field is described as genomics. It is expected that these developments will result in an important step forward in knowledge and understanding of the structure of the tomato genome. The challenge will be how to use the information resources effectively.

APPLICATIONS OF MOLECULAR MARKERS

The high-density tomato genetic linkage map enabled a whole series of new applications. Before the advent of molecular markers, only qualitative traits were mapped, usually in populations that segregated for one or two morphological markers. The advantage of DNA markers is the simultaneous analysis of hundreds of markers in any population, allowing the mapping of any gene by cosegregation of such a gene with DNA markers (Lindhout, 2002) (Fig. 2.5). Generally, the following three mapping strategies can be applied.

Associations between traits and DNA polymorphisms among cultivars

Polymorphisms, observed between obsolete susceptible lines and modern resistant cultivars, are in many cases due to introgressed genome fragments harbouring disease-resistance genes. As a consequence, DNA markers may be present on the introgressed fragment, carrying the resistance gene, and hence are associated with resistances. Van der Beek *et al.* (1992) applied this idea to the tomato and found an association of the RFLP marker TG301 and the *Cf-9* gene for resistance to *Cladosporium fulvum*. As the map position of TG301 is known, *Cf-9* could be mapped on the short arm of chromosome 1.

Segregation analysis of interspecific populations

Interspecific populations, which are obtained when susceptible cultivated varieties are crossed with resistant wild relatives, offer many opportunities for

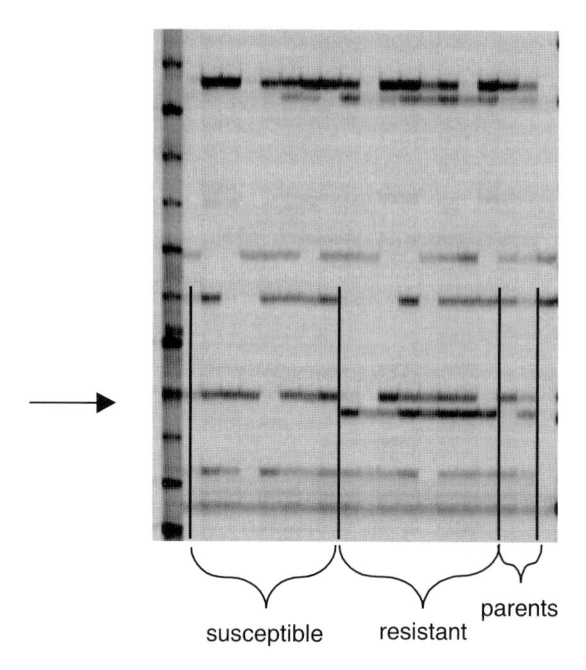

Fig. 2.5. Mapping of a resistance gene by cosegregation with AFLP markers. Arrow indicates the marker that is linked to resistance.

genetic disease analysis. Linkage analysis within such populations often reveals cosegregation of the resistant phenotype with the presence of specific marker bands. For instance, many DNA markers on chromosome 6 were closely linked to the *Ol-1* gene involved in resistance to *Oidium lycopersici* (Huang *et al.*, 2000) (Fig. 2.6). Alternatively, the map position of the *Verticillium* resistance gene *Ve* was determined through linkage analysis of interspecific recombinant inbred lines obtained from a cross between a cultivated tomato accession and a *Lycopersicon cheesmanii* accession (Diwan *et al.*, 1999).

QTL mapping

As the genotype of a plant cannot be directly deduced from the plant phenotype, a different approach is needed when interest lies with 'field' or quantitative resistance. A segregating population (with regard to both marker polymorphisms and observed disease severity) is screened and correlations between marker genotypes and disease severity are calculated along the linkage map. In this way, for each position on the genome, the probability that a quantitative trait locus (QTL) is present is determined. When significant

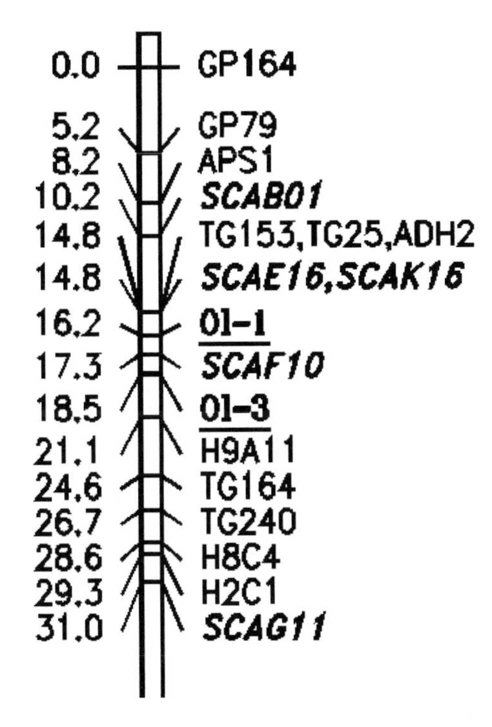

Fig. 2.6. Integrated map of chromosome 6 with the position of two genes, *Ol-1* and *Ol-3*, that confer resistance to *Oidium lycopersici* (Huang *et al*., 2000).

correlations are found, the genome is assigned confidence intervals for QTLs that influence the disease of interest. Alternatively, repeated backcrosses can be made and the progeny are selected for the presence of only one introgression fragment from the donor parents. Such a population of backcross inbred lines (BILs) is very useful, as any difference in phenotype can immediately be linked to the presence of the small introgressions and hence the trait can be mapped (Eshed and Zamir, 1994). Knowledge of the position of these QTLs can be applied in a marker-assisted selection approach, and the markers flanking the QTL confidence intervals can serve as a starting point for genome walking and gene cloning procedures. Numerous tomato QTLs have now been identified and mapped for traits such as cold and salt tolerance, soluble solid content and other traits involved in fruit quality as well as for resistance genes to pests and pathogens (Van Heusden *et al*., 1999) (Fig. 2.7).

After the identification of these QTLs, the corresponding character(s) can be introgressed into tomato using the linked molecular markers that are diagnostic for the trait, instead of selecting for the difficult scorable trait itself (Huang *et al*., 2000). Breeding and selection of quantitative traits as well as

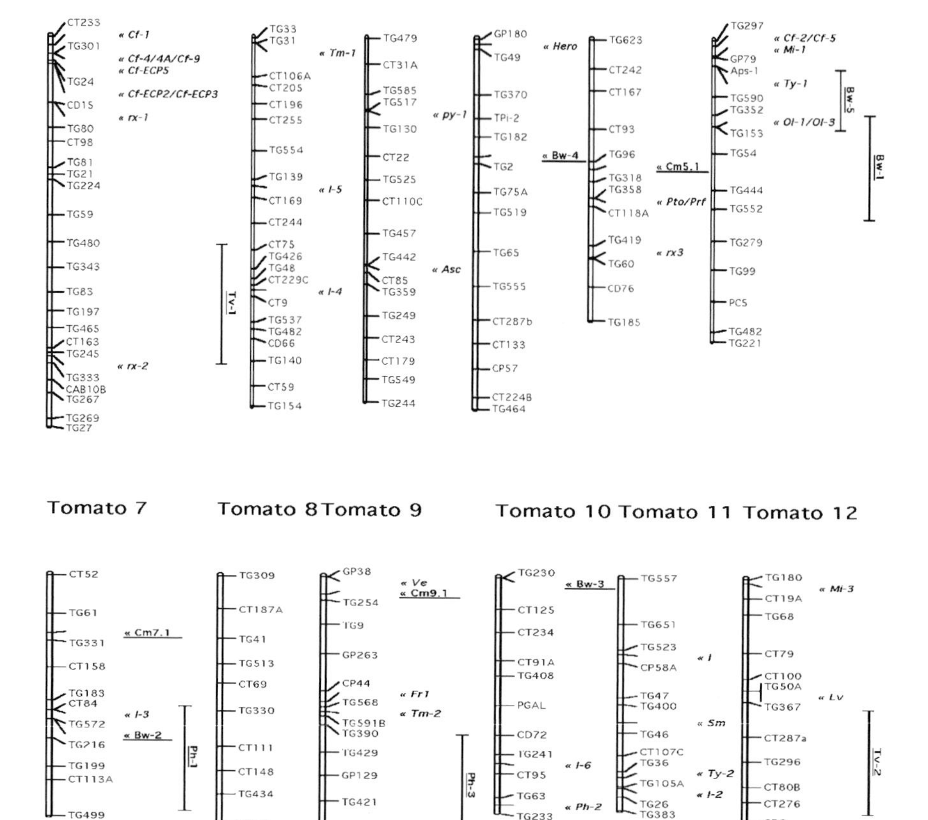

Fig. 2.7. Overview of mapped resistance genes on the tomato genome (Van Berloo and Lindhout, 2001). Marker loci are taken from the core RFLP map (Tanksley *et al.*, 1992). The gene symbols refer to resistance gene loci. Monogenic loci are printed in italics, quantitative loci are underlined and for some loci an estimated interval is indicated.

qualitative traits are called marker-assisted breeding (MAB) and marker-assisted selection (MAS).

The development and application of PCR markers enables the study of traits on seedlings, which requires much less greenhouse space and provides much faster results. It is expected that these techniques of QTL mapping and MAS with the aid of molecular markers will increase the efficiency of breeding programmes considerably.

MAP-BASED CLONING

A high-density linkage map also enables isolation of individual genes through chromosome walking or chromosome landing. Since the cloning of *Cf-9* in 1994, more than a dozen resistance genes have been isolated by using a map-based cloning approach. Remarkably, most resistance genes share common motives such as nucleotide binding sites (NBS) and leucine-rich repeats (LRR) and often are located with several homologous genes in a cluster on the tomato genome (Dangl and Jones, 2001). The genetic organization of *Cf* genes on the tomato genome is even more complex, as five clusters of *Cf* genes have been identified on the short arm of chromosome 1, while each cluster contains several *Cf* homologues (De Kock *et al.*, 2005) (Fig. 2.8). More recently, genes that are involved in quantitative characters such as fruit size and yield have also been cloned.

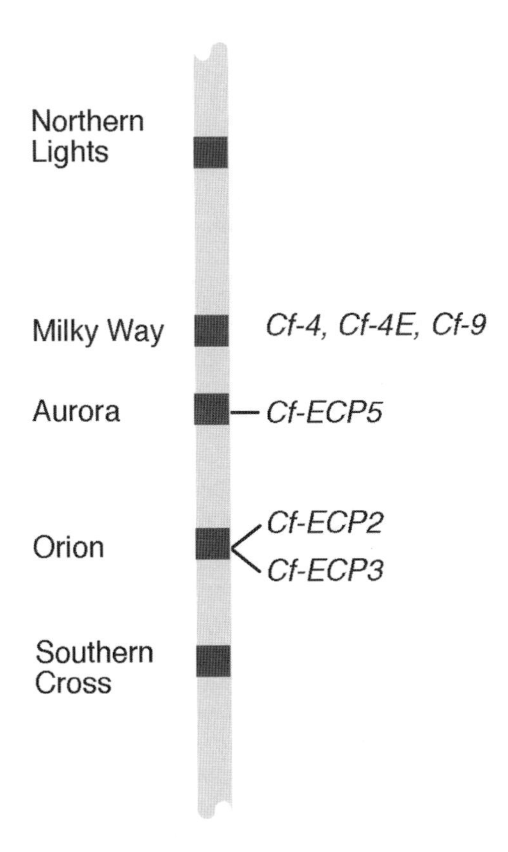

Fig. 2.8. The short arm of chromosome 1 of tomato, with the position of five *Cf* gene clusters and the functional resistance genes at these cluster loci.

HISTORY OF TOMATO BREEDING

After the tomato was brought to the Old World, and its edibility was trusted, its cultivation area spread from southern Europe to the more northern countries. Immigrants took the tomato to North America. In the same period professional seed sellers started to spend more effort in breeding activities and developed into companies that combined seed merchandizing with seed propagation and selection. In the first half of the 20th century these companies applied simple selection methods to develop their own varieties. As the production acreage increased, pathogens and pests also appeared more frequently, causing an increasing demand for resistant varieties. Where resistance was available within the cultivated tomato, it could be easily exploited and introduced into new varieties. If needed, crosses with resistant accessions of wild species were carried out to introgress wild disease-resistance genes. One of the first examples was the exploitation of *Cladosporium fulvum* resistance from *L. pimpinellifolium* in 1934. To protect their intellectual properties, breeders could obtain breeding rights for their cultivars. The seed price remained low, as growers could still propagate seeds on their own nursery if seed prices were high.

During and after the Second World War, national governments funded research institutes to support breeding activities. The acquired genetic knowledge supported breeding activities that rendered breeding programmes more scientific. In 1946 the first hybrid tomato cultivar, 'Single Cross', was released. Hybrids combine good characters from both parents that will segregate in the progeny, discouraging seed propagation by growers. The advantages of hybrid varieties over true-breeding varieties were so great that the growers would buy hybrid seeds at higher prices. The production costs of hybrid seeds were also high, causing a gradual increase in the seed prices of hybrids. Eventually, nearly all new tomato cultivars for the fresh market and an increasing number of cultivars for processing were hybrids. Breeding companies could invest more in professional breeding programmes and new varieties were continuously developed that combined several resistance genes with other good agronomic characters such as high yield, resistance to extreme conditions and high fruit quality.

Breeding goals

The goals of public and private tomato breeding programmes vary widely, depending on location, need and resources. To be successful, growers must produce high yields of high quality fruit while holding production costs as low as possible. Therefore, many of the breeding goals focus on characteristics that reduce production costs or ensure reliable production of high yields of high quality fruit. These can be grouped into the following sections: resistance to biotic stress; resistance to abiotic stress; and fruit quality.

Breeding for resistance to biotic stress

The tomato is host to more than 200 species of a wide variety of pests and pathogens that can cause severe losses. Often, these pests and pathogens have to be controlled by using chemical compounds, such as fungicides or pesticides. These methods may not be fully effective; they raise production costs; and they require compliance with laws governing the use of chemicals. They also cause concern regarding potential risk for the growers, the consumers and the environment. Therefore, one of the most prominent issues in tomato breeding is breeding for resistance to the most destructive pests and pathogens.

Nature has provided a great wealth of resistances that are available in the genus *Lycopersicon* (Van Berloo and Lindhout, 2001; Lindhout, 2002) (Fig. 2.7). Many of the resistances are simply inherited, and tomato breeders have had remarkable success in transferring disease-resistance genes into cultivated tomatoes. As a consequence, modern cultivars can carry more than a dozen resistance genes. These resistances can be grouped according to the pathogen groups against which they can protect the crop.

VIRUS RESISTANCE

- The Tm_2^{2} gene for control of the tomato strain (ToMV) of tobacco mosaic virus (TMV) was obtained from *L. peruvianum* in the 1950s. It is used in nearly all greenhouse tomatoes and some field-grown fresh-market tomatoes, and is still effective after more than 50 years.
- A devastating virus in tropical areas is tomato yellow leaf curl virus (TYLCV), which is transmitted by the whitefly *Bemisia tabaci* and can only partially be controlled by controlling the whitefly. Polygenic resistance to TYLCV has been identified in *L. peruvianum* and an incomplete resistance has been found on chromosome 6 that may give a sufficient level of resistance.
- A less important but well-spread virus in greenhouses is tomato spotted wilt virus (TSWV), which has a very large host range and is transmitted by *Thrips* spp. The TSWV resistance gene *Sw-5* has been identified in *L. peruvianum* and has already been introduced into modern varieties.
- Another virus, cucumber mosaic virus (CMV), is mainly present in subtropical and tropical tomato production fields and is transmitted by aphids or mechanically. No good level of natural resistance is yet available, but transgenic plants carrying a viral gene may have high levels of resistance (see later under biotechnology and transgenic plants).
- Pepino mosaic virus (PepMV) is a new virus that was first recorded in Europe in 1999. PepMV has spread rapidly and is now one of the major threats for greenhouse production. All cultivars are very susceptible to this mechanically transmitted virus and no resistance has been reported in

wild species. Though often no obvious symptoms are visible, the virus may cause reduced growth and development of infected tomato plants and can hence also show reduced yield, especially under suboptimal conditions (http://www.gov.on.ca/OMAFRA/english/crops/facts/01–017.htm).

BACTERIUM RESISTANCE

- Bacterial wilt is one of the most devastating bacterial diseases that threatens the tomato crop in tropical areas. The causal pathogen is *Pseudomonas solanacearum*. No good levels of resistance are yet available, but some genes have been identified that cause a partial resistance to this pathogen.
- Similarly, partial resistance to *Clavibacter michiganensis* has been identified in *L. peruvianum*.
- A relatively new resistance gene is the *Pto* gene that confers resistance to bacterial blight caused by *Pseudomonas syringae*. The tomato–*Pseudomonas* interaction is a well-studied model system for plant defence responses (Pedley and Martin, 2003).

FUNGUS RESISTANCE The dominating group of pathogens that infect the tomato are fungus species. For most pathogens, resistance genes have been identified. Breeders have preferentially used complete or qualitative resistances that are monogenic and race specific. However, these resistance genes are often easily rendered ineffective by adaptation by the pathogen, forcing breeders to introduce a new gene.

- A good example is *Cladosporium fulvum*, where breeders have used at least four resistance genes (*Cf-2*, *Cf-4*, *Cf-5* and *Cf-9*) and yet none is completely effective (Lindhout *et al.*, 1989) (Table 2.2). Still, numerous additional genes are available and can be used for breeding (Haanstra *et al.*, 2000).
- Another pathogen that has been more successfully controlled by breeding is *Verticillium albo-atrum* or *Verticillium dahliae*. For more than 50 years the *Ve* gene has effectively conferred resistance to *Verticillium* race 1.
- Fusarium wilt is controlled by the genes *I*, *I-2* and *I-3*, which all confer resistance to specific races. As *Fusarium* is a soil-borne fungus, the spread of new virulent races is rather slow and consequently resistance *I*-genes may remain effective for decades.
- *Alternaria solanii* is a widespread pathogen that causes early blight disease. Resistance that protects the stems, and to a lesser extent the leaves, has been transferred to fresh-market tomato lines and is starting to appear in hybrid varieties.
- There are two main powdery mildew species that attack the tomato crop: *Leveillula taurica* only occurs in subtropical and tropical areas and can be controlled by the *Lt* gene; *Oidium neolycopersici* occurs mainly in

greenhouse tomatoes but also in the field. This fungus recently spread worldwide but can be controlled by several *Ol*-genes that are available in modern hybrid cultivars.

- As in potato, *Phytophthora infestans* is widespread and can cause the disease late blight, mainly in field tomatoes. Strong resistance has been found in *L. pimpinellifolium* and *L. hirsutum* and resistance lines are close to being released to the market.
- A very common fungus is *Botrytis cinerea*, which is not a major disease but mainly infects plants that are growing weakly (e.g. at the end of the season). No good resistance is yet available, but partial resistance has been described.

INSECT RESISTANCE Resistance to insects has not received as much attention as disease resistance over the years. This is due, in part, to the fact that during the 1950s to 1970s insect pests could be controlled by sprays and monogenic resistance was not available. As a consequence, none of the modern tomato cultivars harbours resistance genes to insects. Yet some wild species show high resistance to insects, which may be caused by toxic compounds, such as acyl-sugars or 2,3-tridecanone in *L. pennellii* and *L. hirsutum*, respectively (Rodriguez *et al.*, 1993).

Resistance to some insect pests has been achieved in plants transgenic for a *Bacillus thuringiensis* construct. Insect resistance is also being introgressed from wild tomatoes, largely *L. pennellii* and *L. hirsutum*. Within the next decade, the release of lines carrying plant-based insect resistance should be accomplished.

Breeding for resistance to abiotic factors

Since the tomato is subtropical in origin, tomato production is suboptimal over large parts of the tomato crop-growing areas, due to unfavourable environmental conditions caused by abiotic factors that include high or low temperatures, excessive water or drought, and soil salinity or alkalinity. Much genetic variation for these traits exists in the tomato germplasm, making breeding for these traits an economically interesting activity.

COLD TOLERANCE Cold tolerance has long been of interest for production in cooler regions with shorter growing seasons. For greenhouse cultivations, the energy crisis of 1973 led to increased energy costs and consequently to more interest in cold resistance, or low-temperature tolerance. It is expected that breeding for low-temperature tolerance will become an increasingly important issue in greenhouse cultivation, due to concerns about global warming and, as a consequence, the agreements on reduction of CO_2 emissions (Fig. 2.9). However, breeding for low-temperature tolerance is not only of particular interest to greenhouse cultivation, since low temperatures may also occur occasionally in subtropical regions.

Table 2.2. Virulence spectrum of some races of *Cladosporium fulvum* from The Netherlands, France and Poland on tomato genotypes (Lindhout *et al.*, 1989).

Tomato genotypes	Resistance genes	Races of *Cladosporium fulvum*[a]											
		Netherlands								France	Poland		
		2	4	5	2.4	2.4.11	2.4.5	2.4.5.11	2.4.5.9.11	2.5.9	4.11	2.4.11	2.4.9.11
Moneymaker	*Cf-0*	S	S	S	S	S	S	S	S	S	S	S	S
Vetomold	*Cf-2*	S	R	R	S	nd	S	nd	S	S	R	S	S
Purdue 135	*Cf-4*	R	S	R	S	nd	S	nd	S	R	S	S	S
Vagabond	*Cf-2, Cf-4*	R	R	R	S	S	S	S	S	R	R	R	R
Ontario 7717	*Cf-5*	R	R	S	R	R	S	S	S	S	R	R	R
Ontario 7818	*Cf-6*	R	R	R	R	R	R	R	R	R	R	R	R
Ontario 7522	*Cf-8*	R	S	R	S	nd	S	nd	S	R	S	S	S
Ontario 7719	*Cf-9*	R	R	R	R	R	R	R	S	S	R	R	S
Ontario 7716	*Cf-11*	R	R	R	R	S	R	S	S	R	S	S	S

S, susceptible; R, resistant; nd, not determined.

[a] The race names refer to the absence of an *Avr* gene that interacts with the corresponding *Cf-* gene (Race 2 is virulent on tomato lines carrying only *Cf-2*).

Note: The genes *Cf-2, Cf-4, Cf-5* and *Cf-9* have been exploited in tomato breeding, but these resistances are ineffective as *Cladosporium* races have occurred that lack *Avr2, Avr4, Avr5* and/or *Avr9*.

Fig. 2.9. Growth of tomato lines at lower temperature and light conditions. The cultivar Premier (left) does not produce fruit and grows vigorously. In contrast, an early breeding line (right) has been selected that sets fruit but grows less vigorously.

Most genotypes are adversely affected at temperatures below 13°C. Prolonged exposure to temperatures below 6°C can even result in plant death. There is a huge variation for low-temperature tolerance in the genus *Lycopersicon* and this variation has been extensively investigated. High-altitude accessions of *L. hirsutum* and *L. peruvianum* that grow at altitudes of over 3000 m show a better adaptation to growth under low-temperature conditions than the cultivated tomato (Venema *et al.*, 2000).

Low-temperature tolerance can be specified into several characteristics, e.g. seed germination, growth rate and fruit set. The last characteristic is mainly determined by poor anther dehiscence and pollination. Cold tolerance is developmentally regulated and there is no clear correlation in cold tolerance at different developmental stages. Although many sources of cold tolerance have been reported and even used in breeding programmes, not much is known about the genetic factors underlying this trait, except that the inheritance is complex. Fruit set seems to be controlled by recessive factors. Also, genetic loci have been identified that affect seed germination at low temperatures. It is expected that more traits involved in cold tolerance will be determined and mapped using molecular marker techniques. For greenhouse cultivation, the best approach is probably to breed for high yields under non-stress conditions. It is likely that higher yields will then also be obtained under suboptimal conditions. This will result in a more efficient usage of energy for heating glasshouses.

HEAT RESISTANCE In most parts of the world, high temperatures during growth of tomato plants in summer can affect production negatively. One of the most sensitive stages is fruit set. Most cultivars lose their blossom after 4 h at 40°C or a day/night temperature regime of 34°C/20°C, though this may vary between genotypes and depends on humidity and soil moisture. Factors involved in heat resistance have been well characterized and heat-resistant germplasm is available. Factors that are involved in heat resistance are photosynthesis, translocation, flower production, flower morphology (stigma exsertion), anther dehiscence, pollen viability and pollen germination. Recessive genes that control a greater flower number have been identified as well as an additive gene for fruit set and partially dominant genes for stigma exsertion with additive components. Considering the large amount of physiological characteristics involved in heat resistance, each controlled by one or several genes, MAS might contribute to the introgression of all these genes into commercially interesting cultivars.

DROUGHT RESISTANCE Drought resistance is not considered to be one of the major issues in breeding. Nevertheless, research has been performed in order to extend the tomato-growing areas. One approach is to breed for drought escape, by breeding for very early or very late varieties (Fig. 2.9). Generally, it is considered that cultivars with a very short life cycle are a good option, so that the drought periods can be avoided. Variation has been observed for this trait between different tomato cultivars and some accessions of the wild relatives *L. pennellii*, *L. cheesmanii* and *L. pimpinellifolium*.

RESISTANCE TO FLOODING Flooding causes anaerobiosis, which in its turn stimulates the production of ethylene, which causes epinasty. Sensitivity to flooding was found to be higher at higher soil temperatures. One indication of sensitivity to flooding is the accumulation of proline. Tolerance to flooding has been found in some tomato cultivars.

SALT RESISTANCE Irrigation may increase the amount of salts present in the soil. Tomato plants are moderately sensitive to salinity and alkalinity. At high levels of salt, a rapid decrease of yield occurs, due to decrease in fruit size (Chapter 4). Several accessions of *L. cheesmanii*, *L. peruvianum* and *L. pennellii* have been found to be salt tolerant. Several developmental stages have been tested, and seed germination was found to be the most sensitive stage. In most cases, salt-tolerant genotypes absorb more Na^{+}.

PHYSIOLOGICAL DISORDERS The tomato fruit is subject to several physiological disorders, such as blossom-end rot, catface, fruit cracking and blotchy ripening (Chapter 5), which affect fruit quality negatively. The inheritance of different forms of cracking is well understood and modern cultivars can be stored for some weeks while maintaining a good fruit quality (high shelf-life, see below). Also, adjustment of nutrition may prevent these physiological disorders (Chapter 6).

Breeding for specific production systems

Since different production systems demand specific characters of traits, breeding has to meet these different characters. The main distinction that can be made in production systems is based upon the difference between processing and fresh-market tomatoes. The processing tomatoes for ketchup, paste, etc. are mechanically harvested all at once, whereas fresh-market tomatoes are picked manually during the whole season.

Several characters are necessary for machine harvesting of tomatoes. The growth has to be determinate and all tomatoes have to be ripe at the same time, i.e. at the time of harvest. Moreover, tomato fruits should break easily from the stem, in order to prevent damage by the stem during transport and storage. As shown above, several fruit quality traits are important for processing tomatoes, such as colour and the content of soluble solids (°Brix × yield).

Most fresh-market tomatoes are indeterminate (Chapter 3). In greenhouses in northern countries, tomatoes are grown year-round and the plants produce marketable fruits from March until November. In recent decades the fruit type has shifted from a uniform size and shape to a wide spectrum of different types, such as cherry, thrush, yellow, beef, pink, orange, pear-shaped, long shelf-life tomatoes and so on. This has put a great constraint on the breeders who breed cultivars for all these markets. However, large variations for these traits are present in the *Lycopersicon* germplasm and so breeding for alternative types remains feasible.

Breeding for quality

There is a growing consumer demand for high quality. Quality is a combination of visual stimuli (e.g. size, shape, colour) and sensory factors (e.g. sugar, acidity, taste) (Fig. 2.10). Consumer perception of quality is also heavily influenced by product appearance and descriptions, such as eco-, sun-ripe, biological or transgenic. To meet consumer demands, large supermarkets put more emphasis on the control of the whole production chain, including cultivar choice, production conditions and handling. Tomatoes are now presented with a brand name that should have a continuous quality, but may comprise different cultivars.

FLAVOUR Flavour comprises mainly sugars, acids and volatile compounds, the latter being of particular importance for fresh fruit (Thakur *et al.*, 1996). Some predictions about flavour can be made by measuring the acidity and

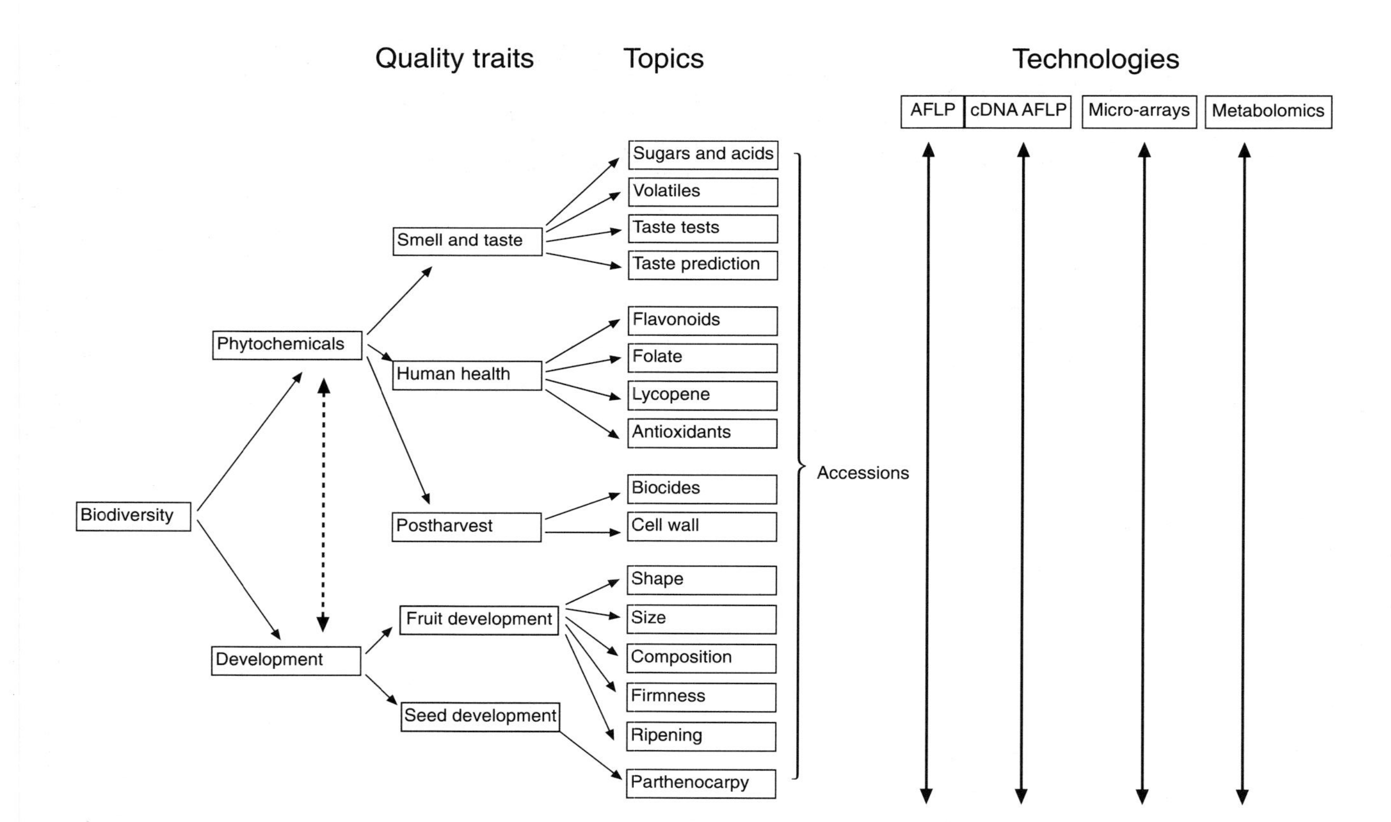

Fig. 2.10. Relationships between various traits, components and technologies applicable in tomato fruit quality genomics (from Arnaud Bovy, Plant Research International, Wageningen, The Netherlands).

refraction, which is equivalent to the soluble solids content. However, taste panels are necessary for a final evaluation of the taste of potential new cultivars. The taste of a tomato is not controlled by a single gene; many genetic loci may play a role and each locus has a partial or quantitative contribution to the trait. These are QTLs and many loci that are associated with quality traits have been identified from *L. pimpinellifolium*, *L. pennellii*, *L. peruvianum* and *L. hirsutum* (Causse *et al.*, 2003; Lecomte *et al.*, 2004; Yates *et al.*, 2004). QTLs for soluble solids contents (°Brix), firmness, stem retention (the ease with which the tomato breaks from the stem at harvesting), pH, viscosity, puffiness (air in locules), colour and fruit shape have been mapped on the tomato genome. Most wild species harbour some QTLs that have positive alleles and others that have negative alleles for a quality trait. For example, QTLs have been observed from the green-fruited *L. hirsutum* that gives a less intense colour, and another QTL has been observed that is involved in more intense red coloration. This has also been observed for traits involved in yield. Wild species with smaller fruits might carry QTLs that give rise to increased fruit size when crossed to tomato, as has been observed for *L. pimpinellifolium*, *L. peruvianum* and *L. pennellii*. QTLs that increase soluble solid contents (°Brix) have been found in all four of the above wild species, as well as QTLs increasing total yield. However, QTLs increasing the total amount of soluble solids, represented by °Brix × yield, are less frequent and have been observed only in *L. pennellii*, *L. pimpinellifolium* and *L. hirsutum*.

The traits that partially determine flavour are °Brix and pH. Breeding and selection have been used to improve these traits. In addition, volatiles are also involved in flavour. Breeding for volatiles has not yet been performed, as much less is known about the relation between flavour and aroma and these volatiles. Moreover, expensive measurements are necessary to determine the amount of these compounds in plants, hampering the screening of material and populations. In ripening tomato fruit, approximately 400 different volatile compounds have been observed. Only a small number of these compounds have been identified as important components of flavour and aroma. These compounds are mainly aldehydes and alcohols. The enzyme alcoholdehydrogenase (Adh) converts aldehydes into alcohols. In the tomato, two Adh enzymes have been identified, of which only Adh2 has been observed in ripening fruit. Transgenic tomatoes have been obtained with elevated levels of Adh2 expression, resulting in increased levels of hexanol and Z-3-hexenol. Plants with higher Adh2 expression were found to have a more intense ripe fruit flavour.

COLOUR The red colour of a tomato is determined by the colour of its skin and of its flesh. Skin colour varies from yellow to colourless, while flesh colour varies between green and red. Flesh colour depends on the amount and type of carotenoid pigments, which are C_{40} isoprenoid derivatives. There are two types of carotenoids in tomato: carotenes and xanthophylls. There is not much variation in xanthophyll levels between green and red fruits, whereas the level of carotenes increases tremendously (Liu *et al.*, 2003). One of the most obvious differences between green and red fruits is the level of lycopene. During ripening, there is a 500-fold increase in the level of lycopene in the tomato fruit. Lycopene comprises 90–95% of the total pigmentation in tomatoes. There has been more interest in lycopene since it was found that this antioxidant is associated with resistance to certain forms of human cancer, in particular cancer of the prostate. The antioxidant activity of lycopene also protects degradation of ß-carotene, which is a source of vitamin A. QTLs for elevated levels of lycopene have been identified in a cross between tomato and *L. pimpinellifolium*.

Several mutants have been identified that are involved in fruit colour. Some of these are affected not only in carotene biosynthesis but also in other fruit-ripening processes. The og^c mutant has elevated levels of carotene, which is at the expense of the level of ß-carotene. Two *hp* mutants are recessive mutations in two different genes that show elevated levels of lycopene as well as ß-carotene and which accompany a general elevated level of chlorophyll. The *hp* mutants also showed higher levels of flavonoids and sucrose, the latter at the expense of the levels of glucose and fructose (Van Tuinen *et al.*, 1997).

RIPENING AND SHELF-LIFE Tomato ripening is a well-studied process (Giovannoni *et al.*, 1999). It is of interest to tomato breeders, since ripening affects several quality traits, some of which have been mentioned above (e.g colour, flavour and soluble solids content).

Another factor that is especially important for fresh-market tomatoes is the shelf-life. During ripening, several processes occur that affect the storage of the fruit negatively. Some genes involved in ripening, such as polygalactoronidase and ethylene synthase, have been cloned and transgenic tomatoes overexpressing these genes have an extended shelf-life.

In the period after the Second World War the focus was mostly on increasing the production of food, including tomatoes. In the 1980s the European tomato market had become saturated. This coincided with increases in consumers' standards of living to a level at which they could afford to be more critical of the quality of the crop. Consumers realized in particular that greenhouse tomato production on non-soil substrates and with a well-controlled environment was highly artificial, and they started to complain about quality. Growers and retailers began to investigate ways of improving quality but this did not lead to practical clues. Variation in samples of the same cultivar were not related with the geographical area, nor with the method of cultivation. It was already known that tomatoes were better if picked at riper stages but this influenced the shelf-life to a great extent. In Israel, Professor Kedar undertook a long research programme to improve the shelf-life of tomatoes by using the mutant 'non-ripening' (*nor*) and 'ripening inhibitor' (*rin*) genes. It took more than 25 years to release a variety, designated 'Daniella', that ripened much more slowly than traditional tomatoes and had a harder skin and texture. The shelf-life was about 2 months, compared with a shelf-life of about 14 days in other tomatoes. This had a great impact on tomato production in Europe. Growers in the subtropical European region (e.g. Spain, Morocco, Tunisia and Italy) turned to 'Daniella', which was shipped to the more northern countries. Growers in these areas suddenly had new competitors who were producing large quantities of tomatoes, of good quality (shelf-life) and at low prices. Noticeably, the taste of 'Daniella' was worse than that of the existing cultivars, but the consumers did not complain: their concern about the other tomatoes was more the way of production than the quality itself. Growers in the more northern countries, such as Belgium, the UK and The Netherlands, gradually developed a new system that was meant to demonstrate to consumers that more attention was paid to the product: they were moving from a bulk product to a high quality product.

Retailers are now gradually controlling the complete production chain, including the choice of cultivars, production and transport systems and selling to the consumers. Tomatoes are merchandized as brands, like 'Gartenaroma' or 'Tasty Tom', and this has been very profitable for growers. This is a good example of how breeders can only marginally improve the crop but that a combined action of all partners in the whole production chain can be economically successful.

PARTHENOCARPY Normally, tomato plants produce seeded fruits upon fertilization. This happens under normal weather conditions, but in extreme conditions (cold, heat or lack of light) no fruit set occurs, due to lack of viable pollen, which prevents pollination and fertilization. This occurs regularly in many cultivation areas. Hence breeders aim at the development of commercially viable parthenocarpic tomato lines for these regions.

Three main sources of facultative parthenocarpy, *pat*, *pat-2* and *pat-3*/*pat-4*, are available for tomato breeding. The development of parthenocarpic fruit in these lines seems to be triggered by a deregulation of the hormonal balance in some specific tissues. Auxins and gibberellins are considered as the key elements in the development of parthenocarpic fruit in these lines, mainly by an increased level of these hormones (in the ovary) that may substitute for the pollen in triggering fruit development. The *pat* genes often show pleiotropic effects such as reduced growth and yield or aberrant fruit size or reduced fruit set. This has hampered the release of commercial cultivars.

Recently, obligate parthenocarpic tomatoes have been obtained by expressing the *rolB* gene from *Agrobacterium rhizogenes*, which enhances auxin sensitivity. This gene was expressed using the TPRP-F1 promoter, which is highly specific for ovaries, young fruits and developing embryos. Though these genetically modified tomatoes have given promising results in terms of quality and quantity of seedless fruit production, consumer concerns still inhibit the release of transgenic cultivars (Gorguet *et al.*, 2005).

Seed production

The tomato is a predominantly inbreeding species. In the field some outbreeding takes place, partly aided by the wind but also by larger insects, such as bees and butterflies. In greenhouses without insects, no pollination occurs. Pollination is stimulated by mechanical shaking of the plants or by bees that are deliberately released into the greenhouse.

Inbreeding results in a decrease of heterozygocity and an increase of homozygocity. After several generations of inbreeding, most if not all loci are homozygous and all seeds of the next generation are genetically identical and also identical to the previous generation.

Until 1946 all tomato cultivars were true-breeding lines that were essentially homozygous and produced progeny that were identical to the parents. Cultivars could easily be maintained and propagated by the growers by collecting seeds from their own crop.

Tomato seed production is simple and efficient. Large quantities of seeds can easily be obtained from a single plant. Each fruit can carry more than 100 seeds, each cluster about ten fruits and each plant five to 25 clusters. Starting with ten plants at the beginning of the season, it is easy to obtain thousands of seeds per plant.

Just as in the production of tomato fruits, seed production is influenced by environmental factors such as light intensity, soil conditions and nutrients. Stress conditions may have an adverse effect on the germination of the seed and on seedling vigour. Harvesting and extracting seeds before full maturity is detrimental to seed viability and further germination.

Fruits can be harvested by hand or mechanically. Seed extraction from the fruits is usually completely mechanized. The extraction results in a mixture of gelatinous seed, which is separated into seeds and debris. After extraction, seeds can be cleaned either by fermentation or by application of 0.7% HCl. Seeds are then filtered, washed with water and air dried for 2–4 days at temperatures of 20–30°C.

To control the seed-borne TMV, seeds can be treated with sodium carbonate. Drying should be uniform and slow, because fast drying shrinks the seed coat around the embryo and reduces seed quality. Tomato seeds can easily be stored for some years at room temperature and for decades at –20°C and at low humidity (< 20%).

Hybrid seed production

Heterosis is a biological phenomenon manifesting itself in hybrids that are more vital, adaptive and productive than their parents (Fig. 2.11). In the

Fig. 2.11. Heterosis in the genus *Lycopersicon*: a hybrid (middle) from an interspecific cross between cultivated tomato (left) and the wild species *L. parviflorum* (right).

generations after F_1, heterosis effects decrease progressively. Heterosis has been explained by overdominance and by additive effects. However, it is still not clear how much each of these effects contributes to the total heterosis. Breeders soon became interested in F_1 hybrids, not only for heterosis effects and for their uniformity but also for the protection of their cultivars: it is not attractive for growers to propagate F_1 hybrids, since the resulting F_2 will show a segregation of all characters that differ between the parents of the F_1 hybrid; therefore, only the breeder can provide the seeds.

In order to obtain superior F_1 hybrids it is necessary to combine the right parental lines, and so parental lines have to be tested for their combining ability (CA). For this purpose, parental lines have to be inspected for as many characteristics as possible, especially the principles of their inheritance in the F_1. These characteristics can be introduced into a database as a passport of such a line. With the aid of the computer, parental lines can be selected for F_1 hybrid production. Then the F_1 hybrids have to be tested for their performance compared with standard cultivars. Results of these tests can again be attributed to the passport data of the parental lines.

A disadvantage of F_1 hybrids is the higher costs for seed production. Since tomato is a self-fertilizing crop, one of the parental lines has to be emasculated and fertilized with pollen from the other parent. This procedure requires a lot of labour, which is why extensive research has been carried out to obtain self-incompatible or male sterile parents and better cross-pollination techniques. Although several genes have been identified that cause sterility, these could not be used in breeding practice, due to problems of sterility in cross-pollinations as well, or due to the existence of some self-pollination. Recent progress has been made using cytoplasmic male sterility (CMS), which has been observed in a backcross (BC_{10}) of *L. peruvianum* with *L. pennellii*. This CMS-*pennellii* has been backcrossed with an F_1 between *L. esculentum* and *L. pennellii* in order to overcome the incompatibility between *L. esculentum* and *L. pennellii*. In this way, the *L. esculentum* genome can be introduced into a CMS background and this might provide a functional CMS system that can be used in hybrid seed production.

Biotechnology and transgenic or genetically modified organisms

Technologies were developed to regenerate plants from tissue or single cells; transformation technologies enabled the introduction of genes of interest into any cultivar; and, with cell fusion techniques, borders between incongruent species could be broken down. The potentials for breeding were evident. With biotechnology, the exploitable genetic variation for tomato breeding increased to almost unlimited levels. In fact, tomato was the first food crop for which transgenic fruits were commercially available. The first transgenic cultivar 'Flavr-Savr'™ was released in 1994 but it was not a commercial success. This failure was probably due to a combination of the other characteristics of the lines (a classic example that one gene or one trait does not make a successful variety) and also commercial problems in arranging for production and marketing of the product. At that time, there were no widespread social concerns regarding GM produce in the USA, so that was not a major factor in the failure.

Since the tomato is a relatively simple system for transformation by using *Agrobacterium tumefaciens*, tomatoes harbouring other transgenes have also been made, but not yet commercially exploited. These traits include other biosynthesis genes affecting ripening or fruit quality, and constructs for herbicide, virus and insect resistance that were hitherto difficult or impossible to breed using a more traditional approach (Schuch *et al.*, 1991).

As stated above, breeding companies have invested in biotechnology but they have not released numerous transgenic tomato cultivars. The reasons include: the complicated and expensive patent filing that is associated with transgenic crops; consumer concerns; the wealth of the available genetic variation to accomplish the same goals without dealing with the patent; and social issues associated with transgenic crops. In the late 1980s large chemical companies invested considerably in biotechnology, as they expected that this would be a booming business. Some two decades later the use of transgenic tomato crops remained small and that of other crops started to decline in 2000. Profits remained below expectations and consequently chemical companies tended to lose interest in the seed industry.

As regulations in the USA tend to become less strict, it is expected that in the coming decade numerous new cultivars with new traits will be released to meet the demands of growers and consumers. However, if international markets restrict the use of transgenic crops or their products, utilization of transgenic plants in the USA may be restricted by marketing concerns. It is expected that, since the recent debate in Europe, the introduction of transgenic cultivars will be prevented for the next decade or more. The question of acceptance of transgenic plants is a social issue currently under debate. Decisions regarding the use of transgenic plants will be made in the social, political and business arenas, rather than by plant breeders.

EPILOGUE

This chapter has summarized the achievements of tomato breeders in the past century and the developments in fundamental and applied tomato research. It has been shown that a large variation is present and exploitable from wild *Lycopersicon* spp. but the vast majority remains hidden in the tomato germplasm. To explore and exploit this variation, introgression lines that carry small introgressed chromosome fragments from related wild species in a cultivated tomato background are most useful. It is remarkable that until now any wild accession under study has been shown to harbour at least several new genes of agronomic interest. It is virtually impossible to exploit all individual wild accessions by generating genetic libraries of introgression lines. Considering the thousands of *Lycopersicon* accessions in genebanks and probably even more that are still growing untouched in the Andes, such an approach is too laborious.

This is now the eve of the genomics era (Fig. 2.10). Thousands of sequences have already become available. It is expected that in the next decade the complete tomato genome sequence, at least for the gene-dense regions, will be established. The next task will be to identify the function of all these sequences. Thanks to the synteny among solanaceous crops and even the microsynteny with *Arabidopsis*, it may be expected that knowledge about the function of tomato genes will gradually increase. This will allow searching for allelic variation of these genes in the *Lycopersicon* germplasm ('allele mining'). Techniques such as eco-tilling will greatly facilitate the identification of useful genes in the wild germplasm (Comai *et al.*, 2004).

In the end the breeder has genetic stock in the seed store and a large database about the seeds. The breeder's capital will shift from the field to the computer. The best combinations of genotypes will be selected and a programme will be designed to combine the traits in a cultivar. This is called breeding by design, or virtual breeding.

ACKNOWLEDGEMENTS

Jair Haanstra and Martha Mutschler are acknowledged for their critical reading of this chapter and for their useful corrections and additions.

REFERENCES

Ali, S.N.H., Ramanna, M.S., Jacobsen, E. and Visser, R.G.F. (2001) Establishment of a complete series of a monosomic tomato chromosome addition lines in the cultivated potato using RFLP and GISH analyses. *Theoretical and Applied Genetics* 103, 687–695.

Budiman, M.A., Chang, S.B., Lee, S., Yang, T.J., Zhang, H.-B., De Jong, H. and Wing, R.A. (2004) Localization of *jointless-2* gene in the centromeric region of tomato chromosome 12 based on high resolution genetic and physical mapping. *Theoretical and Applied Genetics* 108, 190–196.

Causse, M., Buret, M., Robini, K. and Verschave, P. (2003) Inheritance of nutritional and sensory quality traits in fresh market tomato and relation to consumer preferences. *Journal of Food Science* 68, 2342–2350.

Comai, L., Young, K., Till, B.J., Reynolds, S.H., Greene, E.A., Codomo, C.A., Enns, L.C., Johnson, J.E., Burtner, C., Odden, A.R. and Henikoff, S. (2004) Efficient discovery of DNA polymorphisms in natural populations by Ecotilling. *Plant Journal* 37, 778–786.

Dangl, J.L. and Jones, J.D.G (2001) Plant pathogens and integrated defence responses to infection. *Nature* 411, 826–833.

De Kock, M.J.D., Brandwagt, B.F., Bonnema, G., De Wit, P.J.G.M. and Lindhout, P. (2005) The tomato Orion *Cf-Ecp2* resistance locus comprises three homologous genes encoding leucine-rich repeat proteins and a large duplication. *Plant Molecular Biology* (in press).

Diwan, N., Fluhr, R., Eshed, Y., Zamir, D. and Tanksley, S.D. (1999) Mapping of *Ve* in tomato: a gene conferring resistance to the broad-spectrum pathogen, *Verticillium dahliae* race 1. *Theoretical and Applied Genetics* 2, 315–319.

Eshed, Y. and Zamir, D. (1994) A genomic library of *Lycopersicon pennellii* in *L. esculentum*: a tool for fine mapping of genes. *Euphytica* 79, 175–179.

Giovannoni, J., Yen, H., Shelton, B., Miller, S., Vrebalov, J., Kannan, P., Tieman, D., Hackett, R., Grierson, D. and Klee, H. (1999) Genetic mapping of ripening and ethylene-related loci in tomato. *Theoretical and Applied Genetics* 98, 1005–1013.

Gorguet, B., Van Heusden, S. and Lindhout, P. (2005) Parthenocarpic fruit development in tomato. *Plant Biology* (in press).

Haanstra, J.P.W., Wye, C., Verbakel, H., Meijer-Dekens, F., Van den Berg, P., Odinot, P., Van Heusden, A.W., Tanksley, S., Lindhout, P. and Peleman, J. (1999) An integrated high density RFLP-AFLP map of tomato based on two *Lycopersicon esculentum* × *L. pennellii* F_2 populations. *Theoretical and Applied Genetics* 99, 254–271.

Haanstra, J.P.W., Meijer-Dekens, F., Laugé, R., Seetanah, D.C., Joosten, M.H.A.J., De Wit, P.J.G.M. and Lindhout, P. (2000) Mapping strategy for resistance genes against *Cladosporium fulvum* on the short arm of Chromosome 1 of tomato: *Cf-ECP5* near the *Hcr9* Milky Way cluster. *Theoretical and Applied Genetics* 101, 661–668.

Huang, C.C., Cui, Y.Y., Weng, C.R., Zabel, P. and Lindhout, P. (2000) Development of diagnostic PCR markers closely linked to the tomato powdery mildew resistance gene *Ol-1* on chromosome 6 of tomato. *Theoretical and Applied Genetics* 101, 918–924.

Ji, Y.F., Pertuze, R. and Chetelat, R.T. (2004) Genome differentiation by GISH in interspecific and intergeneric hybrids of tomato and related nightshades. *Chromosome Research* 12, 107–116.

Lecomte, L., Saliba-Colombani, V., Gautier, A., Gomez-Jimenez, M.C., Duffe, P., Buret, M. and Causse, M. (2004) Fine mapping of QTLs of chromosome 2 affecting the fruit architecture and composition of tomato. *Molecular Breeding* 13, 1–14.

Lindhout, P. (2002) The perspectives of polygenic resistance in breeding for durable disease resistance. *Euphytica* 124, 217–226.

Lindhout, P., Korta, W., Cislik, M., Vos, I. and Gerlach, T. (1989) Further identification of races of *Cladosporium fulvum* (*Fulvia fulva*) on tomato originating from the Netherlands, France and Poland. *Netherlands Journal of Plant Pathology* 95, 143–148.

Liu, Y.S., Gur, A., Ronen, G., Causse, M., Damidaux, R., Buret, M., Hirschberg, J. and Zamir, D. (2003) There is more to tomato fruit colour than candidate carotenoid genes. *Plant Biotechnology Journal* 1, 195–207.

Miller, J.C. and Tanksley, S.D. (1990) RFLP analysis of phylogenetic relationships and genetic variation in the genus *Lycopersicon*. *Theoretical and Applied Genetics* 80, 437–448.

Monforte, A.J. and Tanksley, S.D. (2000) Development of a set of near isogenic and backcross lines containing most of the *Lycopersicon hirsutum* genome in a *L. esculentum* genetic background: a tool for gene mapping and discovery. *Genome* 43, 803–813.

Pedley, K.F. and Martin, G.B. (2003) Molecular basis of Pto-mediated resistance to bacterial speck disease in tomato. *Annual Review of Phytopathology* 41, 215–243.

Rick, C.M. (1991) Tomato resources of South America reveal many genetic treasures. *Diversity* 7, 54–56.

Rodriguez, A.E., Tingey, W.M. and Mutschler, M.A. (1993) Acylsugars of *Lycopersicon pennellii* deter settling and feeding of the green peach aphid (*Homoptera, Aphididae*). *Journal of Economics and Entomology* 86, 34–39.

Schuch, W., Kanczler, J., Robertson, D., Hobson, G., Tucker, G., Grierson, D., Bright, S. and Bird, C. (1991) Fruit quality characteristics of transgenic tomato fruit with altered polygalacturonase activity. *Hort Science* 26, 1517–1520.

Tanksley, S.D. and Mutschler, M.A. (1990) Linkage map of the tomato (*Lycopersicon esculentum*) ($2n$ = 24). In: O'Brien, S.J. (ed.) *Genetic Maps*, 5th edn. Cold Spring Harbor Laboratory Press, Cold Spring Harbor, New York, pp. 6.3–6.15.

Tanksley, S.D., Ganal, M.W., Prince, J.P., De Vicente, M.C., Bonierbale, M.W., Broun, P., Fulton, T.M., Giovanni, J.J., Grandillo, S., Martin, G.B., Messeguer, R., Miller, J.C.,

Miller, L., Paterson, A.H., Pineda, O., Roder, M.S., Wing, R.A., Wu, W. and Young, N.D. (1992) High density molecular linkage maps of the tomato and potato genomes. *Genetics* 1324, 1141–1160.

Thakur, B.R., Singh, R.K. and Nelson, P.E. (1996) Quality attributes of processed tomato products: a review. *Food Review International* 12, 375–401.

Van Berloo, R. and Lindhout, P. (2001) Mapping disease resistance genes in tomato. In: Zhu, D., Hawtin, G. and Wang, Y. (eds) *International Symposium on Biotechnology Application in Horticultural Crops.* China Agricultural Scientech Press, Beijing, PR China, 12, pp. 343–356.

Van der Beek, J.G., Verkerk, R., Zabel, P. and Lindhout, P. (1992) Mapping strategy for resistance genes in tomato based on RFLPs between cultivars: *Cf9* (resistance to *Cladosporium fulvum*) on chromosome 1. *Theoretical and Applied Genetics* 84, 106–112.

Van der Hoeven, R., Ronning, C., Giovannoni, J., Martin, G. and Tanksley, S. (2002) Deductions about the number, organisation, and evolution of genes in the tomato genome based on analysis of a large expressed sequence tag collection and selective genomic sequencing. *Plant Cell* 14, 1441–1456.

Van Heusden, A.W., Koornneef, M., Voorrips, R.E., Brüggeman, W., Pet, G., Vrielink-van Ginkel, R., Chen, X. and Lindhout, P. (1999) Three QTLs from *Lycopersicum peruvianum* confer a high level of resistance to *Clavibacter michiganensis* ssp. *michiganensis*. *Theoretical and Applied Genetics* 99, 1068–1074.

Van Tuinen, A., Cordonnier-Pratt, M.M., Pratt, L.H., Verkerk, R., Koornneef, M. and Zabel, P. (1997) The mapping of phytochrome genes and photomorphogenic mutants of tomato. *Theoretical and Applied Genetics* 94, 115–122.

Venema, J.H., Villerius, L. and Van Hasselt, P.R. (2000) Effect of acclimation to suboptimal temperature on chilling-induced photodamage: comparison between a domestic and a high-altitude wild *Lycopersicon* species. *Plant Science* 152, 153–163.

Yates, H.E., Frary, A., Doganlar, S., Frampton, A., Eannetta, N.T., Uhlig, J. and Tanksley, S.D. (2004) Comparative fine mapping of fruit quality QTLs on chromosome 4 introgressions derived from two wild tomato species. *Euphytica* 135, 283–296.

3

Developmental Processes

E. Heuvelink

INTRODUCTION

The interpretation of crop responses to environmental factors is facilitated by making a distinction between growth and development. Growth is defined here as the increase in biomass and/or dimensions of a plant or organ (quantitative aspects). Development can be defined as 'ordered change or progress often (but not always) towards a higher, more ordered, or more complex state' (Bidwell, 1974). Examples are phase transitions (e.g. juvenile to adult) and formation and development of new organs. Usually, growth and development proceed together, but there is no strict correlation. For example, two plants that have both initiated the same number of leaves, which is a measure for developmental stage, may differ strongly in mass or leaf area. Also, a tomato fruit of 50 g may be ripe (fully grown), but could also be only halfway through its development from anthesis to ripeness. During development, the plant as a whole, as well as the individual organs, is subject to irreversible changes over time. Often it is possible to identify markers in the sequence of events related to development. Examples are germination, the formation of leaves, anthesis, fruit set and fruit ripening. When markers are defined, it is possible to quantify development in terms of a stage that is attained. A numerical scheme that provides a detailed description of development is the BBCH scale (Meier, 1997).

Developmental processes such as inflorescence formation, flower development, fruit set and fruit ripening all have a strong impact on tomato yield, which is determined by both the number and mass of individual fruits. These processes, together with germination and leaf, stem and root development, are the subjects of this chapter.

GERMINATION

Germination is the transition period between the resting seed and the time of visible radicle emergence (Bewley and Black, 1994). Quality of seed lots is usually determined by a standard germination test conducted under ideal environmental conditions in a laboratory. The criterion for germination used by physiologists is radicle emergence; however, a seed analyst extends this interpretation to classification of seedlings as either normal or abnormal.

Tomato seed germination has been reviewed in Chapter 3 of Atherton and Rudich (1986) and their review is the main source for the present text. Both total germination percentage and uniformity in germination are important characteristics of a seed lot. High germination percentages do not necessarily mean high emergence in practice, especially under suboptimal, fluctuating conditions in the field. For direct-sown tomato crops in the field, seed lots with a high vigour, i.e. 'those properties which determine the potential for rapid, uniform emergence and development of normal seedlings under a wide range of field conditions' (AOSA, 1983), are essential. Seed vigour differs from germination in that vigour emphasizes the germination rate (rapid and uniform) and the application of the results to forecast field emergence rather than laboratory performance. For greenhouse production, seeds are often germinated in controlled-environment chambers, which guarantee optimal conditions. When seeds are sown into multi-plates or peat pots, a germination percentage of close to 100% normal seedlings is required, as opposed to sowing in seed trays and transplanting manually, which allows for some selection.

The seed of a commercial tomato cultivar is a flattened ovoid, up to 5 mm long, 4 mm wide and 2 mm deep, consisting principally of the embryo, endosperm and testa or seed coat. The testa is covered with large soft hairs, which tend to bind with other seeds. Tomato seeds store easily and will retain their viability for long periods between 5°C and 25°C and a fairly wide range of relative humidities, so that no particular special storage measures need be taken. James *et al.* (1964) reported figures of 90% and 59% germination after 15 and 30 years of storage, respectively.

Germination performance of the seed is influenced by its characteristics, e.g. seed size (more rapid germination is associated with smaller seeds, which is thought to be due to a reduced endosperm thickness). Harvesting and extraction of tomato seed before full maturity is detrimental to seed viability and subsequent germination (Kerr, 1963). In the cultivated tomato, the seeds are considered non-dormant, whereas some studies with wild types suggest that abscisic acid (ABA) plays a role in a slight dormancy. Germination is at least partly under genetic control, as cultivars differ in their ability to germinate at either low or high temperatures. Furthermore, inclusion of *Lycopersicon pimpinellifolium* as parental stock in combination with *Lycopersicon esculentum* has resulted in progeny germinating earlier than the *L. esculentum* parents.

Environmental factors strongly influence seed germination. Tomato seeds characteristically germinate best in the dark, and light will inhibit germination in some cultivars. These responses are dependent on the action of phytochrome (P_{fr}). Far-red light has been reported to inhibit germination, whereas red light (at 37°C) promoted germination, suggesting that the presence of P_{fr} in the seed is a prerequisite for germination. The ability of some light-sensitive cultivars to germinate in the dark has been attributed to the high levels of P_{fr} present in the dehydrated seed. Effects of water, salinity and temperature on tomato seed germination will be discussed below. Germination rate can be improved by presowing seed treatments with growth regulators. Stimulatory effects have been demonstrated for gibberellins and various synthetic auxins.

Water and salinity

Without water, no germination will occur. The absorption of water by seeds is characterized by three phases. In the first phase (imbibition), rapid uptake of water will occur, whether the seed is viable or not. In the following phase, hydration of the cotyledons may take place, but the moisture content of the seed remains apparently constant. A resurgence in water uptake marks the final phase (growth), which is associated with radicle protrusion and subsequent growth.

The rate and degree of water uptake for germination are affected by temperature, water content and salinity of the environment. In general, tomato seeds will germinate at soil water potentials ranging from just above the wilting point to field capacity, though optimum germination is usually obtained at 50–75% of field capacity. Reduced germination at high soil-water content is probably the result of reduced oxygen availability to the seeds, since tomatoes show dramatic reductions in germination under conditions of low oxygen availability.

Water stress is often associated with increased soil salinity, which may affect germination in two ways: (i) by creating an osmotic potential that impedes water uptake; and (ii) by the entry of deleterious ions. High salt levels have been shown to decrease tomato seed germination, though tomatoes seem to be more tolerant of saline conditions during germination than during subsequent growth.

Temperature

Effects of temperature on germination can be described according to the well-known concept of heat units (HU):

$$DVS_t = \sum_{i=1}^{t} (T_i - T_{min}) / HU \text{ for } T_i \geq T_{min}, \text{ else } T_i - T_{min} = 0 \quad (3.1)$$

where DVS_t is the developmental stage after t days, T_i is average temperature at day i, T_{min} is minimum temperature (where development stops) and HU is heat units required to attain a given developmental stage, expressed in degree-days. DVS_t as defined here provides a way to interpolate between the start (DVS_0 = 0) and end (DVS_f = 1) of the developmental process under consideration, in this case germination. At constant temperature T, equation 3.1 may be rewritten as:

$$t = HU/(T - T_{min}) \tag{3.2}$$

where *t* is the time to obtain e.g. 50% germination, HU is heat units required to obtain e.g. 50% germination and T_{min} is the minimum temperature for germination. Note that equations 3.1 and 3.2 implicitly assume a linear relationship between development rate and temperature. This seems unlikely for germination of tomato seeds, since optimum temperatures between 20°C and 25°C have been reported. However, over a range of temperatures (13–25°C) this linearity has been observed by Bierhuizen and Wagenvoort (1974), who reported a T_{min} of 8.7°C for tomato germination and a heat sum requirement of 88 degree-days to achieve 50% germination. In a further paper (Wagenvoort and Bierhuizen, 1977), they considered that 13°C was the 'practical' minimum temperature for germination. Membrane lipid changes have been suggested to be involved in cultivar differences in ability to germinate at low temperatures, whereas failure to germinate at high temperatures may be due to thermodormancy, a condition thought to be related to an interaction between temperature and phytochrome action.

LEAF APPEARANCE AND LEAF GROWTH

Leaf appearance rate, or leaf unfolding rate (LUR), is not necessarily equal to the rate of initiation of leaf primordia at the apex but, in the long term, equality is expected. Leaf appearance rate in tomato, as in many other crops, is predominantly influenced by temperature. This is clearly shown in the work of De Koning (1994), who studied the influence of temperature and other growing factors on tomato truss appearance rate. In indeterminate types, since three leaves are initiated between two successive trusses, LUR rate is simply three times the truss appearance rate. Although LUR shows an optimum response to increasing temperature, a linear relationship is observed over a shorter temperature (T) range. From De Koning's (1994) dissertation the following equation for cv. Counter could be deduced:

$$\text{LUR (leaves per day)} = -0.871 + 0.436 \ln(T) \tag{3.3}$$

In the range 17–23°C this relationship is close to linear and at 20°C three leaves are unfolded each week; at 17°C this is 2.5 and at 23°C this is 3.5 leaves per week.

De Koning (1994) observed no effects of fruit load, leaf removal or plant density on truss appearance rate and LUR. Heuvelink and Marcelis (1996) also reported little influence of assimilate supply on LUR (Table 3.1). Assimilate supply was varied with plant density, fruit pruning and removal of leaves that had just unfolded. For example, reducing the number of fruits per truss from seven to one increased vegetative dry mass by 62–67%, reflecting increased partitioning of assimilates towards vegetative growth. Reduced fruit load did not influence LUR, and final leaf number was increased by only 9% when fruit number per truss was reduced from seven to one (Table 3.1). Higher light interception per plant as a result of a lower plant density hardly affected LUR, whereas assimilate supply was strongly improved. However, several other authors observed faster leaf formation at higher light intensities for young tomato plants (e.g. Fig. 3.1). This contrast may be explained by assimilate supply limiting LUR in seedlings, because of the low leaf area and thus low light interception.

The size of a compound tomato leaf is immensely variable, but leaves of common greenhouse cultivars are typically 0.5 m long and a little less in breadth, with a large terminal leaflet and up to eight large lateral leaflets, which may themselves be compounded. Many smaller leaflets (folioles) may be interspersed with the large leaflets. Stomata are always present on the

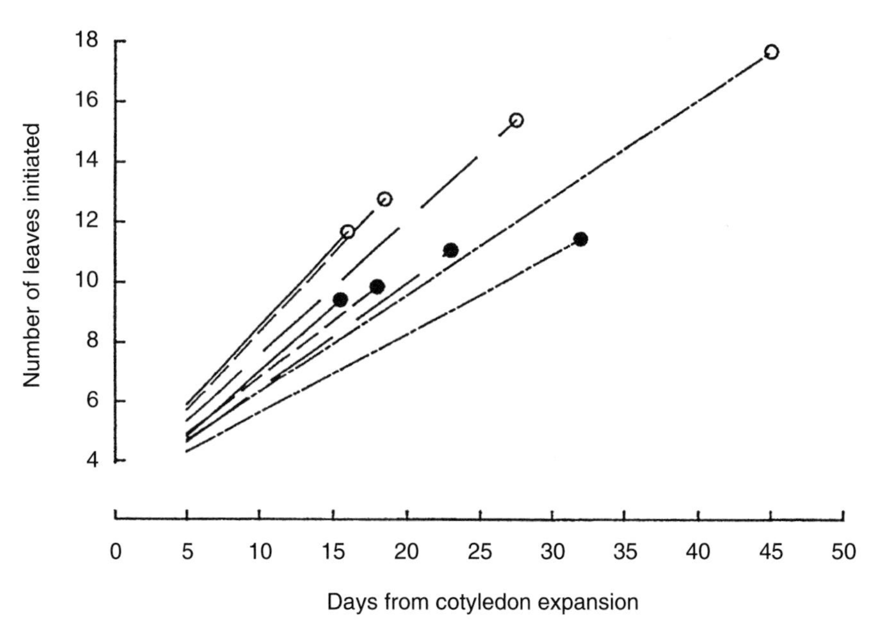

Fig. 3.1. Effect of light and temperature on the rate of leaf production and time of flower initiation. (Reprinted, with permission, from *Journal of Horticultural Science* 34, p. 157, Calvert, 1959.) Key: ● = flowers initiating at 15°C; o = flowers initiating at 25°C. Light levels (W/m²): —— 31.2, - - - 23.4, — — 15.6, —·— 7.8.

Table 3.1. Effect of treatment on the number of leaves visible, total dry weight, vegetative dry weight, average area and dry weight of individual leaves and specific leaf area (SLA) at the end of five experiments with tomato. For the latter three parameters, only the leaves still on the plant were taken into account. Means within an experiment, followed by the same letter, were not significantly different ($P < 0.05$). (Reprinted, with permission, from the *Journal of Horticultural Science* 71, p. 411, Heuvelink and Marcelis, 1996.)

Experiment	Treatment	Number of leaves visible	Total dry weight (g/plant)	Vegetative dry weight (g/plant)	Leaf area (cm^2/leaf)	Leaf dry weight (g/leaf)	SLA (cm^2/g)
1	1.6 plants/m^2	59.5^{a}	636^{a}	256^{a}	467^{a}	3.17^{a}	147^{a}
	2.1 plants/m^2	60.3^{a}	512^{b}	196^{b}	396^{b}	2.36^{b}	169^{a}
	3.1 plants/m^2	57.8^{a}	364^{c}	141^{c}	310^{c}	1.61^{c}	193^{b}
2	1 fruit per truss	68.6^{a}	385^{a}	294^{a}	322^{a}	2.67^{a}	123^{a}
	3 fruits per truss	71.7^{a}	536^{b}	262^{a}	352^{a}	2.53^{a}	140^{a}
	7 fruits per truss	69.4^{a}	519^{b}	176^{b}	329^{a}	1.58^{b}	209^{b}
3	1 fruit per truss	64.8^{a}	414^{a}	341^{a}	494^{a}	3.37^{a}	147^{a}
	7 fruits per truss	59.3^{b}	444^{a}	210^{b}	503^{a}	2.21^{b}	232^{b}
4	50% truss removal	62.7^{a}	471^{a}	252^{a}	419^{a}	2.71^{a}	155^{a}
	Control	62.3^{a}	488^{a}	182^{b}	320^{a}	1.91^{b}	166^{a}
5	50% leaf removal*	37.3^{a}	221^{a}	110^{a}	331^{a}	3.22^{a}	104^{a}
	Control	37.0^{a}	266^{b}	136^{a}	299^{a}	2.54^{b}	118^{a}

*Removal of every other leaf (at a length of 5 cm) above the first truss.

underside of the leaflet and are often present in smaller numbers on the top (Gay and Hurd, 1975).

Leaf size (both area and mass) is influenced by growing conditions (Table 3.1), with an increased assimilate supply resulting in larger leaves of higher mass. However, an individual leaf grown under high irradiance is usually smaller in area, but larger in mass, and the same is observed in enriched CO_2 conditions (Verkerk, 1955; Hurd and Thornley, 1974). Except in low light conditions, a higher temperature usually increases the rate of expansion of individual leaves, but expansion period is decreased relatively more strongly, resulting in a reduced area per leaf. On the other hand, plant leaf area usually increases with temperature, since LUR increases. Leaf expansion is improved by a high turgor pressure, which is favoured by a suppressed transpiration rate. Indeed, tomato leaf area increases with humidity, but the effect is only small. Furthermore, in the long run, high humidity may cause calcium deficiencies, which may lead to leaf area reductions of up to 50% (Holder and Cockshull, 1990).

Leaf thickness increases with assimilate supply (low sink/source ratio), resulting in a lower specific leaf area (SLA) (m^2 of leaf area per unit of leaf dry mass). This is shown in Table 3.1 for wider plant spacing and reduced fruit load, but the same has been reported for increased irradiance or enriched CO_2 concentrations (Verkerk, 1955; Hurd and Thornley, 1974). A higher temperature usually decreases leaf thickness and increases SLA. A further discussion on the modelling of leaf area development and SLA is given in Chapter 4.

Continuous (24 h) light may inhibit tomato leaf growth, but the underlying mechanism is unclear. A reduction in translocation of assimilates out of the leaves under continuous light has been reported. In continuous light, high levels of starch build up in leaves, which develop chlorotic patches and start to die (Bradley and Janes, 1985; Logendra *et al.*, 1990).

Conditions of a long-lasting surplus of assimilates may lead to the so-called short leaf syndrome (SLS), i.e. crisp grey-green to purple leaves, sometimes with necrotic spots and leaf tips. Nederhoff *et al.* (1992) suggested that SLS results from Ca deficiency in the apex and proposed the following underlying theory. The sink/source ratio would indirectly influence the occurrence of SLS, since a low sink/source ratio results in fast-growing organs, fed by assimilates from the phloem. Phloem contains very little Ca, which should reach the organs through xylem flow. However, the Ca-containing xylem sap primarily goes to the transpiring (full-grown) leaves and only small amounts reach the non-transpiring organs. The apex may be considered as such a sink with low transpiration, like fruits. Short leaves can be avoided by increasing plant density, or, in an early-planted year-round crop, by maintaining an extra shoot on part of the plants in spring, thus increasing stem density. In both cases the sink/source ratio is increased.

STEM DEVELOPMENT

At the tip of the main stem is the apical meristem, a region of active cell division where new leaves and flower parts are initiated. It is dome-shaped and is protected by newly formed leaves. Leaves are arranged alternately with a 2/5 phyllotaxy. Between seven and 11 leaves are formed before the apex is transformed into a terminal inflorescence (see section on flowering, below). The lateral bud in the axil of the last initiated leaf becomes the new apex and appears to grow along the initial axis by displacing the inflorescence into a lateral position (Fig. 3.2). This lateral growth is also terminated by an inflorescence and the plant thus continues as a succession of lateral growths. Therefore the tomato stem is a true sympodium. In indeterminate cultivars

Fig. 3.2. Sympodial nature of flowering in tomato. The initial axis terminates in the inflorescence. A, B, C: leaves initiated before flower initiation; D: first initiated leaf on axillary shoot. (Reprinted, with permission, from *National Agricultural Advisory Service Quarterly Review* 70, Calvert, 1965.)

the process is repeated indefinitely, with inflorescences every three leaves. These types are favoured in greenhouses as they produce high yields over an extended period. In determinate types each axis produces a limited number of inflorescences, and strong axillary buds develop at the base of the stem, producing a bushy habit, which is ideal for growing unsupported in the open. Buds in all leaf axils can produce branches, which develop similarly to the original axis. These side shoots are removed when indeterminate cultivars are grown. The side shoot from the leaf axil immediately below an inflorescence is the strongest and can compete seriously with the main apex.

The rate of stem elongation generally increases with temperature, with stem length primarily determined by day temperature (optimum > 28°C) (Langton and Cockshull, 1997). Under a 12 h day (D) and night (N) period, 40-day-old tomato seedlings produced a stem length of 0.52 m when grown at 26/16°C D/N, whereas this was 0.35 m at 20/22°C D/N (Fig. 3.3) and only 0.23 m at 16/26°C D/N, all three regimes having exactly the same average 24 h temperature of 21°C. For plants grown at 16/26°C D/N, leaf production was reduced by only 15%, indicating that observed differences in plant length resulted predominantly from differences in internode extension growth. Reduction in length growth resulting from a so-called negative DIF (difference between day and night temperature) has been reported by many authors and for a large variety of crops. Gibberellin metabolism is almost certainly involved in the regulation of stem extension by temperature (Langton and Cockshull, 1997). The effect of increased root temperature on stem elongation is relatively small.

Fig. 3.3. Tomato plants (8 weeks after sowing, 40 days after start of temperature treatments) grown in climate chambers at a light intensity of 30 W/m^2 at two temperature regimes. D, day temperature; N, night temperature; day length was 12 h.

Under daily total irradiances above about 2 MJ/m^2/day the heights of plants of the same dry mass are constant. Reducing daily irradiance below this level generally increases the rate of stem elongation. The result is a weaker, thinner but taller stem. CO_2 enrichment generally produces taller plants but this is because of faster growth. Plants of similar dry or fresh mass are slightly shorter if grown with additional CO_2. High humidity improves stem elongation, though effects are only small. Holder and Cockshull (1990) did not observe any effect of humidity treatments (0.1, 0.2, 0.4 and 0.8 kPa vapour pressure deficit, maintained continuously for 28 days) on plant height. Salt stress (high EC in the root environment) did not significantly influence stem elongation in the electrical conductivity (EC) range 2–6.5 dS/m; Li and Stanghellini (2001) found a 2% shorter internode length at high EC (6.5 dS/m) compared with a reference EC of 2 dS/m.

Internode length is influenced by sink/source ratio: a reduced fruit load reduces internode length (Heuvelink and Buiskool, 1995). However, Hurd *et al.* (1979) observed an increase in internode length for plants from which two-thirds of the flowers were removed, compared with control plants. Influence of sink/source ratio on internode length may be the result of changes in hormone balances when some flowers are removed.

Shaking both stresses and toughens plants, resulting in reduced stem length, a phenomenon known as thigmomorphogenesis (Picken *et al.*, 1986).

Axillary shoot development is promoted by low temperature or short days. A period of 5 min of far-red light at the end of a light period has been shown to suppress growth of side shoots (Tucker, 1976), which may be of interest for greenhouse crop production, where manual removal of side shoots demands a lot of labour.

ROOT DEVELOPMENT

Tomato seedlings develop a taproot, which can grow longer than 0.5 m but is often damaged in culture. Tomatoes may root as deep as 2 m, a distance reached by lateral roots under favourable soil conditions, but the most active zone of the root system is much shallower. About 60% of the root system lies in the top 0.3 m of soil (Rendon-Poblete, 1980). Adventitious roots, similar in structure to the laterals, develop in favourable conditions from the stem, particularly near the base. They are also initiated on the underside of horizontal portions of stem, enabling the plant to re-root in nature.

Nowadays, many greenhouse tomato cultivations are on artificial substrates, e.g. stonewool or glasswool, expanded clay or sometimes in nutrient solution only (nutrient film technique, NFT). NFT has enabled studies of root growth and development, and both NFT and substrate cultures allow for a much more controlled root environment. In research focused on roots, plants are sometimes grown in aerated containers with nutrient solution. Roots grown (partly) in water have a different anatomy and morphology from roots grown in soil.

Water supply and nutrition, two important factors in the root environment, have profound effects on the growth of plants (Chapter 6). Roots grown at about 14°C are thicker, whiter and less branched than at higher soil temperatures, but nutrition influences this response (Gosselin and Trudel, 1982). Chapter 4 presents effects of root temperature on dry matter production and partitioning. High humidity may encourage aerial adventitious root growth and promote root extension.

FLOWERING

The information in this section is largely a summary of the review presented by Dieleman and Heuvelink (1992), in which references to the original sources are given. An earlier review on flowering in tomato (Picken *et al.*, 1985) has been consulted occasionally. In general the number of leaves preceding the first inflorescence (NLPI) initiated in the tomato plant is at least six to eight. The inheritance of this characteristic was found to be simple; one single gene pair determines the NLPI. Complete dominance is suggested for smaller node numbers over large node numbers to the first inflorescence. The first inflorescence terminates the main axis of growth and subsequent extension growth is made by the lateral shoot in the axil of the last initiated leaf (see section on stem development, above) (Fig. 3.2).

There are two flowering types, which are incorrectly termed determinate and indeterminate (see section on stem development). Botanically, both types are determinate, since each inflorescence terminates the main axis. The tomato does not respond to seed vernalization, i.e. NLPI is not reduced by low temperatures applied to imbibed seeds. The sensitive period, determining the position on the stem at which the first inflorescence will develop, lasts approximately 10 days, starting at cotyledon expansion, though this period can be longer depending on the cultivar. No single environmental factor can be regarded as critical for the control of tomato flowering. Environmental factors such as light, temperature, carbon dioxide, nutrition, moisture and growth regulators directly or indirectly influence flower initiation.

The NLPI can be seen as the result of two processes: the rate of leaf initiation, which determines the number of leaves formed during the vegetative phase, and the time to initiation of the first inflorescence, which determines the end of the vegetative phase. The slopes of the lines in Fig. 3.1 indicate the rate of leaf initiation. The same rate of leaf initiation can result in a different NLPI if the number of days to initiation of the first inflorescence is different. Also a higher rate of leaf initiation combined with a lower number of days to initiation of the first inflorescence may result in the same NLPI (Fig. 3.1).

Factors affecting NLPI

A lower temperature, either by day or by night during the sensitive period for tomato seedlings, causes a smaller NLPI. This effect of mean diurnal temperature on NLPI is mainly determined by the effect of temperature on the rate of leaf production. Temperature does not affect the time to flower initiation if light intensity is high (Fig. 3.1). At low light intensities a higher temperature increases the time to flower initiation compared with a lower temperature. Root temperatures appear to have little or no effect on the NLPI.

The rate of leaf production in a tomato seedling increases with increasing light intensity (Fig. 3.1). On the other hand, the number of days to flower initiation decreases with increasing light intensity, this effect being stronger than the effect on the rate of leaf production. This results in a lower NLPI at higher light intensities. Effects of light and temperature on both the rate of leaf initiation and the time to flower initiation, and hence on NLPI, interact (Fig. 3.1). The small increase in NLPI observed at higher plant densities in the nursery bed (Saito *et al.*, 1963) can be explained by the smaller amount of light available per plant. Flowering in tomato is not usually affected by photoperiod, but some cultivars are regarded as quantitative short-day plants. Several authors have reported a smaller NLPI with shorter photoperiods when the same light integral was maintained.

Carbon dioxide enrichment increases the rate of leaf initiation slightly and decreases the time to flower initiation a little, thus resulting in no clear effect on the NLPI. Removal of parts of the cotyledons of tomato seedlings immediately after they have unfolded increases NLPI, the increase being larger the larger the area removed. Removal of the first two foliage leaves from 6-day-old seedlings reduces the NLPI, whereas NLPI is not significantly altered by the removal of further foliage leaves. Generally, the higher the fertilizer level (NPK), the lower is the NLPI. However, flowering is delayed when salt or water stress is applied in the early stages of development.

The effects that growth regulators have on the NLPI vary with the method of application, the concentration, timing, environment and variety. Application of auxins as well as application of gibberellin as a foliar spray causes an increase in NLPI, but application of a gibberellin solution to imbibed seeds does not influence NLPI. Application of auxin or gibberellin transport inhibitors decreases NLPI; for example, a NLPI as low as 2.4 has been observed after application of 10^{-3} M TIBA (2,3,5-triiodobenzoic acid) foliar spray, which inhibits auxin transport.

Hypothesis on tomato flowering

Most of the observations mentioned in the previous section can be explained by the so-called nutrient diversion hypothesis for flowering (Sachs and Hackett,

1969). According to this hypothesis, the amount of assimilate available to the apex during the sensitive phase has to reach a certain minimum before flower initiation can take place. This level of assimilates is influenced by the rate of assimilate production and also by its distribution. By accepting the nutrient diversion hypothesis, it is expected that a higher production rate of assimilates reduces the time to first flower initiation. When the influence of assimilate production rate on the time to first flower initiation is stronger than the effect on the rate of leaf initiation, a lower NLPI is expected under conditions where the assimilate production is high. Indeed, at higher light intensities the NLPI is reduced and higher plant densities (meaning that less light is available per plant) cause a higher NLPI. Removal of the cotyledons causes a higher NLPI. The cotyledons are a source of assimilates for the young seedling; after their removal more leaves have to be formed until a certain assimilate level is reached. The retarding effect of a reduction in soil moisture or salt stress of flower initiation is probably due to growth reduction. The smaller cotyledons and first foliage leaves under these conditions intercept less light, which results in a greater NLPI.

If the competitiveness of the apex for assimilates is reduced compared with that of other parts, the NLPI will increase as the time to first flower initiation increases. The greater NLPI at high temperatures can be (partially) explained in this way. Leaves excised from plants grown at high temperatures yield more gibberellin-like substances than leaves from plants grown at low temperatures, probably giving these leaves a higher competitive potential for attracting assimilates. In general, effects of growth regulators on tomato flowering can be explained as effects on assimilate distribution. The increased NLPI at extended photoperiod may also be explained in this way. The interaction between light and temperature effects on NLPI (Fig. 3.1) is as expected from the nutrient diversion hypothesis. At high light intensity, high temperature increases the rate of leaf initiation just as under low light intensity. However, at high light intensity the amount of assimilates available in the plant is high and the apex will reach its critical assimilate level at about the same time, and so the initiation date of the first flower is hardly affected. At low light intensity, high temperature delays the first flower initiation, as assimilate supply is limiting and high temperature reduces the amount of assimilate available in the plant (higher respiration) and increases the competitiveness of leaves and stem for assimilates. Thus at low light intensity, temperature has more influence on the NLPI than at high light intensity. The same holds for the influence of light on NLPI at high and low temperatures.

Tomato flowering is complex. Factors that increase the total amount of assimilates in the plant as well as those increasing the competitive potential of the apex decrease the NLPI (Fig. 3.4). A factor might influence both the production and distribution of assimilates. No proof is given here that the nutrient diversion hypothesis is the only way to explain effects of environmental factors and growth regulators on the NLPI in tomato, but no clear contradictions to this hypothesis have been found in the literature.

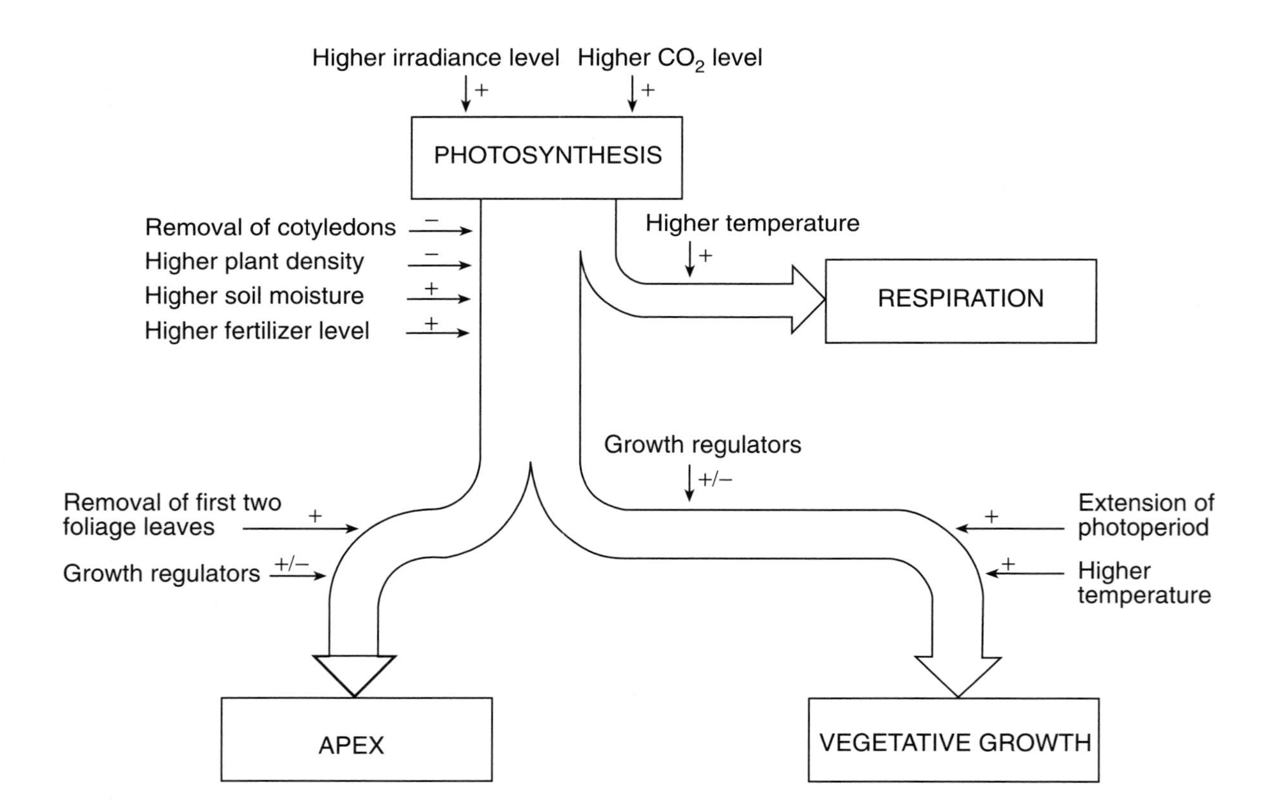

Fig. 3.4. Factors affecting the production and distribution of assimilates in the young tomato plant. Hypothesis: a certain minimum amount of assimilates has to be available in the apex before it can initiate an inflorescence. (Reprinted, with permission, from *Journal of Horticultural Science* 67, Dieleman and Heuvelink, 1992.)

Appearance rate of new trusses

Most tomato cultivars develop 'units', comprising three leaves and internodes terminated by an inflorescence, which build up the sympodial stem (Fig. 3.2). Hence, the appearance rate of new inflorescences (trusses) following the first inflorescence on the sympodial stem, often called the flowering rate, is one-third of the leaf appearance rate (see section on leaf appearance and leaf growth, above). This means that the description of environmental effects on leaf appearance rate also holds for truss appearance rate. Thus, truss appearance rate after the first truss is predominantly determined by temperature. De Koning (1994) observed an almost linear response between temperature (range 17–23°C) and truss appearance rate (0.11–0.17 trusses/day). Truss appearance rate is equally related to day and night temperature and therefore responds to the 24 h mean temperature. No effects of fruit load, leaf removal or plant density on truss appearance rate have been observed (De Koning, 1994). Hence, sink/source ratio has no significant influence on tomato truss appearance rate. Also root temperature, EC in the root environment, air CO_2 concentration and air humidity hardly affect the truss appearance rate.

FLOWER DEVELOPMENT, POLLINATION AND FRUIT SET

Once a tomato plant has terminated its vegetative phase, several processes following the initiation of an inflorescence determine whether eventually fruits will start to grow. Flower buds have to develop into flowers, pollination and fertilization have to be successful and fruit set should occur (Fig. 3.5). All these processes are considered in the present section, mainly based on Chapters 4 and 5 of Atherton and Rudich (1986) and a review on pollination and fruit set by Picken (1984).

Inflorescence size and flower bud abortion

The number of flowers initiated in an inflorescence depends on cultivar and environmental conditions. The inflorescence is a monochasial cyme in which the vegetative axis terminates in the king flower (Picken *et al.*, 1985). Increased irradiance or decreased plant density and decreased temperatures positively influence the number of flowers formed in an inflorescence, with the temperature effect being larger at high irradiance compared with low irradiance. Low air temperatures (< 10°C) during inflorescence initiation promote branching of the inflorescences, usually resulting in more flowers per inflorescence. It is the mean diurnal temperature, rather than day or night temperature alone, that controls branching and flower number (Hurd and

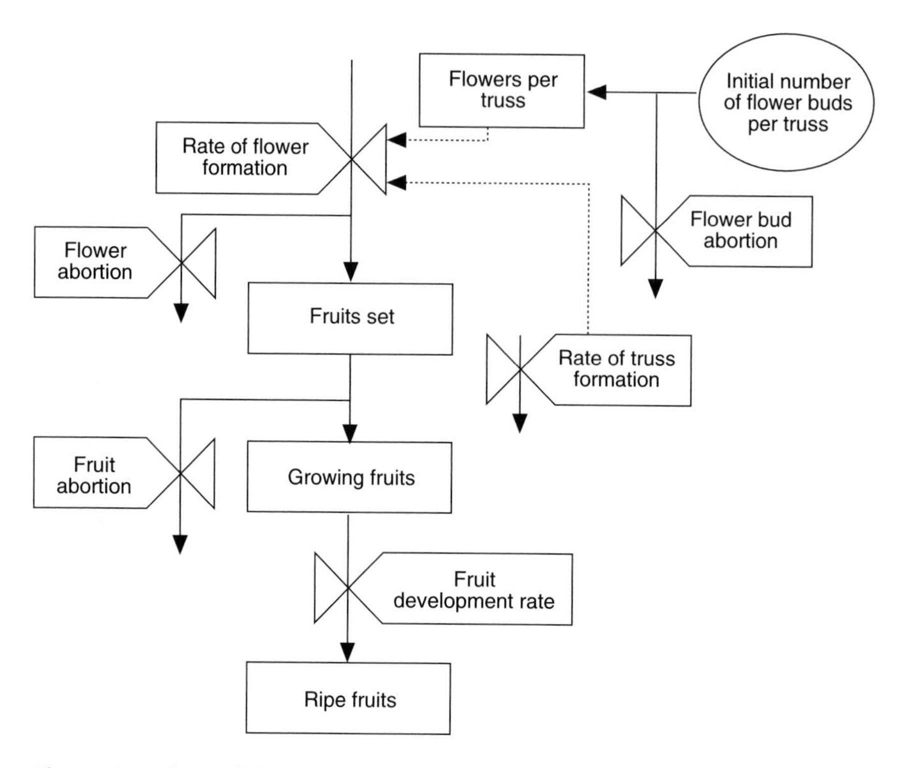

Fig. 3.5. A relational diagram of the number of growing fruits on an indeterminate tomato plant (De Koning, 1994). Boxes are state variables, circles are parameters and valves are rate variables. Solid lines represent number of flowers or fruit and dashed lines represent information flow.

Cooper, 1967). Similar to air temperature, a reduction in root temperature promotes the number of flowers in the first inflorescence. Inflorescence size is reduced when plants are grown under restricted water supply or in small pots containing only a small volume of growing medium. Photoperiod and carbon dioxide levels show no influence on the number of flowers formed in an inflorescence.

Flower development after initiation is primarily influenced by temperature, with higher temperatures resulting in faster development of the flowers. Increased day temperatures are reported to be more effective in promoting flower development than corresponding increases in night temperatures.

Especially under conditions where photosynthesis is low (i.e. low light intensities and/or short days), increased temperatures will stimulate flower bud abortion. Increased photosynthetically active radiation and carbon dioxide enrichment reduce flower bud abortion. High temperatures stimulate flower abortion, but this may be a consequence of failure of fruit set rather than a direct effect of temperature (see below). Overall it appears that low assimilate availability during flower development stimulates abortion of flower buds. Furthermore, when plants are grown at the same total daily irradiance, flower bud abortion is more evident in long days than in short days.

Pollination

Fruit set, defined here as the proportion of open flowers that subsequently set fruit of a marketable size, may fail for many reasons. Flower/fruit abortion is defined as the proportion of flowers that fail to yield fruit of marketable size and hence fruit set is one minus abortion. Tomato flower abortion occurs frequently, whereas tomato fruit abortion is uncommon, although sometimes distal fruits stop growing at a small size and never ripen. Failure of pollen production (amount and viability of pollen) or pollination, pollen germination, pollen tube growth, ovule production, fertilization or fruit swelling may all result in poor fruit set. However, poor fruit set in low light conditions is most frequently caused by failure of pollen production or pollination.

The potential number of pollen grains is genetically determined. For example, some modern tobacco mosaic virus (TMV)-resistant cultivars, containing the *Tm-2* gene, produce less pollen of a slightly lower viability. Pollen development is adversely affected by low light. At severe carbohydrate deficiency, meiosis is abnormal in some pollen mother cells. With less severe carbohydrate stress, pollen development is variable. When microspore development ceases at an early stage, pollen grains are shrunken and irregular, whereas when microspore development proceeds normally until after mitotic division, the pollen grains are morphologically perfect. Abnormal flowers with rudimentary petals, stamens and pistils produced under low light produce sterile pollen. High temperature (40°C) damages pollen, the most critical stage for pollen development being meiosis, which occurs about 9 days before anthesis in plants grown at 20°C. In contrast meiosis appears unaffected by low temperature. Reports in the literature conflict where possible decrease in pollen quality at low temperatures is concerned, possibly because of interactions with light and root temperature.

Pollination – the successful transfer of pollen grains to the stigmatic surface – depends on pollen release, flower characteristics and whether or not pollen grains adhere to the stigma. Although pollen is mature and ready for transfer at the time of anthesis, the stigma becomes receptive about 2 days previously and remains so for up to 4 days or more. All modern cultivars are self-pollinated. Flowers are not frequently visited by insects, since they do not produce nectar. Air movement in the field is enough to stimulate self-pollination, but in greenhouses the wind speed is too low. In greenhouse tomato production, pollination is promoted by vibrating each truss mechanically with an electric 'bee' two or three times a week, or, more recently, by the introduction of bumble bees. It is most effective in winter to vibrate flowers at around midday, maybe as a consequence of the higher temperature and irradiance at this time. However, tomato flowers close at night and open during the day, which may also contribute to the increased effectiveness of vibration at midday.

At high humidities, pollen tends to remain inside the anthers, whereas at low humidities it may not adhere to the stigma. In the range between 50% and 90%, the effects of relative humidity are small. Pollination can be adversely affected by abnormalities in flower structure. For self-pollination, the stigma must lie within the tip of the anther cone. Stigma exsertion beyond the cone enables the pollen to escape before reaching the stigma, and it may also lead to desiccation of the stigmatic surface. The length of the style is both genetically determined and affected by growing conditions. Low light, high nitrogen levels and high temperature have been shown to increase stigma exsertion. In modern glasshouse culture, with careful control of temperature and nutrient supply, stigma exsertion is uncommon. Other structural defects that reduce fruit set are, for example, poor development of the endothecum at higher temperature (which is essential for dehiscence) and splitting of the staminal cone and/or fasciation of the style at low light. Whether or not pollen adheres to the stigmatic surface depends not only on relative humidity (see above) but also on temperature, 17–24°C being optimal.

Germination of pollen grains and fertilization

Pollen germination depends on pollen viability and environmental factors. The time taken for pollen to germinate decreases with increasing temperature. The germination percentage is greatly reduced at temperatures outside the range 5–37°C. The growth rate of the pollen tube increases with temperature between 10°C and 35°C, but is reduced outside this range. Humidity and light only marginally affect pollen tube growth.

Low light may reduce the size of flowers and ovaries and ovule development may cease under such conditions before or soon after formation of the embryo sac. Ovule viability may be adversely affected by high temperature (40°C), 5–9 days before anthesis at about the time of meiosis in the ovule mother cells.

Fertilization will take place once the nuclei from the pollen tubes penetrate viable ovules. Because of an increased tube growth rate, higher temperatures reduce the time to fertilization. Fertilization is not greatly affected by growing conditions, though exposure to high temperature (40°C) can have adverse effects on fertilization and on the processes that immediately precede or follow it. For example, the endosperm can deteriorate at high temperature at 1–4 days after pollination.

Fruit initiation

Fruit swelling does not always follow fertilization, e.g. as a result of adverse environments (high temperatures) or competition with other fruits. Although there is no threshold number of seeds necessary for fruit swelling and parthenocarpy is frequent under certain conditions, ovaries with only a few embryos are found to be more frequently eliminated when there is a heavy fruit load. The precise role of the seeds in the initiation of tomato fruit growth has not been determined, but it has been suggested that they may be sources of auxin, which stimulates fruit swelling. Gibberellins are also involved in the initiation of fruit growth: high levels of endogenous gibberellins in the ovaries have been observed in cultivars that exhibit parthenocarpy. The exogenous application of gibberellins or auxins to flowers results in fruit set in the absence of fertilization.

Fruit set is primarily influenced by assimilate availability, which may be expressed as the source/sink ratio in the plant (Fig. 3.6). De Koning (1994)

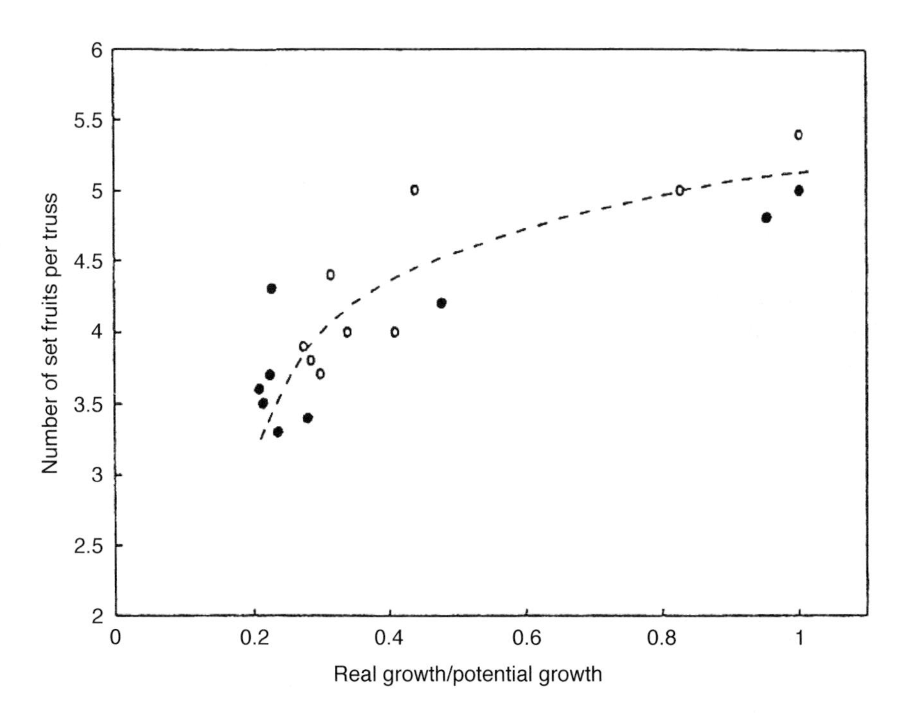

Fig. 3.6. Number of set fruits on the first nine inflorescences of beefsteak tomato 'Capello', grown in CO_2-enriched (o) or non-enriched (•) polycarbonate greenhouses, as a function of the ratio between real and potential fruit growth calculated at fruit set of each inflorescence, pruned to seven flowers. (Reprinted, with permission, from *Annals of Botany* 75, Bertin, 1995.)

used vegetative growth rate as a predictor of fruit formation rate. This is virtually the same as using plant source/sink ratio, because at constant vegetative sink strength the vegetative growth rate is proportional to the source/sink ratio. However, these relationships are not easily generalized. Furthermore, shortage of assimilate supply could not explain the observed fruit abortion in some situations.

FRUIT DEVELOPMENT

After initiation, the fruit will normally develop and grow until the harvestable stage, which is usually earlier than the physiological ripeness of the fruit. In the present section this development rate is addressed. Like most developmental processes, the rate at which an initiated fruit progresses towards ripeness is primarily dependent on temperature. The relationship between fruit developmental stage and fruit growth rate is also presented here. First, fruit morphology and anatomy are considered. The main references for this section are Chapter 5 of Atherton and Rudich (1986) and the dissertation of De Koning (1994).

Fruit morphology and anatomy

Botanically, a tomato fruit is a berry consisting of seeds within a fleshy pericarp developed from an ovary. Fruits of the cultivated species (*Lycopersicon esculentum* Mill.) have two to several carpels and final masses from a few to several hundred grams. Based on the number of carpels, cultivars are grouped as follows: round tomatoes (two to three carpels); beefsteak tomatoes (more than five carpels); and the nowadays popular intermediate types (three to five carpels).

Tomato fruits are composed of flesh (pericarp walls and skin) and pulp (placenta and locular tissue, including seeds) (Fig. 3.7). In general, the pulp accounts for less than one-third of the fruit fresh mass. The pericarp, arising from the ovary wall, consists of an exocarp or skin, a parenchymatous mesocarp with vascular bundles and a single-celled layer of endocarp lining the locules. The pericarp wall may also be divided into the outer wall, radial walls (septa), which separate adjacent locules, and the inner wall (columella). Occasionally the columella is less pigmented than the pericarp and may include large air spaces that cause the tissue to appear white. Most of the cell division in the pericarp takes place during the first 10–14 days after anthesis, but cell division is observed throughout fruit development in *L. pimpinellifolium*.

The fruit skin, or exocarp, consists of the outer epidermal layer plus two to four layers of thick-walled hypodermal cells with collenchyma-like thickenings. Covering the epidermis is a thin cuticle.

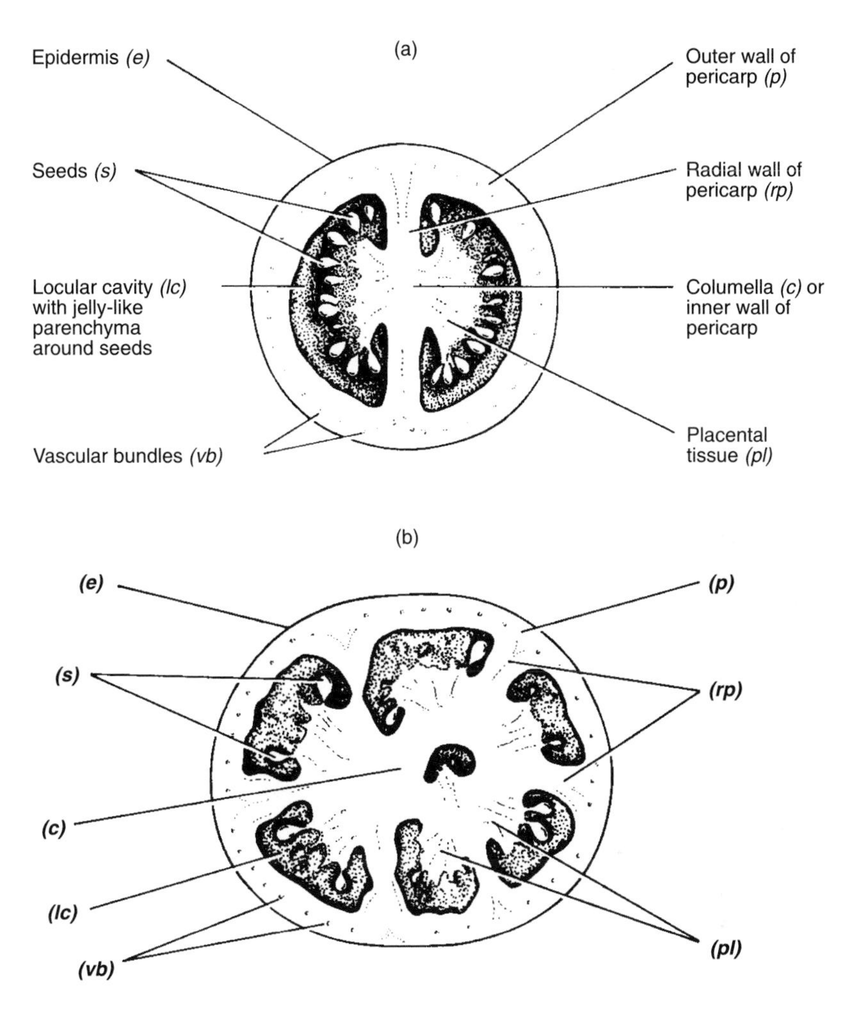

Fig. 3.7. Anatomy of tomato fruits with (a) bilocular or (b) multilocular structure shown as transverse sections. (Glasshouse Crops Research Institute copyright.)

Early in the development of the fruit, the placenta starts to expand into the locules to engulf the seeds within the first 10 days and fills the entire locular cavity in the following few days. In immature fruits the placental tissue is firm, but as the fruits mature the cell walls begin to break down and the locular tissue of a mature-green fruit is jelly-like. At later stages, intracellular fluid may accumulate in the locules. Despite this degeneration, protoplasts usually remain intact.

There are two major branches of the vascular system in fruits: one extending from the pedicel through the outer wall of the pericarp; and the other passing through the inner and radial walls to the seeds. Generally the vascular system is a closed network with very few 'blind' endings.

Fruit development rate

Tomato fruit growth period (the time from anthesis of a flower until harvest-ripe fruit) is about 2 months and predominantly depends on fruit temperature. For example, the fruit growth period may vary from about 73 days at 17°C to only 42 days at 26°C. The fruit growth period can be well described by linearly relating its reciprocal (i.e. fruit development rate) to temperature (Fig. 3.8a). This means that for harvest-ripe fruits a certain temperature sum has to be reached. This concept of heat units has been explained in the section on germination (above). For cv. Counter, good predictions of fruit growth period may be made assuming a temperature sum of 940 degree-days and a base temperature of 4°C. De Koning (1994)

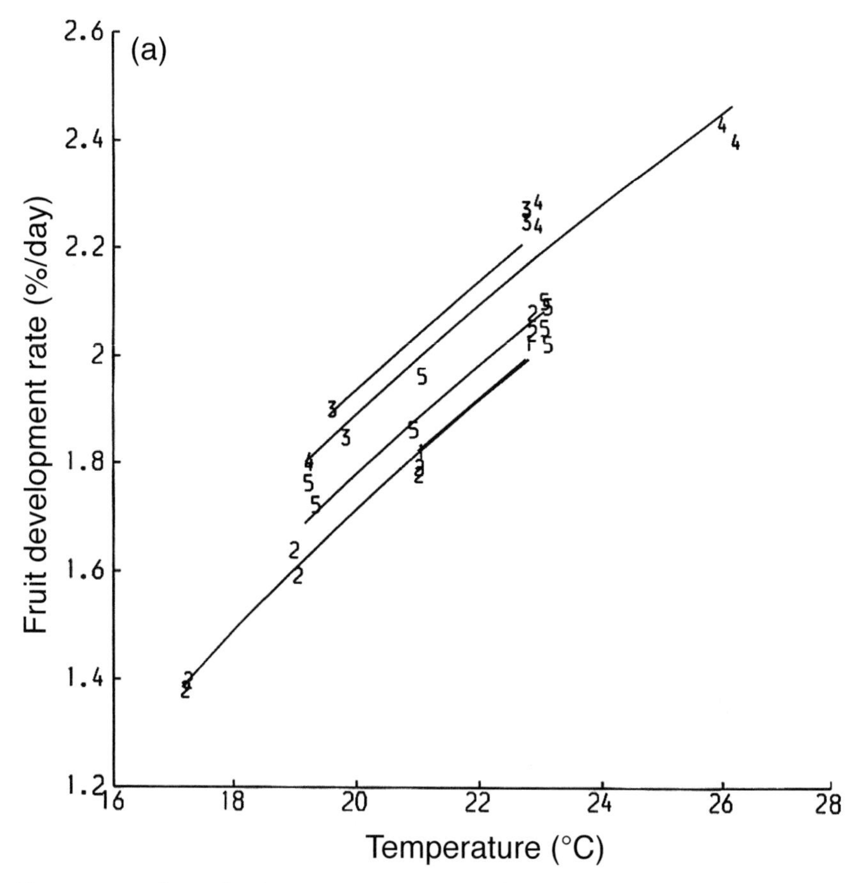

Fig. 3.8. (a) Relationship between temperature and fruit development rate of tomato for five experiments: (1) cv. Counter in spring; (2) cv. Calypso in spring; (3) cv. Counter in summer; (4) cv. Calypso in summer; (5) cv. Liberto in spring (De Koning, 1994).

investigated the interaction between temperature and fruit developmental stage on the duration of fruit growth period by transferring plants for 2 weeks to a compartment with another temperature. The accelerating effects of temperature on fruit development rate vary during fruit development (Fig. 3.8b). In fact, the pattern of sensitivity to temperature appears to be the opposite of that of fruit growth rate, which is at its greatest midway between anthesis and harvest-ripe (see section on relationship between fruit development and fruit growth, p. 76). The course of the temperature response reflects the sensitivity to temperature of successive physiological processes determining the fruit growth period. Just after anthesis, these processes may be cell division and seed growth, while near maturity processes involved in colouring and ethylene production will be accelerated by temperature. At low

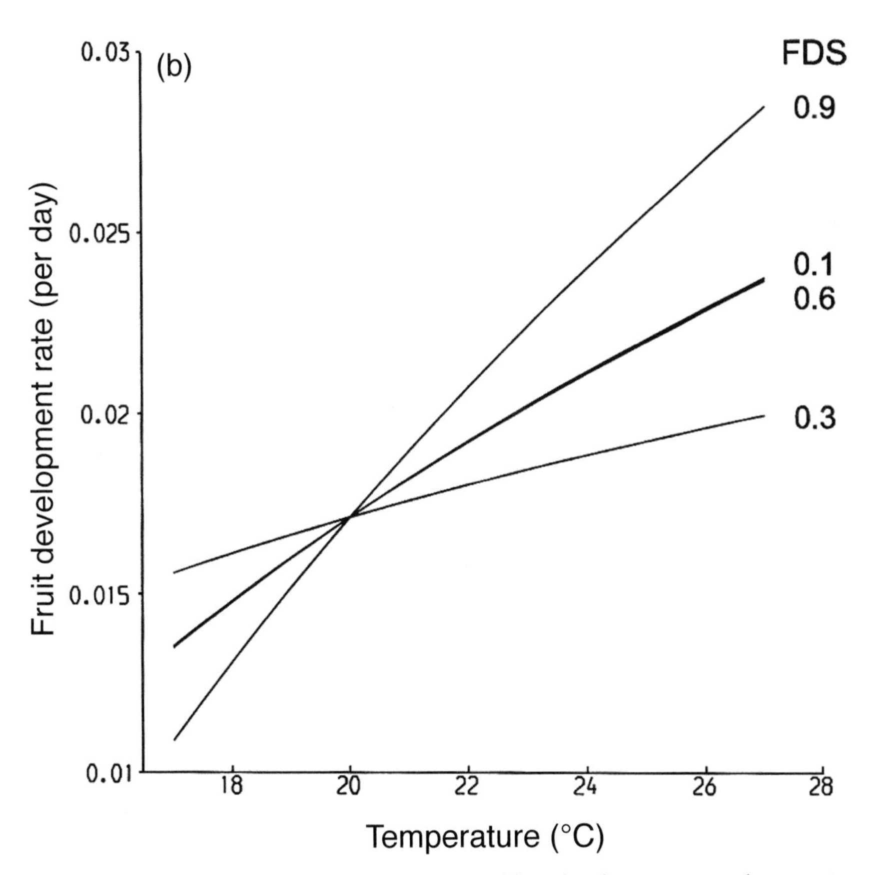

Fig. 3.8. *Continued* (b) Temperature sensitivity of fruit development rate of tomato (cv. Calypso) at 0.1, 0.3, 0.6 and 0.9 fruit developmental stage (FDS) (0 = anthesis, 1 = harvest ripe). (Reprinted, with permission, from *Acta Horticulturae* 519, pp. 92–93, De Koning, 2000.)

irradiance, fruit temperature is equal to air temperature, but at high irradiance, current temperature of exposed fruits may be up to 9°C higher. A difference between measured (air) temperature and fruit temperature may explain observed seasonal effects on the relationship between temperature and fruit growth period (De Koning, 1994).

Differences in fruit growth period between cultivars have been observed, but for related cultivars these differences are very limited. Other factors, such as plant density, light intensity, carbon dioxide, air humidity, fruit load, plant age or salinity in the root environment, have no or only a small effect on fruit growth period. Severe water stress shortens the duration of fruit growth period and fruits affected by blossom-end rot will ripen 1–2 weeks earlier. Fruit ripening is considered in detail in Chapter 5.

Relationship between fruit development and fruit growth

Cumulative fruit growth can be described by a sigmoid curve (Fig. 3.9a). Shortly after anthesis the absolute growth rate is still low and results from both cell division and initial cell enlargement. After about 2 weeks, a 3–5-week period of rapid growth follows, entirely due to cell enlargement. Finally there is a period of slow growth for about 2 weeks in which intensive metabolic changes take place (Chapter 5), but little gain in fruit mass occurs. About 10 days after the first change in colour, assimilate import stops completely; this is caused by the formation of an abscission layer between the calyx and the fruit.

Final fruit fresh mass varies strongly between cultivars, the smallest fruit size being about 15 g (cherry type) and the largest about 500 g (beefsteak type). Mass of individual fruits appears highly correlated with number of seeds and the number of locules (cultivar dependent). Poor pollination reduces seed number and actual fruit mass. Fruit pruning increases fruit size, without influencing seed number, and fruits with many seeds may stay small when grown on plants with a high fruit/leaf ratio. This indicates the influence of assimilate availability on fruit size. Bertin *et al.* (2003b) investigated the link between genetic and developmental controls of fruit size. These authors characterized basal and tip fruits at anthesis and at maturity for two isogenic lines, CF12-C and CF14-L, differing in fruit weight. The influence of competition was studied by removing either basal or tip ovaries at anthesis. On an intact inflorescence, CF12-C fruits grew less than CF14-L fruits, with 1.67 fewer cell layers and similar cell size, suggesting that genes controlling cell division may be responsible for this fruit size variation. Truss thinning masked the difference in fruit size, mainly by reducing the difference in cell number between the two lines and by promoting cell expansion in tip fruits, so that fruit growth was similar at both positions and both lines. Hence, Bertin *et al.* (2003b) found that cell number exerted a control on final fruit

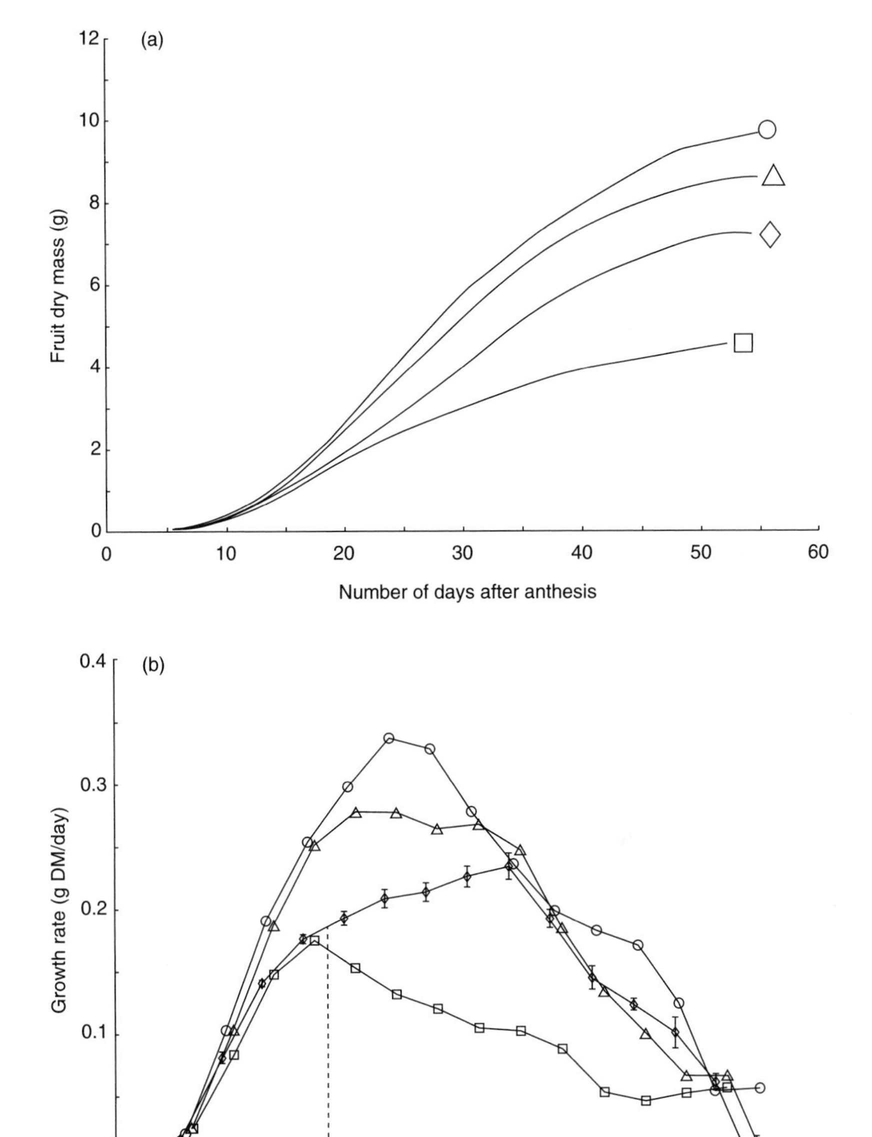

Fig. 3.9. (a) Tomato (cv. Counter) cumulative fruit growth and (b) fruit growth rate as a function of days after anthesis. Fruits were grown with one (circle), two (triangle) or eight (square) fruits per truss or seven fruits were removed from each truss from plants with eight fruits per truss at 18 days after anthesis (diamond). Growth was calculated from diameter measurements on the second and third fruit on the fourth truss. (Heuvelink and Wubben, 1999, unpublished data.)

size only when there was competition among fruits. Seed number is an important factor determining the competitive ability of a fruit to attract assimilates, whereas average fruit mass depends on the availability of assimilates. The prominent role of fruits in assimilate distribution in tomato plants is discussed in Chapter 4.

Under normal conditions, fruit growth in tomato is source limited, as fruit pruning results in an increased growth rate of the remaining fruits. For example, growing eight fruits on a truss resulted in a final fruit dry mass of 4.7 g, whereas in the same glasshouse, fruits growing at one fruit per truss reached 9.9 g (Fig. 3.9a). However, when only two fruits are grown on a truss, fruit growth rate appears to be (almost) at its maximum (potential fruit growth rate), as pruning to one fruit per truss hardly increases growth rate of the remaining fruits (Fig. 3.9a). This so-called potential growth rate, i.e. growth rate under non-limiting assimilate supply, can be used as a measure for fruit sink strength – the competitive ability to attract assimilates (Chapter 4). Fruits grown under source limitation reach the potential growth rate about 2 weeks after the source limitation has been lifted by severe fruit pruning (Fig. 3.9b). This is an interesting observation, as apparently it is the fruit developmental stage that determines its potential growth rate and not the fruit size, which reflects the history of assimilate availability experienced by the fruit. Note that for a fruit that experiences source limitation in the first weeks after anthesis, relative growth rate after lifting source limitation is much higher than for a fruit that has been grown potentially from anthesis onwards, resulting from an equal absolute growth rate after about 2 weeks at a lower cumulative dry mass.

Potential fruit growth depends to some extent on the truss position on the plant and on the fruit position within a truss. Fruits on the first trusses on a tomato plant show a reduced potential mass (Fig. 3.10a). Potential fruit mass at maturity is determined by sink size developed at fruit initiation and sink activity during fruit growth. The main factor of sink size is the fruit's cell number. Changes in cell number during the early period of tomato fruit development were analysed by means of a deterministic model of cell multiplication (Bertin *et al.*, 2003a). The final cell number is reached within 2 weeks after anthesis and depends on the initial cell number in the ovary before anthesis and the rate of cell division thereafter. Hence, the observed increase of final potential fruit mass with higher truss position may be explained by enlargement of the apex and subsequent increase of fruit cell number during ontogeny.

The first trusses on a tomato plant are initiated under poor assimilate availability, since the plant is intercepting only a small amount of light. Assimilate status will be higher under high light levels compared with low light levels. This explains why at low light conditions the effect of truss position on potential fruit size is larger than under high light (Fig. 3.10a).

Distal fruits on a truss often reach lower mass than proximal fruits, and these differences increase with limited assimilate supply. This difference

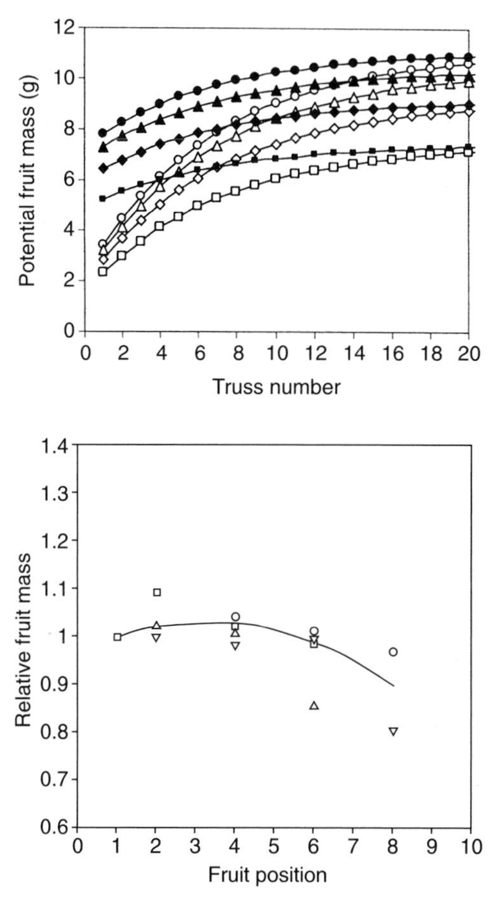

Fig. 3.10. (a) Predicted effect of ontogeny (truss number) and temperature (circle, 17°C; triangle, 19°C; diamond, 21°C; square, 23°C) on potential fruit mass for the first fruit position within a truss of tomato 'Calypso', for winter (open symbols) and summer (closed symbols) planting in north-west Europe. (b) Effect of fruit position within a truss on potential fruit mass relative to the first fruit position (De Koning, 1994).

originates, at least in part, from differences in cell number before anthesis, as proximal ovaries have more cells than distal ones at anthesis, and fruits in both positions show similar cell division activity in the first 10 days of fruit development. When pollination within a truss is synchronized, differences in fruit mass become significantly smaller. Figure 3.10b supports this observation, since potential fruit growth is not much affected by fruit position. Hence, pollination sequence and proximity to the source are significant factors determining the actual growth of fruits in an intact plant. However, these factors are not determined by intrinsic characteristics of fruits and they do not express at non-limiting assimilate supply. Therefore, the effect of fruit position on potential fruit growth is only limited, at least over a range of positions (Fig. 3.10b). This range is expected to be smaller for beefsteak tomato (shorter truss) and (much) longer for cherry tomato (long trusses with more than 20 fruits).

Besides assimilate availability, temperature also strongly influences final fruit size. A higher temperature increases fruit development rate (Fig. 3.8a), whereas the relationship between fruit developmental stage and fruit growth rate is hardly influenced by temperature (Fig. 3.11). Hence, high temperatures result in a reduced final mass per fruit at harvest, as duration of fruit growth is reduced.

Based on studies of carbon import by tomato fruits and translocation speed of photosynthates in the petiole, which are processes promoted by temperature, it is expected that temperature would influence fruit growth rate. However, these observations concern short-term temperature responses of fruit growth rate. Figure 3.11 shows the long-term temperature response. De Koning (1994) presented a possible explanation for differences between the long-term and short-term temperature response. Generally, low temperature reduces the rate of biochemical and biophysical processes, but the apparatus (chloroplast density, amounts of enzymes) enlarges and in this way compensates for the slower reaction rates. In the short term, high temperature immediately increases specific activity, i.e. the activity per unit machinery (e.g. enzyme concentrations). In the long term, the fruit will adapt

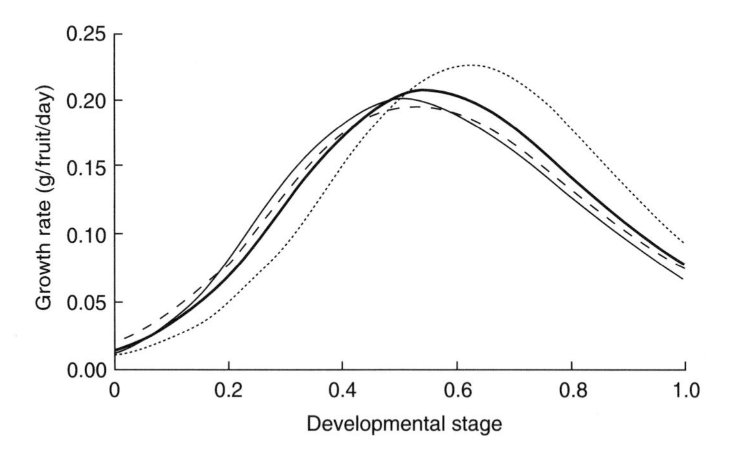

Fig. 3.11. Tomato potential fruit growth rate at 17°C (- - - -), 21°C (— —) and 25°C (—) as a function of fruit developmental stage. Fruit developmental stage was calculated as time after anthesis divided by total fruit growth period, i.e. time from anthesis until harvest-ripe. Bold curve represents average growth curve for the three temperatures. Fruits were grown at one or two fruits per truss in a climate room. (Reprinted, with permission, from *Acta Horticulturae* 260, p. 154, Heuvelink and Marcelis, 1989.)

to this higher specific activity and the amount of machinery will be reduced. The long-term result will be hardly any influence of temperature on growth rate (Fig. 3.11), since effects on specific activity are compensated by effects on amount of machinery.

REFERENCES

AOSA (1983) *Seed Vigor Testing Handbook*. Association of Official Seed Analysts, Contribution No. 32, Las Cruces, New Mexico.

Atherton, J.G. and Rudich, J. (eds) (1986) *The Tomato Crop: a Scientific Basis for Improvement*. Chapman & Hall, London, 661 pp.

Bertin, N. (1995) Competition for assimilates and fruit position affect fruit set in indeterminate greenhouse tomato. *Annals of Botany* 75, 55–65.

Bertin, N., Génard, M. and Fishman, S. (2003a) A model for an early stage of tomato fruit development: cell multiplication and cessation of the cell proliferative activity. *Annals of Botany* 92, 67–72.

Bertin, N., Borel, C., Brunel, B., Cheniclet, C. and Causse, M. (2003b) Do genetic make-up and growth manipulation affect tomato fruit size by cell number, or cell size and DNA endoreduplication? *Annals of Botany* 92, 415–424.

Bewley, J.D. and Black, M. (1994) *Seeds: Physiology of Development and Germination*, 2nd edn. Plenum Press, New York.

Bidwell, R.G.S. (1974) *Plant Physiology*. Collier Macmillan, London, 643 pp.

Bierhuizen, J.F. and Wagenvoort, W.A. (1974) Some aspects of seed germination in vegetables. 1. The determination and application of heat sums and minimum temperature for germination. *Scientia Horticulturae* 2, 213–219.

Bradley, F.M. and Janes, H.W. (1985) Carbon partitioning in tomato leaves exposed to continuous light. *Acta Horticulturae* 174, 293–302.

Calvert, A. (1959) Effect of the early environment on the development of flowering in tomato. II. Light and temperature interactions. *Journal of Horticultural Science* 34, 154–162.

Calvert, A. (1965) Flower initiation and development in the tomato. *National Agricultural Advisory Service Quarterly Review* 70, 79–88.

De Koning, A.N.M. (1994) Development and dry matter distribution in glasshouse tomato: a quantitative approach. PhD Thesis, Wageningen Agricultural University, Wageningen, The Netherlands.

De Koning, A.N.M. (2000) The effect of temperature, fruit load and salinity on development rate of tomato fruit. *Acta Horticulturae* 519, 85–94.

Dieleman, J.A. and Heuvelink, E. (1992) Factors affecting the number of leaves preceding the first inflorescence in the tomato. *Journal of Horticultural Science* 67, 1–10.

Forrester, J.W. (1961) *Industrial Dynamics*. MIT Press, Boston, Massachusetts.

Gay, A.P. and Hurd, R.G. (1975) The influence of light on stomatal density in the tomato. *New Phytologist* 75, 37–46.

Gosselin, A. and Trudel, M.J. (1982) Influence de la temperature du substrat sur la croissance, le developpement et le contenu en elements mineraux de plants de tomate (cv. Vendor). *Canadian Journal of Plant Science* 62, 751–757.

Heuvelink, E. and Buiskool, R.P.M. (1995) Influence of sink–source interaction on dry matter production in tomato. *Annals of Botany* 75, 381–389.

Heuvelink, E. and Marcelis, L.F.M. (1989) Dry matter distribution in tomato and cucumber. *Acta Horticulturae* 260, 149–157.

Heuvelink, E. and Marcelis, L.F.M. (1996) Influence of assimilate supply on leaf formation in sweet pepper and tomato. *Journal of Horticultural Science* 71, 405–414.

Holder, R. and Cockshull, K.E. (1990) Effects of humidity on the growth and yield of glasshouse tomatoes. *Journal of Horticultural Science* 65, 31–39.

Hurd, R.G. and Cooper, A.J. (1967) Increasing flower number in single-truss tomatoes. *Journal of Horticultural Science* 42, 181–188.

Hurd, R.G. and Thornley, J.H.M. (1974) An analysis of the growth of young tomato plants in water culture at different light integrals and CO_2 concentrations. I. Physiological aspects. *Annals of Botany* 38, 375–388.

Hurd, R.G., Gay, A.P. and Mountifield, A.C. (1979) The effect of partial flower removal on the relation between root, shoot and fruit growth in the indeterminate tomato. *Annals of Applied Biology* 93, 77–89.

James, E., Bass, L.N. and Clark, D.C. (1964) Longevity of vegetable seeds stored 15 to 30 years at Cheyenne, Wyoming. *American Society of Horticultural Science Proceedings* 84, 527–534.

Kerr, E.A. (1963) Germination of tomato seed as affected by fermentation time, variety, fruit maturity, plant maturity and harvest date. *Report of the Horticultural Experiment Station and Products Labaoratory, Vineland, Ontario,* 1962, pp. 79–85, in *Horticulture Abstracts* 34, 1909.

Langton, F.A. and Cockshull, K.E. (1997) Is stem extension determined by DIF or by absolute day and night temperatures? *Scientia Horticulturae* 69, 229–237.

Li, Y.L. and Stanghellini, C. (2001) Analysis of the effect of EC and potential transpiration on vegetative growth of tomato. *Scientia Horticulturae* 89, 9–21.

Logendra, S., Putman, J.D. and Janes, H.W. (1990) The influence of light period on carbon partitioning, translocation and growth in tomato. *Scientia Horticulturae* 42, 75–84.

Meier, U. (ed.) (1997) *Growth Stages of Mono- and Dicotyledonous Plants.* BBCH-Monograph, Blackwell, Berlin, 622 pp.

Nederhoff, E.M., De Koning, A.N.M. and Rijsdijk, A.A. (1992) Leaf deformation and fruit production of glasshouse grown tomato (*Lycopersicon esculentum* Mill.) as affected by CO_2, plant density and pruning. *Journal of Horticultural Science* 67, 411–420.

Picken, A.J.F. (1984) A review of pollination and fruit set in the tomato (*Lycopersicon esculentum* Mill.). *Journal of Horticultural Science* 59, 1–13.

Picken, A.J.F., Hurd, R.G. and Vince-Prue, D. (1985) *Lycopersicon esculentum.* In: Halevy, A.H. (ed.) *Handbook of Flowering III.* CRC Press, Boca Raton, Florida, pp. 330–346.

Picken, A.J.F., Stewart, K. and Klapwijk, D. (1986) Germination and vegetative development. In: Atherton, J.G. and Rudich, J. (eds) *The Tomato Crop: a Scientific Basis for Improvement.* Chapman & Hall, London, pp. 111–166.

Rendon-Poblete, E. (1980) Effect of soil water status on yield, quality and root development of several tomato genotypes. PhD dissertation, University of California, Davis, California.

Sachs, R.M. and Hackett, W.P. (1969) Control of vegetative and reproductive development in seed plants. *HortScience* 4, 103–107.

Saito, T., Konno, Y. and Ito, H. (1963) Studies on the growth and fruiting of tomato. IV. Effect of the early environment on the growth and fruiting 4. Fertility of bed soil, watering and spacing. *Journal of the Japanese Society for Horticultural Science* 32, 186–196.

Tucker, D.J. (1976) Effects of far red light on the hormonal control of side shoot growth in the tomato. *Annals of Botany* 40, 1033–1042.

Verkerk, K. (1955) Temperature, light and the tomato. *Mededelingen van de Landbouwhogeschool* 55, 175–224.

Wagenvoort, W.A. and Bierhuizen, J.F. (1977) Some aspects of seed germination in vegetables. II. The effect of temperature fluctuation, depth of sowing, seed size and cultivar, on heat sum and minimum temperature for germination. *Scientia Horticulturae* 6, 259–270.

Crop Growth and Yield

E. Heuvelink and M. Dorais

INTRODUCTION

The yield of a tomato crop is determined by total biomass production, biomass partitioning and fruit dry matter content (Fig. 4.1). Besides affecting the amount of the crop, these attributes also influence product quality (e.g. fruit size and taste) and therefore product price. Yields of field crops usually range between 40 and 100 t/ha, whereas yields from year-round cultivation in greenhouses in north-west Europe or North America easily exceed 500 t/ha and yields as high as 700 t/ha have been reported. The main reasons for these large differences in productivity are: (i) the length of the cultivation period (11–12 months; over 35 trusses harvested) and thus differences in cumulative light interception and biomass production; (ii) the control and optimization of environmental factors (CO_2, temperature, humidity and light); and (iii) intensive cultural practices (hydroponics, fertigation, plant density, additional stems, intercropping, leaf and fruit pruning).

Biomass production is primarily driven by photosynthesis, while photosynthesis to a great extent depends on light interception, which furthermore varies with leaf area. Moreover, a high biomass production does not necessarily result in a high yield, since only the tomato fruit is of economic interest. Partitioning into the fruit may be low because of poor fruit set, and consequently low fruit load, or because of the low capacity of individual fruits to import photoassimilates (fruit sink strength). Finally, fruit dry matter content is of major importance, since this parameter determines which fresh fruit mass results from the dry mass partitioned into the fruit. For processing tomato, dry matter content expressed as soluble solids and sucrose accumulation is of considerable economic importance for the food industry as a major determinant of palatability and processing characteristics (Chapter 10).

The processes that determine crop growth and yield mentioned above will be considered in detail in this chapter, with the emphasis on the influence of environmental factors such as light, CO_2, temperature, humidity and cultural practices. Crop growth shows a delayed response to climatic factors (essentially

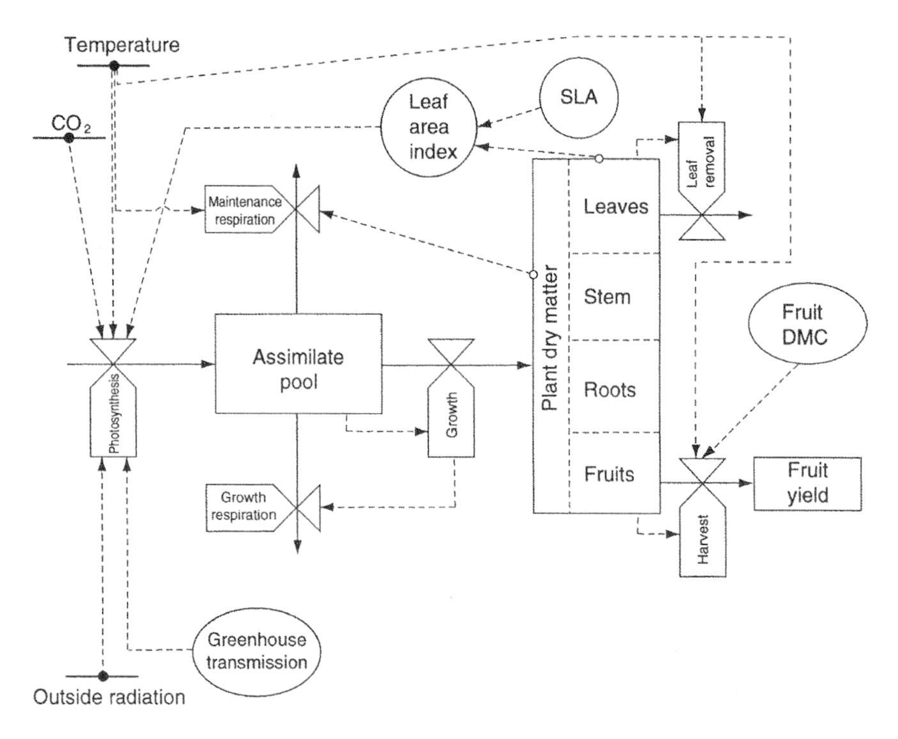

Fig. 4.1. Simplified relational diagram of growth for an indeterminately growing greenhouse fruit vegetable. Boxes are state variables, circles are parameters and valves are rate variables. Solid lines represent carbon flow and dashed lines represent information flow. DMC, dry matter content; SLA, specific leaf area.

with respect to leaf area growth) in addition to the immediate response via crop photosynthesis. For example, crop growth may be increased by a higher temperature in the long term, because increased temperature may increase specific leaf area and thus leaf area development, resulting in higher future light interception. Likewise, maintenance respiration responds immediately to temperature but, since it is also linked to the amount of biomass, the rate of maintenance respiration at a given moment also depends on the integrated (and hence delayed) effect of environmental factors on accumulated dry mass.

BIOMASS PRODUCTION

Cumulative biomass production of a tomato crop basically follows a sigmoid pattern in time (Fig. 4.2). The exponential growth phase, with a constant relative growth rate (RGR) (see section on crop and fruit growth, p. 104),

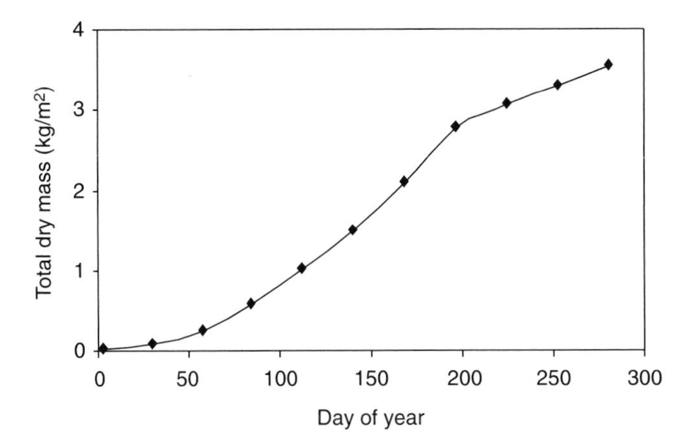

Fig. 4.2. Sigmoid growth pattern of commercially grown greenhouse tomato crop, sown in November and planted in December; first truss at anthesis in second week of January.

gradually evolves into a long phase of linear growth, with a more or less constant absolute growth rate. At ripe fruit stage, growth rate will decrease for determinate tomato types, as a result of leaf senescence and reduction in sink strength negatively influencing photosynthesis (see section on leaf and crop photosynthesis, p. 95). For indeterminate tomato types the linear growth phase continues until the end of the cultivation, resulting in an expo-linear growth pattern. However, this type of growth pattern may not be observed, since towards the end of the cultivation plants are generally decapitated, giving the same response as determinate types. Moreover, low light levels observed in late autumn (October and November) and changes in crop light interception as influenced by leaf area development may also reduce growth rate.

Leaf area and light interception

The amount of intercepted light is a predominant factor in tomato crop growth and biomass production, and depends mainly on leaf area. This relationship (Fig. 4.3) can be described as a negative exponential function of leaf area index (LAI) (m^2 leaf area/m^2 ground area). At an LAI of 3 an indeterminate crop intercepts theoretically about 90% of the incident light. The constant in this relationship is called the extinction coefficient and for tomato its value is about 0.75 for diffuse radiation. For a processing tomato crop, at an LAI of 4.5 the extinction coefficient is around 0.45 (Cavero *et al.*, 1998), while for semi-indeterminate field-grown tomatoes the light extinction coefficient ranges from around 1 early in the season to around 0.2 late in the

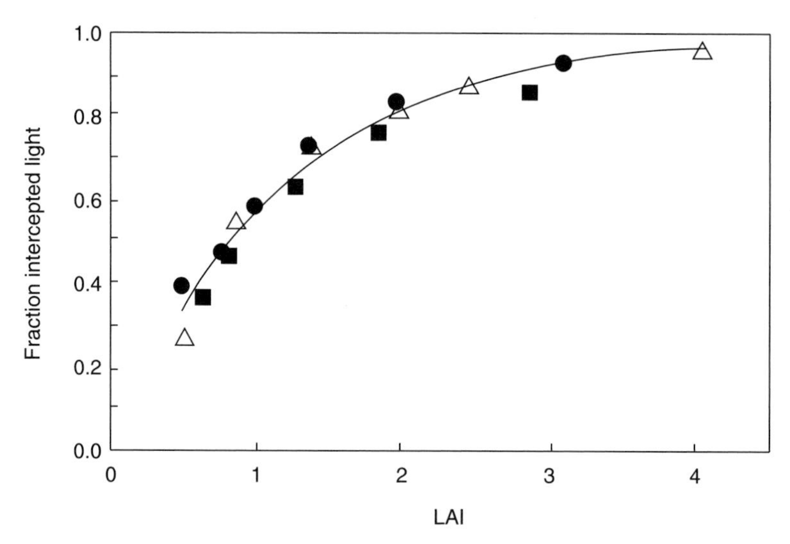

Fig. 4.3. Light interception of young tomato plants arranged at different plant densities in order to vary leaf area index (LAI). Measurements at (●) 26 February, (△) 18 March and (■) 13 July. Solid line is regression equation: $y = 1 - e^{-0.83x}$.

season (Scholberg *et al.*, 2000). Estimated light interception for an LAI of 4–5 is about 50–60% during the period of maximum seasonal fruit development (Scholberg *et al.*, 2000). These values are substantially lower than for greenhouse crops, because of wide row spacing leading to a percentage ground area covered that could be as low as 55% (see below).

Plants in a young crop hardly compete with each other for radiation. At that stage, which is characterized by a low LAI (< 1.5 m^2/m^2), the rate of gross crop photosynthesis (P_{gc}) is almost proportional to LAI (Fig. 4.4a). New leaf area contributes to growth rate to the same extent as existing leaf area, as internal leaf shading is very limited. In a closed canopy, however, leaf area extension is of minor importance compared with a young crop, because most light is intercepted at high LAI (> 3 m^2/m^2) (Fig. 4.3) and further increase of LAI has only a marginal effect on P_{gc} (Fig. 4.4a).

Light interception also depends on the canopy structure, distribution of the leaf area over ground area and cultural practices. For example, under diffuse light conditions, the effect of greenhouse path width on crop photosynthesis becomes important when row height is less than the row distance. In Dutch cultivation, assuming LAI at 3, and with row height being typically 2.25 m and row width 1.25 m at a row distance of 1.6 m, Gijzen (1995) calculated that the daily CO_2 assimilation was on average about 5% less than for a closed canopy at the same LAI. Attempts were made to 'close' the paths when they were not needed for either harvest or crop management, by using movable wires supporting the plants and attached to the top of the

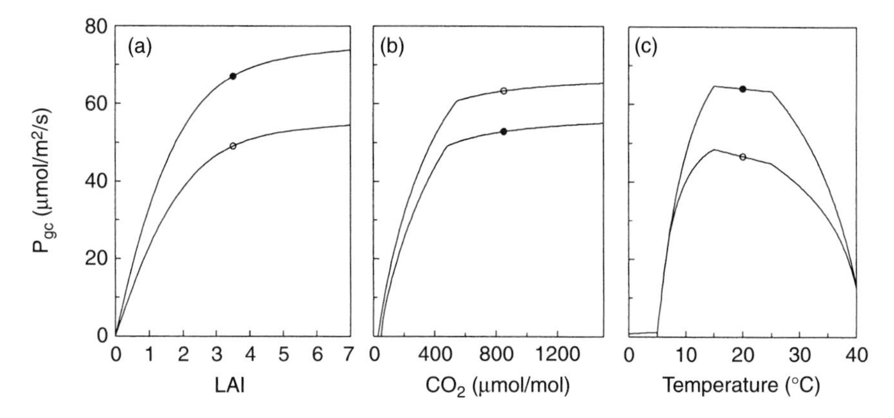

Fig. 4.4. Simulated crop photosynthesis (TOMSIM; see last part of chapter) at 1500 µmol/m²/s diffuse light as a function of: (a) leaf area index (LAI) (○ = at a CO_2 concentration of 340 µmol/mol; ● = at 1000 µmol/mol); (b) CO_2 concentration (○ = at 20°C; ● = at 30°C); and (c) temperature (○ = at a CO_2 concentration of 340 µmol/mol; ● = at 1000 µmol/mol). Effects of temperature and CO_2 concentration are simulated at an LAI of 3 and the effect of LAI is simulated at 20°C.

greenhouse structure. The expected increase in production was not observed, probably because it was counteracted by a negative influence of frequent movements on tomato plant growth. At identical plant density and LAI, a crop grown in four rows will intercept more light in its early stages than a crop grown in two rows, guided according to the so-called V-system. Later in the cultivation, both crops will look the same from above and light interception will hardly differ.

The amount and quality of light intercepted by a crop can be significantly enhanced or modified by the presence of reflecting material on the ground, such as white plastic sheets and mulches. Bare soil reflects 10–20% of the photosynthetic active radiation (PAR) (400–700 nm), whereas white plastic sheets on the soil surface, a common practice in modern greenhouses, may reflect 50–80% of the PAR and increase P_{gc} over the whole season by at least 7% for an LAI of 3 (Gijzen, 1995). This is especially important for the young crop, where a lot of light is transmitted by the crop and reaches the ground.

Greenhouse tomato growers determine the lifetime of individual leaves, since these are removed up to just above the truss that is to be harvested or pruned according to the targeted leaf/fruit ratio. Heuvelink *et al.* (2005) evaluated the influence of leaf area on tomato yield, both by simulations and by experimental work. Measurements at seven commercial tomato farms showed that in the summer season (July–September) light interception was on average 90%, with values varying between 86% and 96%. Simulation of leaf picking based on a desired LAI showed that yield increased up to an LAI of 4, with hardly any effect on yield at higher LAI; the response curve showed a saturation type of curve rather than an optimum response curve. However, both De Koning (1993) and Heuvelink (1999) reported LAI values as low as 1.5 or 2.0 in summer. This results in a reduction of crop photosynthesis, yield (Heuvelink *et al.*, 2005) and fruit quality (Dorais *et al.*, 2004). This could be avoided by using a higher plant density, but early in the season (low light) higher density negatively influences fruit size and yield. Therefore, in modern greenhouse cultivation, a crop starts with a rather low plant density (2.1–2.5 plants/m^2) and in spring a side shoot is retained on one or two of every four plants. In this way, stem density is increased towards the summer and a higher LAI can be maintained, resulting in a significant yield increase without additional planting costs. Increasing stem density towards the summer also results in a more uniform fruit size throughout the season (Cockshull *et al.*, 2001) and better fruit quality (Dorais *et al.*, 2001a). Severe fruit pruning can reduce the length of leaves and consequently LAI and specific leaf area (SLA) (cm^2 leaf/g leaf) because of a reduced sink/source ratio (overproduction of assimilates).

Leaf growth and canopy characteristics for field-grown tomatoes also depend on genetic traits and management practices. Determinate tomato types, which may grow to 2 m in height, are erect and bushy with a restricted flowering and fruiting period. From five to 11 leaves are formed before the apex is transformed into a terminal inflorescence, and further growth is from the leaf axillary buds (Picken *et al.*, 1986). For semi-determinate field-grown tomatoes, LAI increases exponentially with main-stem node number (from 0 to 19–21; average development rate 0.5 nodes/day), and maximum LAI is attained 11 weeks after transplanting (Scholberg *et al.*, 2000). After an initial lag phase of 225°Cd (degree-days), LAI values increased linearly with degree-days after planting (base temperature of 10°C), resulting in an average maximum value of 3.8 (Fig. 4.5a) (Scholberg *et al.*, 2000). Near-optimal light interception within the row for field-grown tomato appears to be attained within 4–6 weeks at intra-row spacings of 45 and 60 cm, respectively (Scholberg *et al.*, 2000). Canopy closure between rows does not occur in practice, and pesticide applications and harvesting operations may become difficult with row spacing < 1.8 m. Thus, based on a maximum observed canopy width of 1.0 m, the percentage ground area covered by the crop would be around 83%, 67% and 55% for row spacings of 1.2, 1.5 and 1.8 m (Scholberg *et al.*, 2000). For optimum light interception and fruit yields of a field-grown tomato crop, the LAI should be around 4–5 with a plant population of 10,000–60,000 plants/ha, according to the cultivars, soil fertility, growing and irrigation systems (raised bed, mulch, drip irrigation, sub-irrigation) and available solar radiation. Lower LAI values would reduce light interception and increase yield losses due to sunburn, while higher values may delay the onset of fruit production and reduce the effectiveness of foliar pesticides (Scholberg *et al.*, 2000).

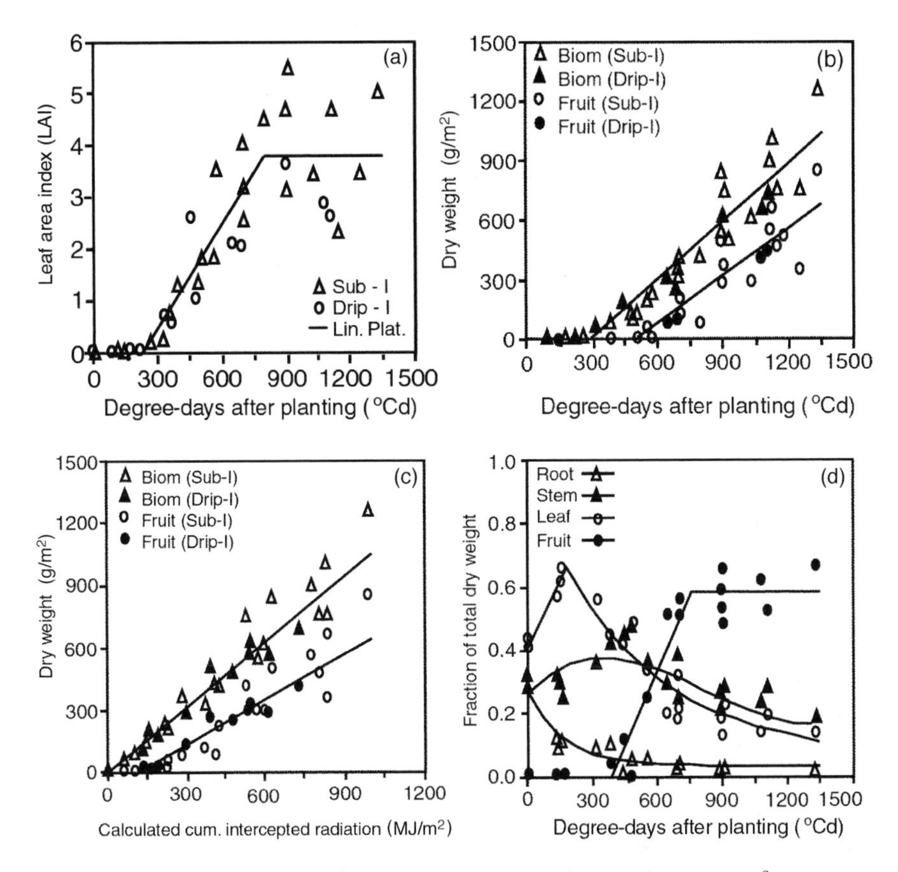

Fig. 4.5. (a) Leaf area index ($r^2 = 0.74$) and (b) total dry weight (Biom) ($r^2 = 0.91$) and fruit dry weight ($r^2 = 0.79$) accumulation of field tomato grown in Florida, USA, as a function of thermal time; (c) estimated cumulative intercepted radiation ($r^2 = 0.92$ and $r^2 = 0.84$, respectively) with subirrigation (Sub-I) and drip irrigation (Drip-I); and (d) fraction of total dry weight accumulated in root ($r^2 = 0.92$), stem ($r^2 = 0.50$), leaf ($r^2 = 0.93$) and fruit ($r^2 = 0.81$). (Reprinted, with permission, from Scholberg *et al.* (2000), *Agronomy Journal* 92, 152–159.)

Light

Environmental factors such as light, CO_2 concentration, temperature and humidity could also influence tomato leaf area. In general, growth responses to environment are influenced by both quantity (cumulative light or light sum or light integral; light intensity × light period; the number of photons intercepted per m^2 per unit of time) and quality (spectral distribution) of light and there may be interaction with temperature and cultural practices. The quantity of light intercepted is affected by a combination of day length, solar angle, atmospheric cover, plant/stem density, canopy structure and, for greenhouse crops, by greenhouse structure and cover materials, while the spectral distribution of light received at a given point depends on solar angle, atmosphere, transmission through leaves, and reflection from nearby plants and other objects, including the soil surface. The phytochrome system (350–800 nm) within the plant acts as a sensor and regulates metabolic events that result in adaptive responses such as stem length, leaf shape and thickness, and carbon partitioning between plant organs (see section on biomass partitioning, below). Cryptochrome (320–500 nm) and ultraviolet-B (280–350 nm) receptors are two other kinds of photoreceptors involved. For example, a high UV-B exposure reduced tomato leaf area after 10 and 19 days of treatment (Hao *et al.*, 1997). High solar radiation may also reduce tomato leaf area and SLA. This so-called short leaf syndrome has been associated with Ca deficiency in the apex when the sink/source ratio is low (high amounts of phloem sap available) and could be prevented by maintaining a higher plant density or an extra shoot on the plant in spring and summer.

The use of supplemental lighting (SL) could influence the light quantity (photosynthetic photon flux (PPF) and light integral; assimilation light) and quality (types of lamps; control light) intercepted by the crop, and then leaf characteristics. Thus, extending the photoperiod to 18 h and 24 h by high-pressure sodium (HPS) lamps (PL 780/N 400, 120 μmol/m^2/s) had no significant effect on the leaf area of greenhouse tomato plants but increased their dry weight. Under a high PPF provided by metal halide lamps (GTE-Sylvania 400 W, 350 μmol/m^2/s), the leaf area of tomato plants grown under 24 h photoperiod was lower than under the 14 h lighting treatment, but had a higher specific weight. In several cases, leaf chlorosis was observed after several days of continuous light treatment with SL. Foliar chlorosis was also reported among tomato plants exposed to a 17 h photoperiod where the 7 h nocturnal period was split into 2 short nights of 3.5 h each (Dorais and Gosselin, 2002). Under 24 h of natural lighting in Finland, tomato plants do not show such negative symptoms.

Carbon dioxide

It has been shown that CO_2 enrichment increases or has little effect on tomato leaf area, while the SLA declines during the development of CO_2-enriched plants (Nederhoff, 1994). High CO_2 increases the dry matter content of leaves and stem, as well as leaf starch concentration. For example, specific leaf weight (SLW) (g leaf per unit of leaf area) increased by up to 51% after 8 weeks of high CO_2 treatment compared with ambient CO_2 level (Yelle, 1988), suggesting that more assimilates were stored per unit of leaf area. For tomato seedlings, CO_2 enrichment increases transplant leaf area and leaf dry weight, with the result that more carbohydrate may be available to the plant for future growth (Tremblay and Gosselin, 1998). CO_2 enrichment induced a 45% increase in SLW in plants with two fruiting trusses, but hardly affected SLW of producing plants (Bertin and Gary, 1998) and it was proposed that structural SLW varies between a maximum and a minimum value according to the ratio of assimilate supply and demand during leaf development. Leaf area was independent of the supply of assimilates when the minimum structural SLW was realized. When the maximum structural SLW was attained, a storage pool of assimilates may have accumulated in leaves during periods of high supply and low demand.

The short leaf syndrome, related to stomatal closure and then leaf calcium deficit, has also been reported under high CO_2 levels (Nederhoff, 1994). Better climate and fertigation management as well as an appropriate fruit/leaf ratio have now solved these side effects of CO_2 enrichment on leaf area.

Temperature

In young crops, temperature plays a role in leaf expansion and hence in the interception of radiation, while in closed canopies its main effect is through maintenance respiration. Growth analysis confirms that in young tomato plants the main effect of temperature on RGR may be attributed to its effect on leaf area ratio (LAR) (leaf area per plant dry weight; cm^2/g), whereas there is only a minor effect on net assimilation rate (NAR) (see section on crop and fruit growth, below) (Table 4.1). LAR is positively correlated with temperature, mainly due to the response of the SLA, whereas leaf weight ratio (LWR) tends to be independent of temperature (Heuvelink, 1989) (see section on crop and fruit growth, p. 104).

LAR and SLA respond not only to the average diurnal temperature, but also to the difference between day and night temperature (DIF). In tomato, SLA was positively affected by increased DIF, while an inverse temperature regime (negative DIF) reduced growth of young plants, primarily because of lower SLA (see Chapter 3). There are indications that the mechanism of DIF relates to the action of phytochrome. Similarly to the DIF, the use of low-temperature pulse (LTP) of 12°C during the end of the 20 h photoperiod reduced the length of the youngest leaves (Dorais *et al.*, unpublished data), while an LTP given at the beginning of the day length of tomato seedlings

Table 4.1. Correlations between growth parameters (average values over a dry weight interval from 20.1 to 2981 mg) of young tomato plants. Data used for correlations are from six treatments (26°C/16°C, 24°C/18°C, 22°C/20°C, 20°C/22°C, 18°C/24°C, 16°C/26°C day/night temperatures; 12 h day length) with an average 24 h temperature of 21°C (Heuvelink, 1989).

	RGR	NAR	LAR	SLA
Net assimilation rate (NAR)	0.469			
Leaf area ratio (LAR)	0.810*	−0.131		
Specific leaf area (SLA)	0.838*	−0.034	0.946***	
Leaf weight ratio (LWR)	−0.354	−0.273	−0.145	−0.455

*$P < 0.05$; ***$P < 0.001$

reduced the number of leaves and their leaf-petiole length (Grimstad, 1995).

Humidity

Although field tomato crops in cool temperate climates rarely experience extremes in humidity, high or low humidity conditions could be found for protected crops during summer or winter, respectively. In some fruit vegetables an effect of air humidity on leaf size has been reported (Bakker *et al.*, 1995). A high humidity could increase the LAR although the NAR is reduced (Grange and Hand, 1987). In short-term studies, humidity had little effect on RGR, LAR and NAR of tomatoes grown under low light conditions (Hurd, 1973), while leaf area increased as humidity increased from 1.0 to 0.2 kPa (Suto and Ando, 1975). Misting could also increase leaf area under high light conditions. Nevertheless, this positive effect can be overruled in tomato by a negative effect of calcium nutrition, which gives rise to smaller leaves (Holder and Cockshull, 1990; Bakker *et al.*, 1995). In a long-term study (7 months), higher humidity (0.4 kPa compared with 0.97 kPa vapour pressure deficit (VPD) and a third treatment with VPD adjusted according to a transpiration rate of 800 ml/day per plant) reduced the leaf dry weight, leaf area and specific leaf weight of the fifth and tenth leaves of tomato plants, due to a low leaf Ca concentration (Iraqi *et al.*, 1997). Under low VPD, the transpiration rate of a leaf might be low, resulting in Ca-related disorders (Grange and Hand, 1987). Low Ca concentrations in tomato leaf laminae were also observed under VPD of 0.1–0.4 kPa (Holder and Cockshull, 1990). Thus, the effect of humidity on plant growth is determined either by the water uptake or through mineral uptake and mineral balance.

Salinity

High water quality is now considered to be the 'blue gold'. Decline in both quantity and quality of irrigation water in many countries is responsible for the use of high-salinity irrigation water (Dorais *et al.*, 2001b). The relationships between salinity and mineral nutrition in horticultural crops are extremely complex. Under high salinity, crop performance may be adversely affected by a water deficit arising from low water potential of the nutrient solution (osmotic effect), and by salinity-induced nutritional disorders associated with excessive ion uptake or nutrient imbalance by nutrient availability, competitive uptake, transport or partitioning within the plant (ionic effect). In the first case, salinity decreases water availability and water uptake and thus reduces root pressure-driven xylem transport of water and solutes. The rate of supply of both water and mineral nutrients to the shoot becomes restricted under these conditions (Dorais *et al.*, 2001b). For example, individual tomato leaf area decreased by 8% per dS/m for electrical conductivity (EC) > 6 dS/m (Li and Stanghellini, 2001). The effect of salinity on SLA is different for different varieties. In the range of 1–6 dS/m, either no effect or a 5% decrease with 1 dS/m increase in EC has been observed.

Leaf and crop photosynthesis

Photosynthetic CO_2 assimilation is a key process in crop production. PAR supplies the energy for crop photosynthesis. The RuBisCO enzyme (ribulose-1,5-*bis*-phosphate carboxylase-oxygenase) catalyses the incorporation of CO_2 (carboxylation), but RuBisCO also has affinity for O_2 (oxygenation), resulting in release of CO_2 (photorespiration). RuBisCO contains as much as 25–40% of the total leaf nitrogen and so it is understandable that leaf nitrogen and leaf photosynthetic capacity are strongly correlated. For example, the maximum rate of photosynthesis ($P_{g,max}$) of a tomato leaf was positively correlated with nitrogen content, while dark respiration was regulated by $P_{g,max}$ and nitrogen content (Osaki *et al.*, 2001). N deficiency decreased $P_{g,max}$ and F_v/F_m ratio as well as stomatal conductance in tomato (Guidi *et al.*, 1998). The relationship between nitrogen status of the tomato plant and photosynthesis will be treated in Chapter 6.

For leaves, the light response curve of gross photosynthesis (P_g) can be generalized (Marcelis *et al.*, 1998) to:

$$P_g = P_{g,max} \times f(x) \tag{4.1}$$

where $P_{g,max}$ is the maximum rate of leaf photosynthesis and x is the dimensionless group $\epsilon H/P_{g,max}$, ϵ is the initial light use efficiency and H is the absorbed radiation per leaf area. Note that ϵ is expressed per unit of absorbed radiation, whereas incident radiation is frequently used for photosynthesis–light response curves. A tomato leaf absorbs about 85% of incident PAR, hence ϵ for incident PAR will be 15% lower than for absorbed PAR. The function value $f(x)$ ranges from 0 in darkness to 1 at saturating light intensities, when $P_{g,max}$ is reached. One of the most general functions for this relationship is the non-rectangular hyperbola:

$$f(x) = \{1 + x - \sqrt{((1 + x)^2 - 4\, x\, \Theta)}\}/(2\, \Theta) \tag{4.2}$$

The parameter Θ regulates the shape of this function and Θ varies between 0 (leading to the softly rounded rectangular hyperbola) and 1 (leading to the sharp-shouldered Blackman response). For the intermediate value of 0.7, the shape of the function is very close to the widely used negative exponential function:

$$f(x) = 1 - e^{-x}$$

Crop photosynthesis can also be described with equation 4.1, but with different parameter values, by applying the total intercepted radiation by the crop for H.

Under the same measurement conditions, tomato leaves grown in different environments may show different photosynthetic rates. For example, it is well known that leaves adapted to low irradiance show a low $P_{g,max}$, whereas ϵ is not or hardly affected (Ludwig, 1974; Hurd and Sheard, 1981).

This means that the oldest leaves in a tomato canopy, which experience low light conditions as compared with younger leaves above them, show a reduced $P_{g,max}$. For indeterminate tomato plants, it has been observed that $P_{g,max}$ of the tenth, 15th and 18th leaves from the apex was reduced, respectively, to 76%, 37% and 18% of $P_{g,max}$ of the fifth leaf, while the estimated net photosynthetic rate of those leaves was 50%, 21% and 7% of the fifth leaf (Xu *et al.*, 1997). The lower $P_{g,max}$ of older leaves is attributed to a reduced RuBisCO content rather than a lower activity. The net photosynthesis rate reaches saturation faster in lower leaves than in upper leaves in response to increasing light. However, as these leaves operate at low light levels this reduced $P_{g,max}$ (which is the light-saturated rate of gross leaf photosynthesis) hardly influences the actual photosynthesis of the whole crop. In a tomato canopy of leaf area index 8.6, Acock *et al.* (1978) observed that the uppermost 23% of the total leaf area assimilated 66% of the net CO_2 fixed by the canopy. In contrast to individual leaves, the light saturation for $P_{gc,max}$ (the maximum rate of crop gross photosynthesis) does not occur for a tomato crop, as part of the canopy does not reach light saturation.

In conditions where the rate of photosynthetic CO_2 assimilation exceeds the capacity of sink organs (e.g. high light and CO_2 concentrations), there is an accumulation of carbohydrates in the leaves which might trigger a down-regulation of photosynthesis involving a repression of several photosynthetic genes, namely RuBisCO and thylakoid proteins. Leaf carbohydrate metabolism such as sucrose-metabolizing enzymes like sucrose phosphate synthase (SPS) and phloem loading could also limit photosynthesis under high light and CO_2 levels. Under these conditions, photosynthesis rate is largely determined by the rate of regeneration of ribulose bisphosphate (RuBP) and the capacity of sucrose synthesis to regenerate inorganic phosphate (P_i). Low P_i levels may affect the transport of triose-phosphates out of the chloroplast, resulting in a high starch concentration that is not completely mobilized overnight. Evidence of the role of carbohydrate metabolism and phloem loading in the photosynthetic limitation of tomato is provided by SPS-transformed plants. Tomato plants over-expressing SPS showed a higher rate of photosynthesis (at CO_2 concentrations of 550 and 750 μmol/mol) and had 2–3 times less starch accumulation in the leaves than non-transformed plants at midday. At the end of the day, leaves of SPS-transformed plants had 30–83% less starch than those of non-transformed plants. Over-expression of SPS in the leaf also increased the phloem loading, SPS and sucrose synthase activity (27%) of the fruit (sink strength), and the sucrose unloading into the fruit (70%), which could explain the higher yield of transformed plants (up to 30%) and the higher fruit soluble solid content (Dorais *et al.*, unpublished data).

Truss peduncle, pedicels, calyxes and green fruit may be quite active photosynthetically and therefore contribute significantly to plant growth. A correlation has been reported between chlorophyll content of peduncle, stem and fruits, and $P_{g,max}$ (Xu *et al.*, 1997). At light saturation, the photosynthetic rate in young truss peduncles reached 5 mg CO_2/h/g dry weight (DW) compared with 3.5 mg CO_2/h/g in older ones, while the gross photosynthetic rate per green fruit surface area was 15–30% of that in the leaves (Czarnowski and Starzecki, 1990). For small fruit (10 g fresh weight), $P_{g,max}$ reached 0.0043 μmol CO_2/kg FM/s, which was similar to a 3 cm^2 leaf (Xu *et al.*, 1997). As the fruit size increased, $P_{g,max}$ decreased to a negligible value and no net photosynthesis was found in a 100 g fruit. Moreover, photosynthesis in orange fruits did not balance the endogenous catabolic processes and even under optimal conditions (high CO_2 concentration and high PPF), no photosynthetic CO_2 assimilation was observed (Czarnowski and Starzecki, 1990). For immature, red-turning, mature and over-ripe tomato fruits, even though no sign of CO_2 fixation was observed, the effective electron transport, expressed by the values of chlorophyll fluorescence parameters in green fruits, and the RuBisCO activity of immature fruits were similar to those determined in the leaves (Carrara *et al.*, 2001). At PPF of 185 μmol/m^2/s, 29% of photosynthetic electron transport activity on a surface area basis was located in tissues other than leaf laminae, with fruit accounting for 15% (Hetherington *et al.*, 1998). The levels of photosynthetic activity found in the calyx, green shoulder, pericarp and locular parenchyma suggest that all of these tissues have significant roles in CO_2 scavenging and the provision of carbon assimilates, and could influence the ratio of fruit acid to sugar and hence fruit quality (Smillie *et al.*, 1999). Except for young fruit, respiration always exceeds gross photosynthesis in tomato fruit, even under high light. The accumulation of dry matter in a tomato fruit is thus dependent on the import of leaf assimilate.

Light use efficiency

For broad categories of plant species, light use efficiency (LUE) (expressed as mol CO_2 fixed/mol photons absorbed) can respond to variation in atmospheric CO_2 concentration and humidity, the composition of solar radiation, and soil moisture content. Model results indicate that a 10% increase of LUE increased daily crop photosynthesis by 5% on a clear day and by 8.8% on a cloudy day (Gijzen, 1995). At ambient CO_2 concentration, leaf LUE is about 0.05 mol/mol, with a maximum of 0.08 mol/mol without photorespiration (e.g. very high CO_2 levels). LUE of tomato was increased by about 6% to 15% per 100 μmol/mol increase in CO_2 concentration (Nederhoff, 1994). A combination of CO_2 enrichment and SL for greenhouse crops grown in winter in northern regions could have a synergistic effect in increasing LUE, the LAI combined with biomass having a great influence on the LUE (De Koning, 1997). A high temperature, however, increases (photo)respiration and consequently decreases LUE. At ambient CO_2 concentration, LUE decreases by about 15% when temperature rises from 15°C to 30°C. In greenhouse-grown tomato plants, LUE may be higher than in field-grown plants because of the UV-protective epidermis of field plants.

Supplemental lighting and photoperiod

There is small variation in the photosynthetic effectiveness of photons of different wavelengths. Nevertheless, the LUE of SL may differ from that of natural light, due to the different spectrum and the wavelength dependency of quantum efficiency. The efficiency of SL at 50 μmol/m^2/s of lighting equals 0.052 and 0.051 mol/mol for sun and shade leaves, respectively, and decreases when natural PPF is added, but increased by 19% when CO_2 concentration was increased to 700 μmol/mol (Gijzen, 1995). Theoretically, the relative photosynthetic yield of HPS lamps should be 34% higher than that of natural light (Bakker *et al.*, 1995).

For a winter tomato crop (seedlings in June/July) grown under SL, photosynthetic rates of the fifth and tenth leaves from the apex followed the variations of incident light from November to January and then became saturated in March when light at the canopy level reached 900 μmol/m^2/s. In May, the photosynthetic rate was maximum early in the morning and then declined after only 4 h of light period even though the incident light was increasing. These changes in photosynthetic activities were accompanied by large accumulations of carbohydrate in leaves, such as starch and hexoses. The F_v/F_m ratio, which is a quantitative measure of photochemical efficiency of photosystem II (PSII), was constant during the winter, while in spring the F_v/F_m ratio decreased at midday and then increased to its initial value. This response to daily changes of light was related to the carbohydrate accumulation and suggests that tomato leaves were subject to photoinhibition. However, this photoinhibition was reversible, as F_v/F_m recovered at the end of the day. So in winter, tomato crops are source limited by the low natural light even if SL were used. In spring, however, down-regulation of photosynthesis and daily accumulation of carbohydrate suggest a sink limitation of the crop. On the other hand, when a winter tomato crop (seedlings in June/July) was compared with a spring crop (seedlings in November/December), CO_2 assimilation rate of the fifth leaf from the apex of the winter crop declined during the day whereas the CO_2 assimilation rate of the spring crop remained constant during the day. This decrease in CO_2 assimilation rate was related to a reduction of the stomatal conductance accompanied by a decrease in the internal CO_2 concentration, suggesting a stomatal limitation of photosynthesis in winter tomato culture. Nevertheless, the maximum photosynthetic rate of the winter tomato crop leaf was similar to that of the spring tomato crop leaf when measured in the presence of high CO_2 concentration that relieves any stomatal limitation. Both tomato crops had identical effective quantum yield of PSII electron transport and fluorescence quenching coefficient (q_p). Under non-limiting conditions, the leaf of the winter crop appeared to be as efficient as the leaf of the spring crop. From a commercial point of view, these physiological responses are associated with a lower yield: in spring the yield of a winter tomato crop under SL could be 50% lower than that of a spring tomato crop. This physiological response could result from important limitations of the leaf water status (less efficient root system) combined with carbohydrate accumulation, which both reduced photosynthetic efficiency of the tomato leaf (Dorais and Gosselin, 2002). Nowadays, the use of grafted plants (Chapter 9) with better root systems solves part of this problem. Abnormal photosynthetic responses to irradiance, such as hysteresis, photosynthetic depression under high radiation and oscillation, induced or promoted by high salinity (Dorais *et al.*, 2001b), could also be reduced or prevented by grafting.

It is important to take into account that a tomato plant grown under SL needs a minimum dark period of about 6 h. Extended photoperiod > 12 h resulted in a decrease in net photosynthesis (26% and 29% under 18 h and 24 h photoperiod, respectively, compared with a 12 h photoperiod) and the translocation efficiency was only 54–69%. Hence, carbohydrates are accumulated in the leaves, as loss by dark respiration is low or nil. In fact, starch concentration in tomato leaves increased by up to 45% under extended photoperiod, while hexose concentration increased by up to 15% compared with natural light (Dorais and Gosselin, 2002).

Foliar chlorophyll (Chl) content and ratio of Chl *a*/*b* of tomato leaves was negatively correlated to lengthening of the photoperiod with HPS lamps. However, neither PSII nor PSI activities have been related to the loss of photosynthetic efficiency of tomato plants exposed to extended photoperiod, and no photoinhibitory damage was observed. Tomato thylakoid membranes adapt to an extended photoperiod provided by HPS lamps by increasing their PSII/PSI ratio, decreasing the amount of Chl associated with PSI light-harvesting Chl *a*/*b* protein complexes and increasing the amount of Chl associated with PSII light-harvesting Chl *a*/*b* protein complexes. No significant effect of SL by extended photoperiod on the reaction centres of either PS was observed for tomato, and the light saturation point of PSII remained unchanged (Dorais and Gosselin, 2002).

Carbon dioxide

CO_2 is the substrate for photosynthesis. Higher concentrations increase the rate of diffusion of CO_2 into the leaf and therefore gross leaf and crop photosynthesis. Just as for light intensity, gross photosynthetic rate shows a saturation type of response to CO_2 (see Fig. 4.4b). The CO_2 concentration at which net photosynthetic rate equals zero is called the CO_2 compensation concentration (Γ). The initial slope (at Γ) of the curve relating net photosynthesis to CO_2 concentration is called the CO_2 use efficiency (τ). At low CO_2 concentration, net photosynthesis is limited by the affinity of RuBisCO for CO_2 as O_2 competes with CO_2 for its carboxylation site (photorespiration). At high CO_2 concentration, photorespiration is reduced or inhibited and the curve saturates because of the limitation of RuBP regeneration resulting from the increased demand for RuBP. Furthermore, net photosynthesis does not respond any more to higher CO_2 levels, and can even decrease at higher levels, as a result of end-product synthesis limitation. This response can be observed for tomato plants grown under high light and CO_2 levels.

Until quite recently, it was widely believed that the effect of CO_2 enrichment was negligible at low light intensities. This resulted from the misconception that only one factor, either CO_2 or light, could be the limiting factor, restricting photosynthesis and yield. Hence, it was believed that at low light intensities, light was limiting and so CO_2 enrichment would be useless. However, because of inhibition of photorespiration by CO_2, gross photosynthesis increases with CO_2 concentration, even at low light intensities.

Simulated gross crop photosynthesis (P_{gc}) increased by 24% when CO_2 concentration was doubled from 350 to 700 μmol/mol at a PAR intensity of 500 μmol/m^2/s, and by 32% at 1500 μmol/m^2/s. An increase in CO_2 concentration from 350 to 1000 μmol/mol increased crop photosynthesis by 33% and 43% at 500 and 1500 μmol/m^2/s, respectively (Gijzen, 1995). For a tomato crop, the calculated value of P_{max} at 1000 μmol/mol of CO_2 and incident PAR of 1350 μmol/m^2/s was 6.5 g/m^2/h (Nederhoff, 1994). Using regression equations, the relative increase of photosynthesis (X, % per 100 μmol/mol) caused by additional CO_2 at certain CO_2 concentration (C, μmol/mol) can be roughly estimated for indeterminate tomato plants grown under normal greenhouse conditions by the following rule of thumb $X = (1000/C)^2 \times 1.5$ (Nederhoff, 1994). The light compensation point decreases with increasing CO_2. An increase of 100 μmol/mol in CO_2 concentration could reduce tomato leaf conductance by about 3% and it has a small effect on crop transpiration. However, no significant effect of CO_2 concentration on average leaf temperature was established, except under very low radiation (Nederhoff, 1994).

Although photosynthesis often shows an initial higher short-term response to CO_2 enrichment, the long-term response often declines markedly (Cure and Acock, 1986; Besford *et al.*, 1990; Besford, 1993). For a wide range of plants and climatic conditions, Cure and Acock (1986) concluded that net CO_2 exchange initially increased by 52% when doubling ambient atmospheric CO_2 concentration, but the plants acclimated to elevated CO_2 and the stimulation of photosynthesis rate declined to about 29%. Biochemical and molecular changes of the photosynthetic apparatus under high CO_2 atmosphere could explain the acclimation of tomato. For indeterminate tomato types it was shown that, after 1 week of CO_2 enrichment (900 μmol/mol), the carbon exchange rate of the leaf increased by 37% compared with ambient concentration, but then declined to the rate of the ambient concentration after 9 weeks of treatment (Yelle, 1988). The partial closure of stomata did not affect the carbon exchange rate since substomatal carbon dioxide concentration (C_i) was not reduced by CO_2 enrichment (Yelle, 1988), but decreased transpiration and could lower nutrient uptake. Moreover, neither carbohydrate accumulation nor modification of chloroplast ultrastructure by starch accumulation under high CO_2 explains the loss of photosynthetic efficiency of tomato plants grown in high CO_2 concentrations. As RuBisCO has a crucial role in the fixation of atmospheric CO_2 into the pentose sugar, RuBP, decline of photosynthesis and decrease in RuBisCO with extended use of CO_2 enrichment have been reported. When measured in the growth CO_2 concentration, Besford (1993) observed that P_{max} of the high CO_2-grown leaf was similar to P_{max} of the low CO_2-grown leaf at full leaf expansion, while the P_{max} of the high CO_2-grown younger leaf (25% to around 75% of full expansion) was higher than the low CO_2-grown younger leaf (Fig. 4.6). P_{max} was obtained at 0.25 leaf expansion for low CO_2-grown leaf and at around 0.50 leaf expansion for high CO_2-grown leaf. Similarly, Van

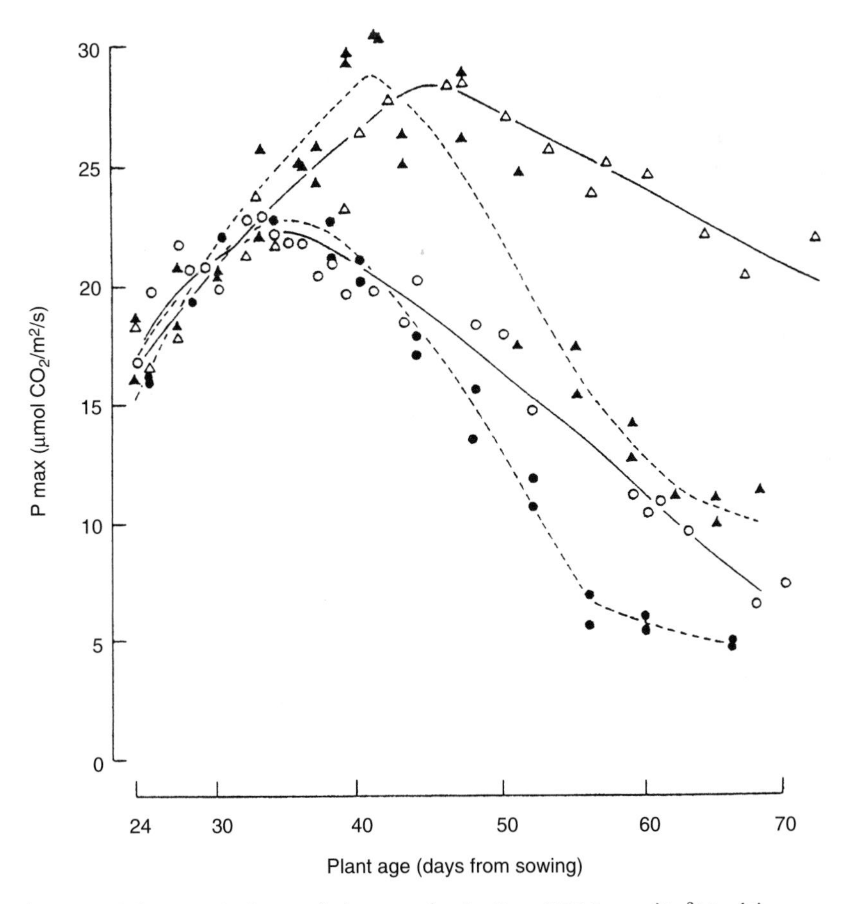

Fig. 4.6. Light-saturated rate of photosynthesis, P_{max} (1500 μmol/m²/s) of the unshaded fifth leaf of tomato plants at various stages of development. Plants grown in 340 μmol CO_2/mol (○, △) or in 1000 μmol CO_2/mol (●, ▲) and measured in 300 μmol CO_2/mol (○, ●) or in 1000 μmol CO_2/mol (△, ▲). (Reprinted, with permission from Kluwer Academic Publishers, from Besford (1993), *Vegetatio* 104/105, 444.)

Oosten and Besford (1995) showed that at 60% and 95% leaf expansion, P_{max} of high CO_2-grown plants measured at growth CO_2 was greater than the P_{max} of the plants grown at ambient CO_2. However, in the fully mature leaves, P_{max} at growth CO_2 declined linearly as growth CO_2 concentration increased. In those leaves, acclimation to elevated CO_2 involved a decline in RuBisCO, Chl, the ratio of RuBisCO/Chl and Chl *a*/*b*. PSI core protein, D_1 and D_2 of the PSII core complex and cytochrome *f* were reduced by elevated CO_2 only in the fully mature leaves (31 days), whereas the large and small subunits of RuBisCO and RuBisCO activase proteins had already declined after 22 days of exposure. In addition, dark respiration measured at growth CO_2 was smaller for the high CO_2-grown plants compared with the control plants. They concluded that the effects of CO_2 on changes in photosynthesis, A/C_i curves, RuBisCO activity, various chloroplast proteins, mRNA and carbohydrate content are leaf-age dependent, and could involve hexose repression of nuclear photosynthesis-associated genes (Van Oosten *et al.*, 1994) via the hexokinase (Jang and Sheen, 1994; Van Oosten and Besford, 1995; Dai *et al.*, 1999). Nevertheless, more studies have to be done to prove the role of hexokinase in CO_2 acclimation.

Other carbohydrate metabolism enzymes, such as SPS, can also play a key role in CO_2 acclimation. For example, over-expression of SPS in tomato leaves (up to sixfold) and fruit (up to 2.5-fold) led to an increase in photosynthesis at high light and elevated CO_2 (800–1500 μmol/mol) (Foyer *et al.*, 1999). The CO_2-dependent increase in foliar starch accumulation was much lower in the leaves of the SPS transformants than in those of the untransformed controls in the same conditions. However, maximal extractable RuBisCO activity was also reduced for SPS transformants grown at high CO_2. Thus, an increased foliar SPS activity does not prevent acclimation of photosynthesis in tomato plants grown with long-term CO_2 enrichment (Murchie *et al.*, 1999).

Temperature

Over a wide range, temperature has only a minor effect on gross photosynthesis of tomato (see Fig. 4.4c). Temperature affects mainly the assimilate demand by the fruit (see section on biomass partitioning, below) without changing photosynthesis rate in the range of 15–25°C (De Koning, 1994). Only under conditions of high light and high CO_2 can P_{gc} be significantly affected by temperature. At decreasing temperature, the photosynthetic apparatus enlarges and in this way the plant compensates for slower enzymatic rates (Berry and Björkman, 1980). Full acclimation of the photosynthetic apparatus of mature leaves to altered temperature requires several days at least. Stomatal conductance increased exponentially with temperature over the range of 15–35°C, even when assimilation rate decreased at high temperature (Bunce, 2000). Below these ranges, photosynthesis increases with temperature because of the increased rates of the light reaction and carboxylation, but net CO_2 assimilation is depressed by the increase in dark respiration and photorespiration. The optimal temperature is higher under high PPF and CO_2 concentrations.

The initial slope of assimilation rate vs. C_i exhibits an optimum temperature which ranges from 20°C to 30°C, while this optimum temperature at high C_i ranges from 25°C to 30°C. For C_3 plants grown at one temperature, photosynthesis does not respond to elevated CO_2 as strongly at low compared with high temperature (Laing *et al.*, 2002). Temperature can also modulate stomatal responses to water deficits and to ABA (Correia *et al.*, 1999) and prevent tomato leaf chlorosis damage under extended photoperiod provided by SL. No leaf symptoms were observed when LTP was used under 20 h photoperiod with HPS lamps (Dorais *et al.*, unpublished data).

The expression of the light-harvesting complex genes can also be influenced by temperature. For field crops, low-temperature exposure in combination with high irradiance causes rapid, often very severe inhibition of photosynthesis in a broad range of plants, including tomato. Several elements that contribute to this inhibition have been identified and all may ultimately arise from the photosynthetic production of oxygen radicals. For chilling-sensitive species such as tomato, the circadian rhythm controlling the activity of SPS and nitrate reductase (NR), key control points of carbon and nitrogen metabolism in plant cells, is delayed by chilling treatments (Jones *et al.*, 1998). The mistiming in the regulation of SPS and NR, and perhaps other key metabolic enzymes under circadian regulation, underlies the chilling sensitivity of photosynthesis in tomato. Byrd *et al.* (1995) observed that photosynthetic rate, RuBisCO activation state and RuBP concentration were reduced after exposing tomato plants to low temperature (4°C for 6 h). These authors explained this response to low temperature by interference at thioredoxin/ferredoxin reduction, which limits bisphosphatase activity, and at RuBisCO activase, which reduces the RuBisCO activation state. Moreover, it was observed that low temperature (6°C for 5 days) altered the membrane lipid composition of tomato chloroplast and decreased the number of granal thylakoids, but had no effect on the total content of protein (Novitskaya and Trunova, 2000). The possible role of variation in membrane lipid composition as a causal factor in the chilling sensitivity of several species has been commented on by Somerville (1995).

Humidity

Little is known about the response of tomato photosynthetic rates to different humidities. However, in many other species, humidity has either little effect in the range of 0.5–2.0 kPa or increases photosynthesis with increasing humidity. Humidity can influence leaf photosynthesis through its effect on stomatal conductance: low humidity may result in water stress and (partial) stomatal closure. Acock *et al.* (1976) observed that, at a CO_2 concentration of 400 μmol/mol, ϵ was 18% larger at a VPD of 0.5 kPa than at 1.0 kPa. At 1200 μmol/mol, ϵ was only 5% larger. Carbon assimilation *per se* appears not to be greatly affected by humidity in the range of 0.2–1.0 kPa unless leaf or plant transpiration exceeds the water supply and water stress results (Grange and Hand, 1987). Photosynthesis of the fifth and tenth leaves from the apex of tomato plants grown under high VPD (adjusted according to a transpiration rate of 800 ml/day per plant, or grown under 0.97 kPa), expressed in terms of net maximum oxygen production, was significantly higher in comparison with plants grown under 0.4 kPa VPD. Apparent quantum yield of the fifth and tenth leaves increased by 38–41% and 42–48%, respectively, compared with the high humidity treatment, while total pigment content of the fifth leaf increased by 34–40% (Iraqi *et al.*, 1997). Consequently, the dry matter accumulation by tomato leaves decreases with increasing humidity from 0.8 to 0.1 kPa (Adams and Holder, 1992). On the other hand, high humidity has been used to partially mitigate the negative effect of high salinity in the root environment, by reducing the transpiration rate and thus the water demand (Li *et al.*, 2001). Hence, photosynthesis was decreased by low humidity only in high EC- and/or water-stressed plants.

CROP AND FRUIT GROWTH

Young plants usually show exponential growth with constant RGR, as mutual shading of leaves is limited (Hunt, 1982):

$$W_{t2} = W_{t1}\, e^{\mathrm{RGR}(t2-t1)}$$

where W_{t1} and W_{t2} are plant dry mass at times $t1$ and $t2$, respectively. RGR can be separated (Fig. 4.7) into a 'photosynthetic term', net assimilation rate (NAR), and a 'morphogenetic term', leaf area ratio (LAR) (Hunt, 1982):

$$\mathrm{RGR} = \mathrm{NAR} \times \mathrm{LAR}$$
$$\mathrm{NAR} = 1/A_t\, \mathrm{d}W_t/\mathrm{d}t$$
$$\mathrm{LAR} = A_t/W_t$$

where A_t is plant leaf area at time t.

An evaluation of experimental design and computational methods in plant growth analysis is given by Poorter and Garnier (1996). LAR may be further separated into the leaf weight ratio (LWR), which is the ratio of leaf dry weight over total plant dry weight, and the specific leaf area (SLA), which is the ratio of leaf area over leaf dry weight. Variations in LAR are usually primarily caused by variations in SLA (Hunt, 1982).

Growth analysis has proved highly effective in studying a plant's reaction to environmental conditions. However, it has its limitation, especially when analysing crop growth, where mutual shading of leaves begins within a few weeks of emergence and is the major reason for a rapid decline of both RGR and NAR. Therefore, attempts to correlate these quantities with changes in

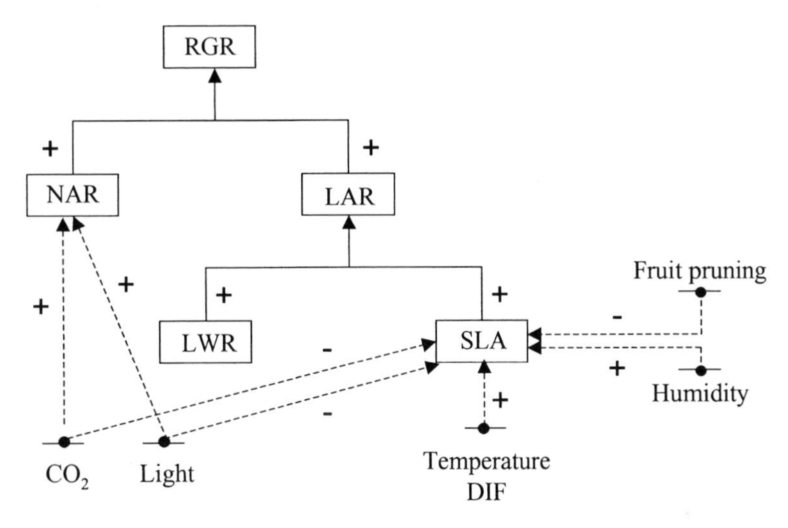

Fig. 4.7. Relational diagram between growth parameters and the influence of climatic and cultural factors of a tomato crop. Solid lines represent relationship between growth parameters and dashed lines represent the positive (+) or negative (–) influence of climatic factors and fruit pruning. LAR, leaf area ratio; LWR, leaf weight ratio; NAR, net assimilation rate; RGR, relative growth rate; SLA, specific leaf area.

the environment during the growing season have rarely been successful (Goudriaan and Monteith, 1990). Furthermore, NAR and LAR are often mutually dependent, showing a negative correlation (Thornley and Hurd, 1974; Bruggink and Heuvelink, 1987). Finally, a comparison based on time-averaged growth parameters may be misleading and averaging over a dry weight interval is preferred, as growth rate is size dependent (ontogenetic effect; Heuvelink, 1989).

Light

A minimum light requirement of 4 mol/m^2/day (1 MJ/m^2 natural PAR = 4.6 mol/m^2 = 71.9 kilo lux hours (klxh) = 6640 foot candles (ft-c), while for HPS 1 MJ/m^2 PAR = 5 mol/m^2 = 118 klxh = 10,970 ft-c; 1 J = 1 W/s; 1 kWh = 3.6 MJ) is reported for tomato fruit set and growth. For tomato seedling production the minimum requirement is lower, whereas a light integral of 4.8–6.0 mol/m^2/day is generally favourable for seedling production, which corresponds to a light intensity of 83 μmol/m^2/s during a photoperiod of 16–20 h. The NAR of tomato seedlings was maximum when the mean daily light integral was 400 J/cm^2 or more, and RGR was maximum at 300 J/cm^2/day. At high radiation, the NAR response of young tomato plants to changes in radiation was partly compensated for by adaptations in LAR (Bakker *et al.*, 1995). An increase in NAR of 10% resulted in summer in a decrease in LAR of 4% and therefore RGR increased by only 6%.

As for many other crops, a linear relationship between cumulative intercepted PAR and tomato dry mass (DM) production has been reported (see Fig. 4.5c) or can be deduced from published data (Heuvelink, 1995b). The slope of this relationship is the LUE, expressed as g DM/MJ intercepted PAR. Heuvelink and Buiskool (1995) observed LUE values varying between 2.8 and 4.0 g/MJ intercepted PAR, in experiments where no CO_2 enrichment was applied. As CO_2 enrichment increases P_{gc} (see Fig. 4.4b) and crop growth at each light intensity, it will increase LUE. On the other hand, De Koning (1993) observed a hyperbolic relationship between PAR intercepted by a tomato crop (range 2–7 $MJ/m^2/day$) and crop growth rate ($g/m^2/day$), with an average LUE of about 2.5 g/MJ. The observed non-linearity may be attributed to the high radiation and low CO_2, the negative correlation between radiation level and the fraction of diffuse radiation (lower LUE with direct radiation), and the low LAI values prevailing in summer. For field-grown tomatoes, an LUE of 2.1 g/MJ (PAR) has been observed (Scholberg *et al.*, 2000), while 2.4 g/MJ is generally used for processing tomatoes (Cavero *et al.*, 1998).

Cockshull (1988) reported a linear relationship between tomato cumulative yield and cumulative solar radiation at crop height within a greenhouse. In a shading experiment (6.4% and 23.4% reduction of solar radiation incident on tomato plants), Cockshull *et al.* (1992) observed over the first 14 weeks of harvest (February to May) that, regardless of treatment, 2.0 kg fresh weight of fruit were harvested for every 100 MJ of global radiation incident on the crops from the onset of harvest (Fig. 4.8). A rule of thumb estimating the effects of light on production, and often used in practice, is the one per cent rule, stating that 1% reduction in light will reduce production by 1%. Despite its simplicity, the one per cent rule often gives good estimates of the consequences of light loss on tomato yield. Cockshull (1988) discussed the problems associated with comparing cumulative yields with cumulative light integrals. If the linear relationship passes through the origin, 1% less light will always give 1% less yield, but other ratios may be obtained if this is not the case. Accurate estimates of the expected yield increase (supplementary assimilation light) or decrease (e.g. double glazing, polycarbonate panels as greenhouse covers) are necessary for sound economic decision making on possible investments. Here, mechanistic models such as TOMSIM (see section on crop growth models, below) are a great help, as effects of simultaneous adjustments in cultivation practices (e.g. narrower spacing under supplementary light or retaining lower numbers of fruit per truss under polycarbonate panels to obtain the same average fruit weight) can also be estimated.

Under low light conditions, initiation of the first inflorescence is delayed as more leaves are initiated prior to the inflorescence, and reduced flower development and flower abortion are observed (Ho, 1996). In commercial practice, fruit set could temporarily be improved under low light by removing

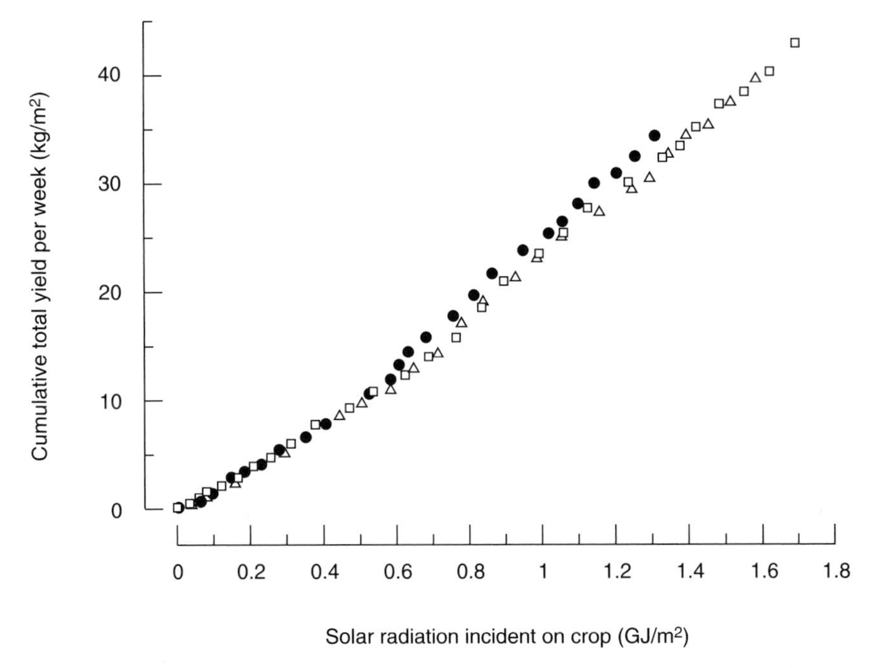

Fig. 4.8. The relationship between cumulative total yield (kg/m^2) and cumulative solar radiation incident on the crop (GJ/m^2) in successive weeks in unshaded (□), light shade (△, −6.4%) or heavy shade (●, −23.4%) treatments. (Reprinted, with permission, from *Journal of Horticultural Science* 67, 18, Cockshull *et al.*, 1992.)

some young leaves (competing sink organs for assimilate). The use of SL for greenhouse crops to secure flowering and fruit set in the winter is successful when there is surplus assimilate after the demand of more competitive organs is met (see section on biomass partitioning, p. 115).

Consequently, SL may be used to increase tomato crop growth and yield in greenhouses, as is done in Canada (Dorais and Gosselin, 2002), Finland and Norway and by a few growers (in total about 90 ha) in The Netherlands. For such an energy- and capital-intensive system to be profitable, high yields must be obtained. On an experimental basis, year-round production can reach more than 90 kg/m^2 with SL of 118–175 μmol/m^2/s (HPS light: 1 μmol/m^2/s = 85 lx = 0.2 W/m^2 PAR = 7.9 ft-c). It has been shown that SL during the winter could increase yield by 70–106% compared with natural light and could sustain a minimal weekly yield of 1 kg/m^2 from November to February (Dorais and Gosselin, 2002). Increasing SL from 100 to 150 μmol/m^2/s gave an additional 20–36% increase in yield, which has been related to a higher number of fruit instead of a higher fruit size (Dorais and Gosselin, 2002). In Canada, the maximum income return has been established at between 100 and 150 μmol/m^2/s, depending on location, which represents 10–45% of the total PAR. In The Netherlands, depending of the time of the year, SL of 100 μmol/m^2/s with HPS lamps provided for 16 h (light integral of 5.76 mol/m^2/day) contributes to 19–71% of the total light integral inside the greenhouse (45% of PAR in the sunlight and 60% of light transmissivity) as the natural PPF decreases from 24.5 mol/m^2/day in June to 2.3 mol/m^2/day in December (1 μmol/m^2/s = 56 lx = 0.217 W/m^2 PAR = 5.2 ft-c). Supplemental lighting is quite often used in tomato seedling production in north-west Europe, but is little used in tomato production for economic reasons. However, for the past few years SL for tomato production has been expanding quite rapidly in The Netherlands and in other European countries such as France, Hungary, Germany and Russia.

Sequential intercrops trained in a V-shaped system have shown that annual yield of 72 kg/m^2 could be obtained with a SL of 150 μmol/m^2/s and a mean weekly yield >1.3 kg/m^2 during winter months was reached (Dorais and Gosselin, 2002). Under this growing system, increasing SL from 100 to 150 μmol/m^2/s increased yield, but did not have a significant effect on fruit quality expressed by the percentage of the crop with blossom-end rot (BER) and misshapen fruit. However, the percentage of Class I fruit increased significantly (larger fruit). Lower initial plant density (2.3 compared with 3.5 plants/m^2) under SL increased the incidence of fruit with BER. On the other hand, tomato fruit sugar content and ascorbic acid concentration increased with PPF, while titratable acid content decreased.

Nevertheless, reduced growth was observed following several days of uninterrupted light. Extending the photoperiod of tomato plants up to 24 h by SL (HPS) of 120 μmol/m^2/s increased shoot dry weight by 10–33% and leaves were thicker, as shown by a lower SLA, but had no effect on total fruit weight (Dorais and Gosselin, 2002). Logendra *et al.* (1990) showed that starch accumulated in the leaves of young tomato plants grown for 7 days under 20 h light, whereas no such accumulation took place for plants grown at 8 h or 16 h (Fig. 4.9). This suggests that chlorotic symptoms in tomato plants grown under 20 h or 24 h light periods results from excess starch accumulation in the leaves, which may disrupt the chloroplast. However, even if starch grains were much larger in size than under natural light, they did not disturb chloroplast structure and thylakoid integrity (Fig. 4.10) (Dorais and Gosselin, 2002). The inability of tomato plants to use SL in an extended photoperiod (18–24 h) was not the result of photoinhibitory conditions, but seems rather to depend on the balance between the production of carbohydrate in source leaves, their loading into and unloading from the phloem, and their utilization or storage in sink organs.

Carbon dioxide

The estimated relative effect of doubling atmospheric CO_2 concentration on production increase varies between 11% and 32% for fruit vegetable crops. For several species, including tomato, relative increases in RGR at elevated

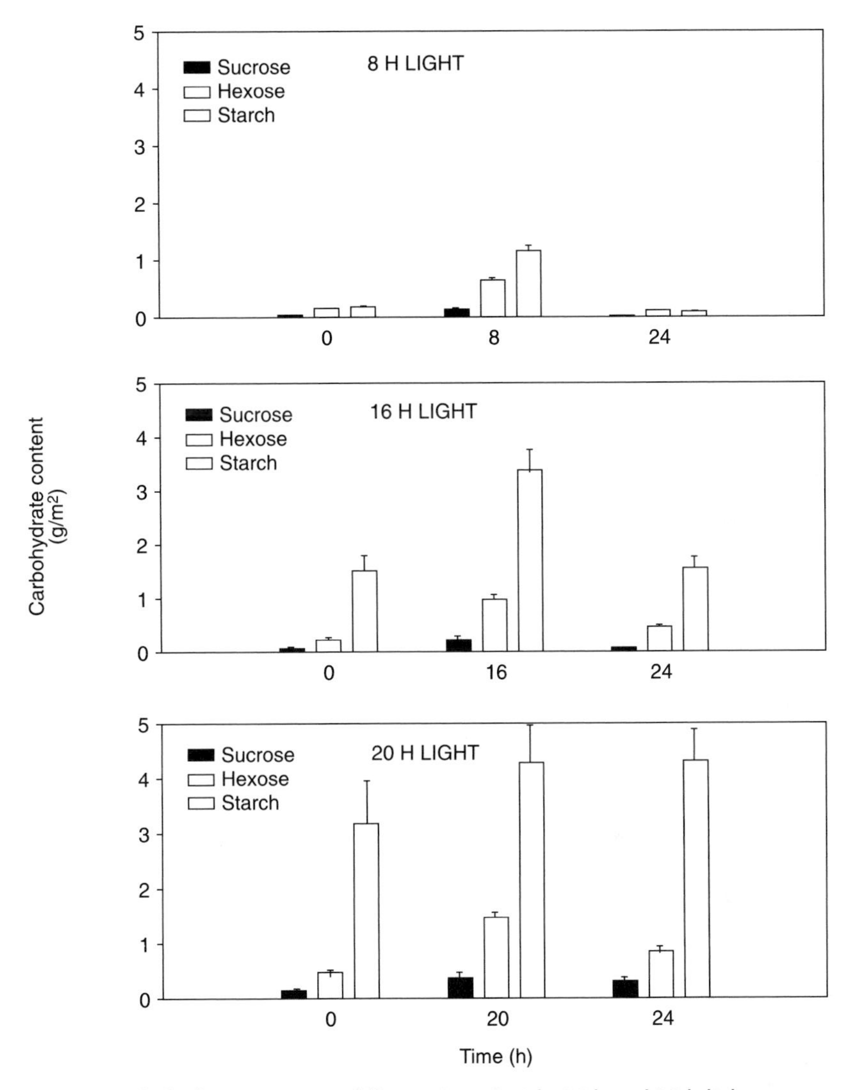

Fig. 4.9. Carbohydrate content at different times in 8 h, 16 h and 20 h light treatments at the beginning of light period (0 h), beginning of dark period (8, 16 or 20 h), and end of dark period (24 h). Vertical lines represent standard errors. (Reprinted, with permission of Elsevier Publishers, from *Scientia Horticulturae* 42, 80.)

CO_2 are correlated with the relative increase in NAR and partly counteracted by a decrease in LAR, because of decreased SLA. After 4 weeks of high CO_2 concentration (330–900 μmol/mol), tomato yield increased by 21% (Yelle, 1988), while an increase of 16% was observed after 20 weeks of harvest of plants grown under 450 μmol/mol compared with 350 μmol/mol

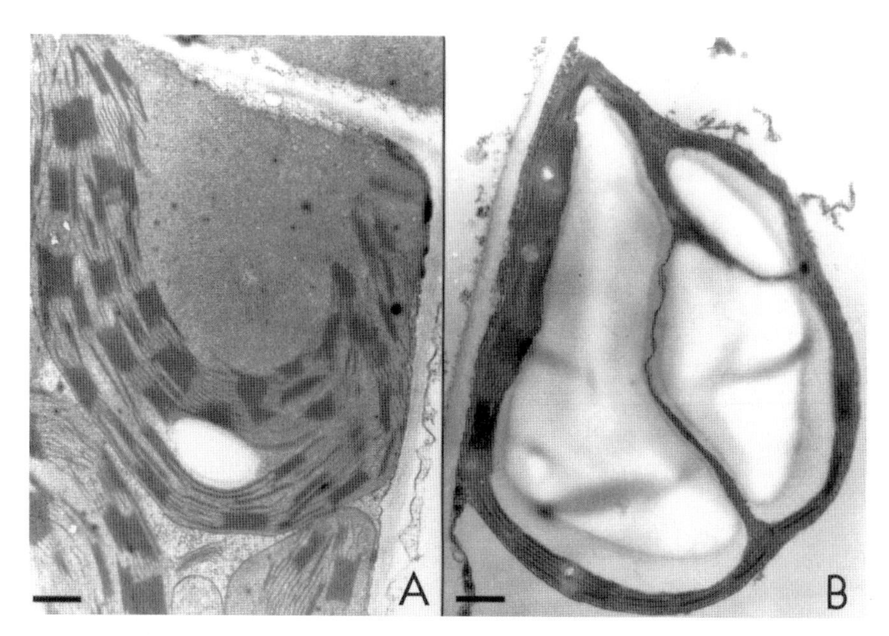

Fig. 4.10. Electron micrograph of tomato chloroplasts (*Lycopersicon esculentum* cv. Trend) after 8 weeks of (a) natural light condition and (b) natural light with 24 h of supplemental light of 120 μmol/m²/s (HPS, PL 780/N 400). Scale bar = 0.05 mm.

(Nederhoff, 1994). Under poor light conditions, CO_2 supply increases tomato fruit set. However, CO_2 concentration has no effect on rate of appearance of leaves and trusses or on the dry matter allocation, as sink strength rather than source strength determines the assimilate partitioning (see section on biomass partitioning, below).

Temperature

Temperature has a direct influence on plant metabolism and thus affects crop growth and yield. The effect of temperature on the growth of closed canopies is mainly through maintenance respiration. The rate of maintenance respiration has been described as an exponential function of temperature with a temperature coefficient (Q_{10}) of about 2, meaning that its rate is doubled with a rise in temperature of 10°C. In the long term, the effect of temperature on crop maintenance respiration rate is small. At cool temperatures a larger biomass will compensate for a lower respiration rate per unit of biomass, whereas at warmer temperatures biomass is reduced, which in turn compensates for the increase in respiration rate per unit of biomass (De Koning, 1994; Heuvelink, 1999).

In an indeterminate tomato plant, temperature affects floral initiation, floral development, fruit set and fruit growth simultaneously (Chapter 3) and its effect is closely associated with light conditions. Under optimal temperatures but insufficient light, floral development may suffer or even result in abortion (Ho, 1996). The optimum temperature for vegetative growth is 18°C to 25°C, while the flowering rate increases almost linearly when temperature increases from 17°C to 27°C (Chapter 3). The optimal temperature for anthesis is 19°C, and fruit set is interrupted above day/night temperatures of 26/20°C (El Ahmadi, 1977). High temperatures may reduce pollen quality, increase floral anomaly and consequently reduce the fruit number (Dorais *et al.*, 2001a), while a low night temperature (10–13°C) can induce the seedling to produce a higher flower number (Ho, 1996). A low temperature at the beginning of truss initiation may also cause split trusses and malformed flowers. Fruit set can be inhibited under low temperature as a result of low pollen viability at 10°C (Ho, 1996). The influence of temperature on partitioning, both direct and indirect through fruit set and abortion, will be discussed in the section on biomass partitioning.

Fruit growth rate (FGR) is strongly related to the ambient temperature and to the water supply (Pearce *et al.*, 1993). FGR increases rapidly at the beginning of the day, peaks at midday, and then decreases by the end of the day (Ehret and Ho, 1986a). The increase in growth rate during the morning may well be related to temperature increase when plant water status is not limiting. Pearce *et al.* (1993) reported significant correlation of temperature and FGR with temperature sensitivities of approximately 5 μm/h/°C. Average fruit size decreases with temperature, being a consequence of increased truss appearance rate and accelerated fruit development (including reduction of the ripening time). Hence, at higher temperature an almost similar amount of assimilates has to be distributed over a larger number of fruits, resulting in a lower average fruit weight. Thus, the potential fruit weight at 23°C is about 40% lower than that at 17°C (De Koning, 1994). Moreover, higher temperature can enhance canopy transpiration to induce water stress, resulting in lower fruit volume growth (Pearce *et al.*, 1993). Reduction of duration of fruit growth might also be expected if the appearance of the peroxidase activity responsible for growth termination were increased at high temperatures (Thompson *et al.*, 1998).

The long-term average temperature, rather than the temperature regime, determines crop growth and yield in tomato crops (De Koning, 1988, 1990). A higher day temperature can compensate for a lower night temperature and vice versa (De Koning, 1988). De Koning (1990) studied growth and development of tomato plants receiving low- and high-temperature regimes alternating every 3, 6 or 12 days. During each cycle of 6, 12 and 24 days, respectively, the temperature sum was the same as that for a control kept constant at a moderate temperature. Two temperature amplitudes (difference between high and low temperature level), 3°C and 6°C, were applied. Only the combination of 6°C amplitude and 24-day cycle was excluded. When treatments started 10 days after anthesis of the first truss, no reductions in growth or development were observed. When treatments started 10 days before anthesis of the first truss, only temperature alternation with a 6°C amplitude combined with a 12-day integration period slightly reduced total growth with respect to the control.

This great ability of a mature tomato crop to integrate (compensate within a certain period) temperature offers good possibilities for saving energy in greenhouse cultivation. When using a thermal screen during the night in winter, it is cheaper to heat at night, since the greenhouse is better insulated than during daytime. Thus, using less energy, the same average temperature can be reached. In spring, energy can be saved by allowing the greenhouse to heat up above the desired long-term average temperature by solar radiation during daytime, which compensates for reduced heating during the night. Also, in periods with high wind speed (high energy loss), the temperature set-point may be below the desired average, to be compensated for by a higher-than-average set-point in periods with low wind speed, reducing total heating costs. Depending on the actual conditions and on the temperature fluctuations a grower wishes to accept, annual heating costs may be reduced by 10–15%, using temperature integration. Nowadays, these control possibilities are included in greenhouse-climate computer software in modern facilities.

In greenhouse tomato production, energy input is an important aspect with regard to economics and environmental concerns (CO_2 emissions). Within the cultivated tomato, there is only a little genetic variation in response to temperature, which hampers breeding for equal production and quality at lower temperatures. However, low-temperature tolerance is present in wild tomato species (e.g. *L. pennellii, L. hirsutum*) (Fig. 4.11). RGR in 'Moneymaker' is reduced by 17% at 16°C, compared with 20°C, whereas no reduction is observed in *L. pennellii*. In both species NAR was not influenced, but 'Moneymaker' showed a 21% reduction in LAR at the lower temperature, whereas in *L. pennellii* LAR was equal at 16 and 20°C. The decrease in LAR in

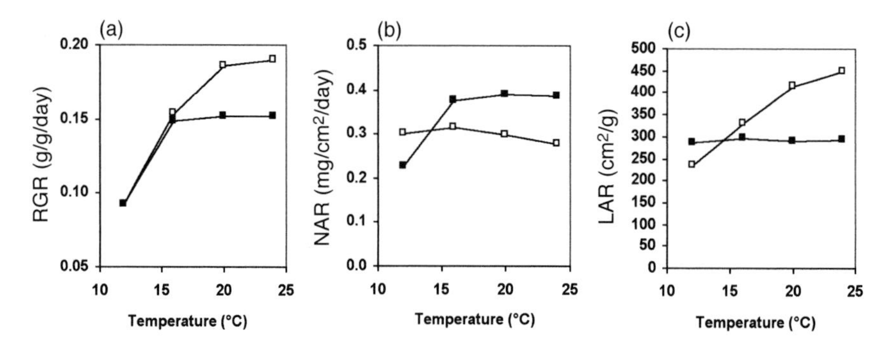

Fig. 4.11. Effect of temperature on (a) relative growth rate (RGR), (b) net assimilation rate (NAR) and (c) leaf area ratio (LAR) of seedlings of *L. esculentum* 'Moneymaker' (□) and *L. pennellii* LA716 (■). (J.H. Venema, A. Van der Ploeg and E. Heuvelink, unpublished.)

'Moneymaker' at decreasing growth temperatures could be wholly attributed to a decrease in SLA. Venema *et al.* (1999a,b) showed that the decrease in SLA in 'Moneymaker' was related to the accumulation of non-structural carbohydrates (soluble sugars + starch), which was much less pronounced in more cold-tolerant related wild *Lycopersicon* spp. The interspecific differences in carbohydrate accumulation in leaves at low root temperatures could be to a great extent ascribed to differences in inhibition of the sink capacity of the roots (Venema, Dijk and Van Hasselt, unpublished results). Through the creation of backcross inbred lines (BILs), low-temperature tolerance may be introduced into commercial cultivars in the near future. A set of near-isogenic and backcross recombinant inbred lines containing most of the *L. hirsutum* genome in an *L. esculentum* genetic background has been developed as a tool for gene mapping and gene discovery by Monforte and Tanksley (2000).

Humidity

Grange and Hand (1987) reported that humidities between 1.0 kPa and 0.2 kPa VPD have little effect on growth and development of horticultural crops, while an increase of VPD from 1 to 1.8 kPa could reduce plant growth of several crops (Hoffman, 1979). Tomato pollination, fruit growth and development are generally not influenced in the range of 0.2–1.0 kPa VPD (Picken, 1984). Under Mediterranean conditions, however, VPD > 2 kPa during the hottest hours of the day reduces fruit growth and fruit fresh weight, probably due to a reduction of the fruit-stem water potential gradient. Under extreme conditions of low humidity (1.5–2.2 kPa or 10–14 g/kg), there is a reduction of growth due to stomatal closure and therefore reduced photosynthesis. Total yield is under those conditions also reduced by a decrease in fruit size (higher dry mass content). Moreover, there is an increase in the number of fruit affected by BER due to a high foliage transpiration rate, thereby limiting the supply of xylem sap (calcium) to the fruit.

For VPDs varying from 0.2 to 0.8 kPa, Bakker (1990) reported low yields, reduced fruit size and short fruit shelf-life under high humidity (low VPD). Yield reductions between 18% and 21% under a VPD of 0.1 kPa (high humidity) have been observed compared with a VPD of 0.5 kPa. Such yield reductions are probably related to a diminution of leaf due to calcium deficiency in the foliage (Holder and Cockshull, 1988) and reduced fruit growth rate (Bakker, 1990). Under sustained high humidity, reduced growth or death of the apex, leaf damage and stem fasciation of tomato could also be observed. There is also a reduction in the ovule fertilization rate, due to more difficult release of the pollen. Conversely, calcium intake into tomato fruit is greater under low VPD or when nights are humid rather than dry.

The lower sensitivity of greenhouse tomatoes to salinity compared with field tomatoes could be attributed to a higher atmospheric humidity under protected cultivation. In the Mediterranean spring/summer growing season, greenhouse misting of salinized tomato plants increased total plant leaf area by 50%, dry matter production by 80% and yield by 100% (Romero-Aranda *et al.*, 2002). Thus, a humid atmosphere may modulate the salinity effect of the nutrient solution. At high salinity (10 dS/m), depressing plant transpiration by 35% under the same solar radiation reduced the incidence of BER from 20% to 2% of the total yield, and might have enlarged fruit size, thereby improving fruit quality and yield value (Stanghellini *et al.*, 1998). Moreover, lower VPD that reduced potential tomato transpiration could alleviate the salinity effect (9.5 dS/m) on fruit growth rate, but could not modulate its effect on fruit development period (Li, 2000). Under moderate salinity, however, higher humidity (VPD 0.2 kPa compared with 0.4 kPa) did not offset the reduction in fruit size when EC increased from 1.8–3.5 to 2.9–5.6 dS/m (Dorais *et al.*, 2000). Low VPD did not affect the enhanced organoleptic quality of fruit grown under these high EC levels.

Salinity

The effect of salinity on tomato crops and yields has been reviewed recently by Dorais *et al.* (2001b). In general, high salinity can reduce the rate of fruit growth and the final fruit size by an osmotic effect. High salinity lowers water potential in the plant, which will reduce the water flow into the fruit and therefore the rate of fruit expansion. Depending on temperature, tomato cultivar, severity of salinity and duration of salinization, the number of trusses per plant as well as the number of flowers per truss might be influenced by extreme salinity conditions, which are not normally found under field or greenhouse soilless growing conditions. For example, ECs of 4.6–8 dS/m reduced fruit yield because of a reduction in fruit size, whereas an EC of 12 dS/m reduced both the number and size of fruit. At very high ECs, decline of fruit number explains the main proportion of yield reduction. According to different studies and growth conditions, salinities > 2.3–5.0 dS/m result in an undesirable yield reduction, while ECs of 3.5–9.0 dS/m improve tomato fruit quality (Dorais *et al.*, 2001a,b).

The effects of high salinity on fruit yield vary according to the cultivar and not all tomato cultivars reduce their fruit size to the same extent. In general, the yield and individual fruit weight of cultivars with smaller fruit size would be less affected by high salinity than cultivars with larger fruit. Thus, growing smaller-fruit cultivars could be advantageous when available water is highly or moderately saline. For field cultivation, models of yield response of tomato plants as a function of root-zone salinity have been made but they should be adapted and validated to soilless cultivation (inert substrate and different cultivars) because there is a buffer soil effect in the field and thus a delay in the build-up of salinity in the soil (Dorais *et al.*, 2001b). In soilless cultivation (rockwool, perlite), root-zone salinity is generally similar to the salinity of the nutrient solution because of the high rate of daily leaching (10–40%), and the adjustment of the salinity according to solar radiation and consequently plant water needs.

BIOMASS PARTITIONING

Total dry matter production and partitioning within the tomato plant may differ considerably. For example, in tomato plants at anthesis of the first truss, Heuvelink (1989) observed that 13% of total dry matter was distributed to the roots (data not presented), while a 4% distribution to the roots in producing tomato plants has been observed by Khan and Sagar (1969). For a long-season crop, the total yield is determined by the balance between vegetative and reproductive growth, and consequently the assimilate partitioning. Usually in a tomato plant, assimilate availability (source strength) is lower than assimilate demand (sink strength), shown by increased fruit size when some of the fruits are removed at an early stage. As shown climatic factors, light intensity and CO_2 concentration influence source strength, whereas sink strength primarily depends on temperature. Based on a simulation model, De Koning (1994) estimated total assimilate demand to be twice the assimilate availability, averaged over a whole growing season. Bertin (1995) calculated a ratio between tomato source and sink strength that was even lower, being about 0.3. (This difference may have a genetic background, as these authors did not use the same cultivar.) Hence, competition between sinks becomes the determining factor for control of biomass allocation. Competition occurs between vegetative and generative plant organs, among trusses and among fruits within a truss, and can be cultivar dependent.

Only biomass allocated to the fruits contributes to yield (see Fig. 4.1); therefore biomass allocation has a direct impact on tomato crop production. For tomato plants grown in spring, summer or autumn for a period of about 100 days after anthesis of the first truss, 54–60% of total dry mass produced was located in the fruits when trusses were pruned to seven fruits (Heuvelink, 1995b). The fraction of dry mass partitioned into the fruits was reduced in winter to 35–38% as a result of flower and/or fruit abortion under low light. De Koning (1993) reported for a year-round greenhouse crop that 72% of total dry mass was allocated to the fruits (Table 4.2), while Cockshull *et al.* (1992) determined a harvest index (HI) of 69% (plants without side shoots), compared to 53–71% with an average of 58% for field-grown semi-determinate tomato plants (Scholberg *et al.*, 2000) (see Fig. 4.5d) and 57–67% for processing

Table 4.2. Total performance of an early tomato crop (De Koning, 1993).

	Fresh weight		Dry weight	
	kg/m^2	%	kg/m^2	%
Fruits	51.7	84.2	2.96	71.5
Leaves	6.6	10.7	0.76	18.2
Stem	3.1	5.1	0.43	10.3
Total	61.4		4.15	

tomato (Hewitt and Marrush, 1986; Cavero *et al.*, 1998). High-yielding crops typically have HI values of about 65%. Higher and lower values may be observed according to the number of trusses harvested and crop management. In an indeterminate plant, fresh weight gain by fruit accounts for almost 85% of the plant gain, because fruits accumulate more water than other organs (Table 4.2). A certain balance between vegetative (future production potential) and generative growth (short-term productivity) should be maintained, as sufficient but not too much new leaf area has to grow for future light interception and biomass production. Besides its influence on total fruit mass, partitioning also influences individual fruit mass in tomato.

Although the mechanism by which a plant partitions its resources between the different organs is of considerable theoretical and practical interest, it is still only poorly understood. It is generally agreed that sinks play an important role in partitioning and in a tomato plant the fruits are the most important sinks. Tomato biomass allocation is both dynamic and complex. It involves the transport of assimilates from sources to sinks. Hence, the sources, the transport path and the sinks may influence allocation.

Source strength

Source strength (availability of assimilates) has no direct influence on assimilate partitioning, as different assimilate availabilities affect growth rate of all organs to the same relative extent. For example, despite a large reduction in assimilate availability at high plant density (plants grown at 1.6, 2.1 and 3.1 plants/m^2 reached a total dry mass of 596, 478 and 340 g/plant, respectively), biomass allocation was not influenced (Heuvelink, 1996b). Indirect effects, through the number of sinks (sink strength), were excluded in this experiment by fruit pruning. However, at very low assimilate availability, priority of the apex over an initiating inflorescence has been found. Under such conditions the inflorescence, or part of its flowers, may abort, while the growth of young leaves continues.

Transport path

Supply of assimilates from leaves to trusses in a multi-truss tomato plant is localized, the three subtended leaves of a truss being the principal suppliers. A truss together with the three leaves immediately below it has been regarded as a sink–source unit, although this relationship is not absolute. As fruit attract assimilate locally, the supply of assimilate to the apex and the roots may be mainly confined to a few upper leaves and the bottom leaves, respectively (Ho, 1996). Within each truss, the fruits on one side tend to receive more assimilate from leaves on the same side of the stem, even if this distribution is not absolute. Slack and Calvert (1977) showed that removing a truss at anthesis resulted in yield increases on some of the remaining trusses closest to the one removed (Fig. 4.12). These observations seem to suggest that the transport path (phloem resistance) plays an important role in biomass allocation in tomato plants. However, a study involving double-shoot tomato plants where, on half of the plants, no trusses were removed from one shoot and all trusses were removed at anthesis from the other shoot (100–0), whereas on the other plants every other truss was removed from both shoots (50–50), showed that biomass allocation was the same for both treatments (Fig. 4.13). No significant influence of distance between source and sink on

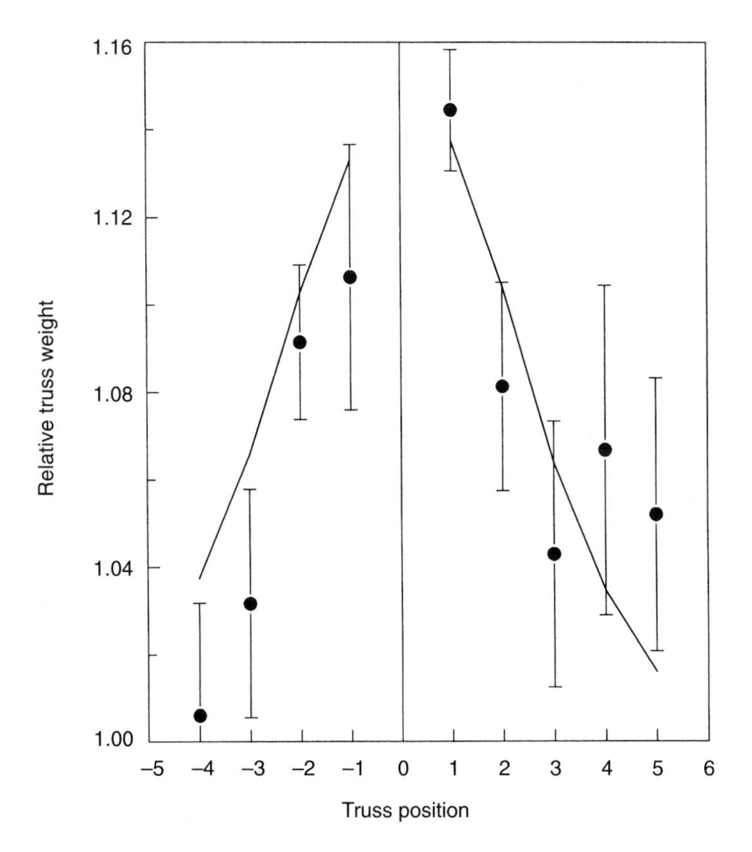

Fig. 4.12. Relative truss weight over controls for trusses above (positive numbers) and below (negative numbers) an excised truss (0) on tomato plants. Measurements (●) (reprinted, with permission, from *Journal of Horticultural Science* 52, 312, Slack and Calvert, 1977) and simulation lines (TOMSIM, see later in this chapter). No influence of position on the plant on sink strength of a truss or vegetative unit was assumed. Vertical bars indicate standard error of mean.

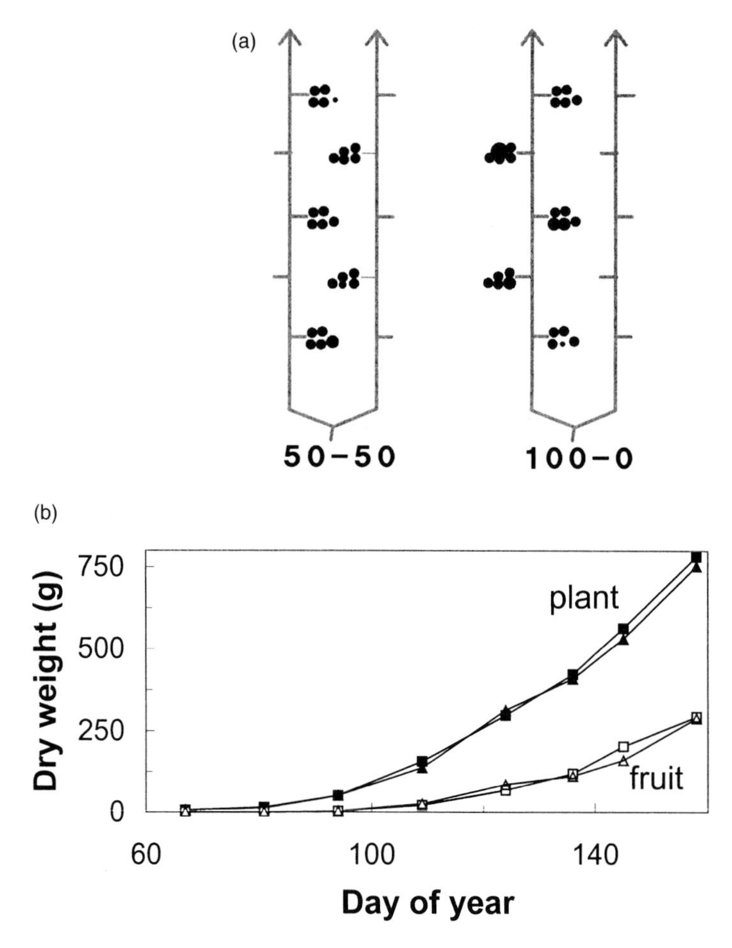

Fig. 4.13. (a) Schematic presentation of two pruning treatments on tomato plants with two equal stems: removal of every second truss at anthesis on each of the two stems (50–50) or removal of all trusses at anthesis from one stem and no truss removal from the other stem (100–0). (b) Total (closed symbols) and fruit (open symbols) dry weight increase for pruning treatment 50–50 (squares) and 100–0 (triangles).

partitioning was found. Hence, biomass allocation may be considered to originate from one common assimilate pool. This does not exclude the possibility that, in intact plants, trusses are fed predominantly by the leaves nearby, nor does it conflict with the observations of Slack and Calvert (1977), which can easily be explained without assuming a 'distance effect' on assimilate partitioning. Trusses closest to the one that has been excised get the highest yield increase. Earlier-initiated trusses have a shorter growth period remaining in which to profit from removing a truss; and later-initiated trusses miss a larger part of the period where removal of the truss plays a role. After one fruit growth period (about 60 days at 20°C) after the expected anthesis date of the excised truss, its removal no longer plays a role. Furthermore, trusses closest to the excised truss exhibit highest sink strength in the period when excision has the largest influence on total sink strength, i.e. the period when the highest sink strength of the excised truss would have occurred. In conclusion, transport path plays only a minor role in tomato biomass allocation, which agrees with observations made for many other plant species (Marcelis, 1996).

Sink strength

Partitioning may be analysed according to the concept of sink strengths. The term sink strength is used to describe the competitive ability of an organ to attract assimilates. Sink strength can be quantified by the potential growth rate of a sink, i.e. the growth rate under conditions of non-limiting assimilate supply (see Fig. 3.9), and depends on sink activity and size. Whereas sink activity is determined by processes such as phloem transport, metabolism and compartmentation, sink size is determined by the cell number (Ho, 1996).

The order of priority in assimilate partitioning changes from the order of root > young leaf > flower in the flowering plant to that of fruit > young leaf > flower and root in the fruiting plant (Ho, 1996). The low sink strength of flowers might be due to low cell division activity, as the enhanced cell division activity in the ovary caused by hormonal treatment (cytokinin and gibberellic acid) may generate a sink strength greater than that of the apex. Under low assimilate supply, cell division is a main limiting factor for fruit growth, but cell enlargement during further fruit development is also affected (Bertin *et al.*, 2002).

The sink strength of a fruit is primarily determined by the fruit developmental stage. The fraction partitioned into the fruits shows a saturation-type response with time (Figs 4.5d and 4.14): in the early stages only a small proportion is partitioned into the fruits, as only a few small fruits are competing with vegetative growth. Once fruit harvest has commenced, a steady-state constant partitioning is reached, if stem growth below the harvest-ripe truss is ignored, and assuming equal numbers of fruit for each truss. Fruits are competing with vegetative parts as a whole, as partitioning within the vegetative parts is fairly constant.

Xiao *et al.* (2004) studied the effect of removal of the second leaf of each vegetative unit between two successive trusses when the leaf was only 1–3 cm long. As expected, the reduced vegetative sink strength by leaf pruning increased partitioning to the fruits from 66% to 74%. However, yield was hardly affected, as leaf pruning resulted in a lower LAI and hence lower light interception and total biomass production. When leaf pruning was combined

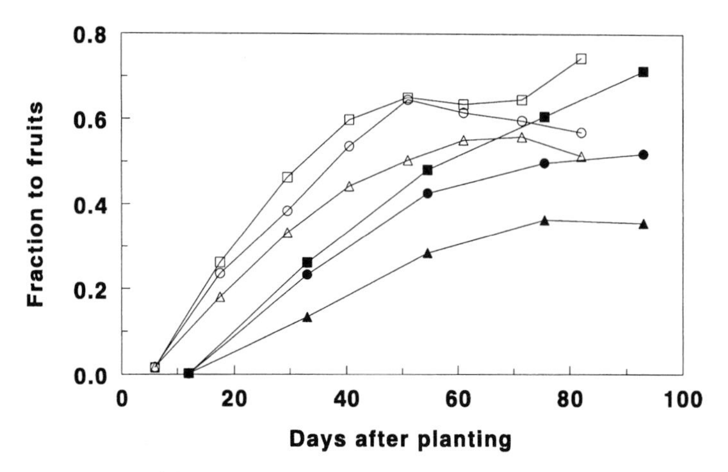

Fig. 4.14. Fraction of dry matter partitioned into the fruits as a function of days after planting, for several pruning treatments: two (▲), three (△), four (●), five (○), six (■) and seven (□) fruits per truss, for a crop planted in January (▲,●,■) or April (△,○,□).

with delayed picking of older leaves to obtain the same LAI as in the control, a yield increase of 13% was predicted. Hence, a tomato cultivar with two instead of three leaves between trusses would have improved yields, when combined with measures to keep LAI sufficiently high.

In the developing tomato fruit, sucrose-metabolizing enzymes may regulate sucrose unloading and sink strength and thus accumulation of fruit dry matter. It has been shown that acid invertase may not play an important role in the regulation of assimilate importation by the tomato fruit (Dorais *et al.*, 1999). Over-expression of SPS in field tomato fruit increased sucrose synthase (Susy) activity by 27%, and 70% more sucrose was unloaded in transformant fruit (20 days after anthesis) compared with the untransformed control (Dorais *et al.*, 1999). Acid invertase and ADPGlc Ppase activities remained at similar levels or were slightly lower than in the untransformed control. Unexpectedly, the repression of Susy activity in the fruit (antisense cDNA of the tomato fruit-specific Susy, TOMSSF) did not affect the unloading capacity when compared with the untransformed plants. Four sucrose turnover cycles (I, sucrose degradation and resynthesis in the cytosol; II, sucrose degradation in the vacuole and resynthesis in the cytosol; III, sucrose degradation in the apoplast and resynthesis in the cytosol; and IV, starch degradation and synthesis in the amyloplast) may control sink activity of tomato fruit (Dorais *et al.*, 1999). Thus, sink utilization (i.e. respiration, cellular structure, growth and storage) determines the sucrose import rate.

Sucrose-metabolizing enzymes may affect the unloading rate but they are not the main regulatory factors. For indeterminate tomato plants, it has recently been observed that fruit dry weight of over-expression SPS plants (maize SPS cDNA with the RuBisCO SSU promoter) was 15–18% higher than for untransformed plants under enriched CO_2 atmosphere and SL, while their fruit soluble solids were 13–20% higher (Dorais *et al.*, unpublished data). Approaches to the manipulation of sink strength in order to improve harvest index and thereby crop yield and quality were discussed by Herbers and Sonnewald (1998).

If sink strength determines dry matter distribution, the fraction of dry matter partitioned to the fruits (F_{fruits}) may be calculated as generative sink strength (SS_{gen}) divided by total plant sink strength, the latter being the sum of generative and vegetative sink strength (SS_{veg}):

$$F_{fruits} = SS_{gen} / (SS_{gen} + SS_{veg}) \quad (4.3)$$

If the number of fruits on a plant is varied without influencing the age distribution of the fruits, it may be assumed that generative sink strength is proportional to the number of fruits on the plant (N_f):

$$SS_{gen} = N_f \times SS_{fruit} \quad (4.4)$$

where SS_{fruit} is the average sink strength of a fruit on the plant. Equations 4.3 and 4.4 can be combined as:

$$F_{fruits} = N_f / [N_f + (SS_{veg}/SS_{fruit})] \quad (4.5)$$

If N_f is divided by the number of trusses on the plant and each truss has the same number of fruits, this results in:

$$F_{fruits} = n_f / [n_f + (SS_{veg.unit}/SS_{fruit})] \quad (4.6)$$

where n_f is the number of fruits per truss and $SS_{veg.unit}$ is the average sink strength of a vegetative unit (three leaves and internodes between two trusses). It is implicitly assumed that every fruit on a truss has the same sink strength. However, at high fruit numbers per truss, the generative sink strength is no longer proportional to the number of fruits per truss: the distal fruits show a reduced sink strength (De Koning, 1994). According to Fig. 4.15a, the ratio between the average sink strength of a vegetative unit and the average sink strength of a fruit has a constant value of 2.96. This agrees with the constant in equation 4.5 (24.2 according to Fig. 4.15b), representing the total vegetative sink strength relative to the average sink strength of a fruit, as the number of growing vegetative units on a tomato plant is about 8 (De Koning, 1994).

Shoot/root ratio

A concept that helps in understanding the partitioning of assimilates between shoot and root is that of functional equilibrium (Brouwer, 1962). According

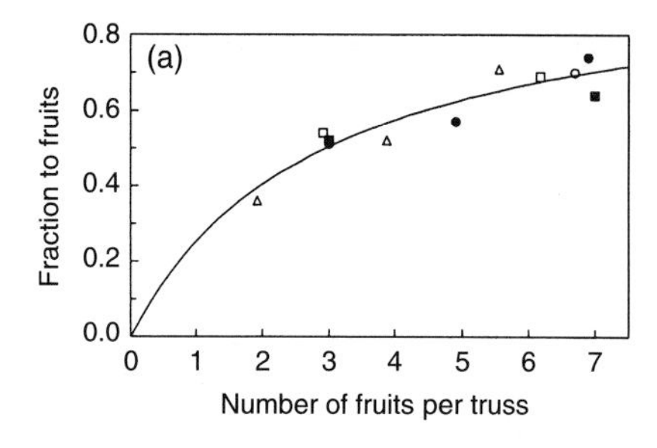

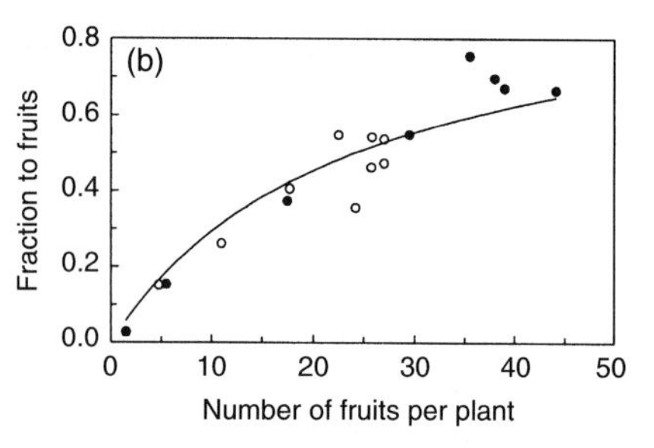

Fig. 4.15. (a) Fraction of dry matter partitioned into the fruits as influenced by fruit pruning just after fruit set for each truss (achieved average number of fruits per truss). Different symbols reflect different experiments. Solid line represents regression curve (equation 4.6: $y = x/(x + 2.96)$, $n = 12$, $r^2 = 0.87$ and standard error of regression is 0.04). (b) Fraction of total dry matter production distributed to the fruits for each time interval between two destructive harvests as a function of number of fruits on the plant averaged over the same time interval: (●) no truss pruning or (○) every other truss removed at anthesis, seven fruits per truss. Solid line represents regression curve (equation 4.5: $y = x/(x + 24.2)$, $n = 18$, $r^2 = 0.89$ and standard error of regression is 0.07).

to this theory, partitioning between shoot and root is regulated by an equilibrium between root activity (water or nutrient absorption) and shoot activity (photosynthesis):

$$W_s / W_r \propto S_r / S_s$$

where W_r is root mass, W_s is shoot mass, S_s is the specific photosynthesis rate of the shoot and S_r is the specific absorption rate of the root. Observations of environmental effects on the tomato shoot/root ratio are well explained by this theory. For example, the shoot/root ratio is lower for plants grown in higher irradiance, which increases specific shoot activity. Similarly, the shoot/root ratio is higher for plants grown at higher root temperature or improved nutrition, which improves specific root activity. Increasing the temperature of the root zone from 14 to 26°C increased the quantity of water absorbed during a day by 30%, the rate of absorption of N, K and Mg by 21–24% and the rate of absorption of Ca and P by 45% and 64%, respectively (Dorais *et al.*, 2001a).

The functional equilibrium theory is useful in understanding partitioning in young, vegetative tomato plants. However, it is not easily applied to fruiting plants. The shoot/root ratio increases as plants grow, and root growth often ceases completely in the early fruiting phase. It later resumes and thereafter the ratio of vegetative shoot to root fresh weight remains essentially constant (4:1).

Influence of growing conditions and cultural practices

While the priority of assimilate partitioning may be determined by the intrinsic potential sink strength of all the sink organs, the actual sink strength may still be affected by growing conditions.

Light can play an indirect or direct role in regulating source–sink relationships involved in allocation of photoassimilate within the growing plant. The light intensity received by the plant affects the quantity of assimilates available to the plant organs and thus their degree of sink competition. Guan and Janes (1991) showed that light also directly stimulates tomato fruit growth, due to mechanisms other than photosynthesis. This can be interpreted as a direct influence of light on tomato fruit sink strength. In field tomato crops grown in summer sunlight over mulches, it has been reported that an upwardly reflected FR:R ratio higher than the ratio in incoming sunlight would signal the growing plant to partition more of its new photosynthate to shoot and fruit growth, while an FR:R ratio lower than that in incoming sunlight would favour partitioning to the root (Kasperbauer and Kaul, 1996).

CO_2 concentration has almost no direct effect on dry matter allocation in summer (Nederhoff, 1994). Indirectly, however, CO_2 enrichment in tomato plants increases fruit set and thus dry matter allocation to the fruits. Similarly, the dry matter allocation into tomato fruits is not influenced by air humidity, although interaction has been observed between air humidity and plant fruit load (Leonardi *et al.*, 2003). The higher the leaf/fruit ratio, the greater are the responses to an increase in air humidity (Gautier *et al.*, 2001).

In greenhouse commercial practice, temperature seems to be the most suitable variable to control biomass partitioning, as sink strength of individual fruits is immediately affected by temperature. High temperature enhances early fruit growth at the expense of vegetative growth. At high temperature, the rate of plant development (increase in new leaves and trusses) is higher (Chapter 3). Therefore, growing young plants (when only a few trusses have appeared) at high temperatures will increase partitioning into the fruits, as it increases the number of growing fruits on the plant. The strong assimilate demand by the growing fruits at the higher temperatures causes not only reduced leaf growth, but also delay of growth of newly set fruits and even flower abortion. As a consequence, after some time total sink strength of the fruits is low and the plant recovers vegetatively and healthy flowers develop. Subsequently, these flowers form strong sinks, resulting in the onset of a second cycle of strong fruit growth. This leads to a more or less pronounced alternation of fruit growth and vegetative growth (Fig. 4.16). These effects of temperature should not be confused with a direct effect of temperature on dry matter allocation.

Published data are not conclusive on a possible direct effect of temperature on partitioning in tomato. Heuvelink (1995c) grew tomato plants in two

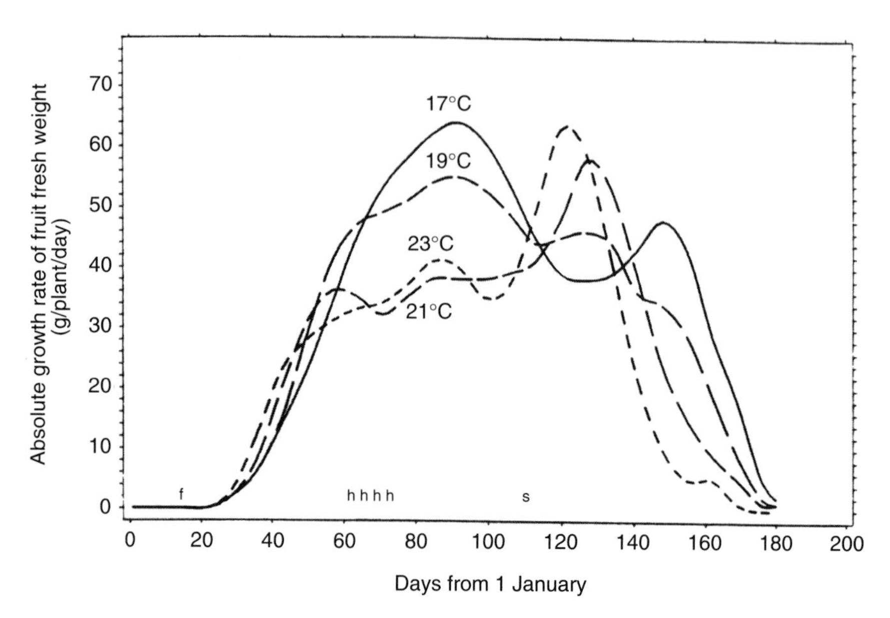

Fig. 4.16. Absolute fresh weight growth rates of fruits (n = 4) at different temperatures from day 14 (14 January) until day 70 (11 March): f, first flowering; h, first harvest at 23, 21, 19 and 17°C, respectively; s, stopping of the plants. (Reprinted, with permission, from *Acta Horticulturae* 248, 335, De Koning, 1989.)

glasshouse compartments. Once harvest had commenced, one compartment was kept at about 24°C, whereas in the other compartment temperature was maintained at about 19°C during a period of 3 weeks. Partitioning towards the fruits was only a little improved at the higher temperature. De Koning (1994) conducted a similar experiment, starting the temperature treatment for a period of 2 weeks at anthesis of the sixth truss, and reported that the fraction of dry matter partitioned to the fruits was 0.68 and 0.80 at 19°C and 23°C, respectively. The use of a low-temperature pulse for a similar 24 h average temperature had no effect on dry matter partitioning of tomato (Dorais *et al.*, unpublished data).

Fruit number appears to be the main determinant for dry mass partitioning between fruits and vegetative parts in tomato plants, whereas temperature (19–23°C) and irradiance (8–15 $MJ/m^2/day$) hardly influence partitioning (Heuvelink, 1995b). Thus, a higher fruit number increases partitioning to the fruits; the fraction partitioned into the fruits is almost doubled when fruit number increases from two to seven per truss (Fig. 4.15b). Conversely, individual fruit mass increases with decreased fruit number per plant, as competition for assimilates among fruits is reduced. This increase in fruit mass results from a higher average growth rate of individual fruits, as fruit growth period (time from anthesis until harvest ripe) is hardly affected by fruit load

(Chapter 3). Fruit pruning can be applied to obtain a certain desired average fruit weight. Expected total fruit growth, mainly influenced by the amount of intercepted light, is important here, as well as temperature. Higher temperature will reduce the fruit growth period (Chapter 3) and, under otherwise equal conditions, result in a lower fruit weight. As shown in Fig. 4.17, it can be deduced that the number of fruits per truss needed for a desired average fruit weight is proportional to the expected total fruit growth rate, and inversely proportional to expected truss appearance rate (temperature influence), plant density and desired fruit weight.

In practice, water and salinity stresses are believed to favour generative development in tomato. Young plants are often stressed to stimulate fruit growth on the first truss. There is no proof that water and salinity stresses directly influence partitioning, but indirect influence by improving fruit set is well known. As shown in Fig. 4.15b, improved fruit set leads to more fruits, and a higher fruit number on the plant will increase partitioning towards the fruits. Under an EC of 6 dS/m, biomass and dry matter partitioning among fruits, vegetative parts and roots were not affected by salinity. Nevertheless, an EC of 10 dS/m reduced plant dry weight by 19% as compared with an EC of 2 dS/m, but did not influence dry matter partitioning. An EC of 17 dS/m slightly reduced dry matter distribution to the fruits (Dorais *et al.*, 2001b).

FRUIT DRY MATTER CONTENT

Fruit dry matter content (dry matter as percentage of fruit fresh weight) decreases during fruit development. Generally, 5–7.5% of tomato content is dry matter; this could reach 17% for wild tomato species (Dorais *et al.*, 2001a). De Koning (1993) reported variations in dry matter content of harvest-ripe tomato fruits from 5.2% in spring to 6.0% in summer. This

(a) Truss appearance rate × truss growth period = number of trusses on plant
(b) Number of trusses on plant × plant density = number of trusses per m^2
(c) Total fruit growth rate / number of trusses per m^2 = average truss growth rate
(d) Average truss growth rate × truss growth period = final truss weight at harvest
(e) Final truss weight / desired fruit weight = number of fruits per truss needed

Substitution of (a), (b), (c) and (d) in formula (e) results in

(f) Number of fruits per truss needed =

$$\frac{\text{Total fruit growth rate}}{\text{Truss appearance rate} \times \text{plant density} \times \text{desired fruit weight}}$$

Fig. 4.17. Formula to determine number of fruits per truss needed to obtain a desired average fruit weight (De Koning, unpublished).

variation in dry matter content affects fresh yield by as much as 15%. E. Dayan (1995, personal communication) observed much larger variations under Israeli conditions: the dry matter content of tomato fruits varied between 4% in winter up to 8% in summer. The rate of dry matter accumulation by a tomato fruit varies among cultivars (Chapter 2), and dry matter content positively correlates with a better taste, as normally sugars represent about 50% of the dry matter in fruits and a higher sugar concentration correlates with a better taste (Chapter 5). Dry matter content of harvest-ripe fruit is slightly higher for fruits grown at non-limiting assimilate supply, probably due to extra storage of carbohydrates. It also increases with temperature (De Koning, 1994).

Dry matter content of the fruit increases linearly with the salinity of the root environment (De Koning, 1994; Dorais *et al.*, 2001a). However, neither the fruit sink strength nor the quantity of assimilates imported by the fruit was affected by high salinity and by a reduction in water absorption (Dorais *et al.*, 2001b). Generally, measures that affect water relationships change only the fresh weight, not the dry weight of the fruits (Ho, 1996) and, therefore, only alter the water import into the fruit. Nevertheless, starch accumulation in the fruit and the activity of the sucrose synthase enzyme were intensified when plants were cultivated under high salinity (Ho, 1996). Ehret and Ho (1986b) reported that carbohydrate partitioning between starch and soluble sugars can be influenced by the osmotic potential of young fruit. Under saline conditions, the starch concentration in the fruit can account for as much as 40% of dry matter. Gao *et al.* (1998) showed that NaCl salinity (0, 50 or 100 mM NaCl) enhanced the transport of ^{14}C-assimilates from the pulse leaf to adjacent fruits and the diversion of ^{14}C label to the starch fraction of the fruit. It also prolonged the period of starch accumulation in developing fruits. These authors suggested that sucrose unloading to the developing fruits is enhanced under high salinity by increased starch, which will later be hydrolysed back to soluble hexoses during fruit maturation, with a resulting improvement of fruit quality. Thus, under saline conditions, both a higher concentration of sucrose in the leaves (higher activity of SPS, lower acid invertase activity) and a faster rate of starch synthesis in the immature fruit (higher activity of ADPG-Glc Ppase) may constitute part of a mechanism responsible for a higher sugar content in the mature fruit.

Unfortunately, there is often an inverse relationship between fruit size, total yield and fruit total solids content, even though a higher solids content of tomato fruit for processing has been the goal for plant breeders.

CROP GROWTH MODELS

As seen in the previous sections, tomato crop growth is a complex phenomenon. The use of simulation models seems appropriate here, as these models have proved to be successful tools for predicting and explaining the behaviour of such a complex, dynamic system as a growing crop.

Marcelis *et al.* (1998) provided an overview of published photosynthesis-based tomato models. Two of these are TOMGRO (Dayan *et al.*, 1993; Gary *et al.*, 1995) and TOMSIM (Heuvelink, 1996b, 1999), which include many of the processes already described (e.g. leaf and crop photosynthesis, light interception, maintenance respiration, dry mass partitioning) and the developmental processes (e.g. truss appearance rate, fruit growth period) presented in Chapter 3. A comparison between the two models has been published (Bertin and Heuvelink, 1993; Heuvelink and Bertin, 1994).

In this chapter a general description of the TOMSIM model is given, as this model will be used in some case studies aimed at an improved understanding of tomato crop growth and yield formation. For a detailed description, see Heuvelink (1999) and references therein. The modular (explanatory) structure and the thorough sensitivity analysis and validation, also at submodel level, make TOMSIM a valuable tool for many purposes. It enables the study and analysis of the tomato crop, a very complex system, as a whole. Crop growth models, such as TOMSIM, may contribute significantly to computerized management support in greenhouse cultivation (Lentz, 1998) and may be used for optimization in greenhouse climate control (Challa, 1990).

TOMSIM: a model for tomato crop growth, development and yield

The potential crop growth is simulated, i.e. the dry mass accumulation under ample supply of water and nutrients in a pest-, disease- and weed-free environment under prevailing climatic conditions. Assimilation rates are calculated separately for shaded and sunlit leaf area at different cumulative leaf areas in the canopy and the daily crop gross assimilation rate ($P_{gc,d}$) is computed by integration of these rates over total crop leaf area and the day. Crop growth results from $P_{gc,d}$ minus maintenance respiration rate (dependent on temperature, crop weight and crop relative growth rate as a measure for metabolic activity) (Heuvelink, 1995a), multiplied by the conversion efficiency of carbohydrates to structural dry matter.

Dry matter partitioning is primarily regulated by the sinks, i.e. individual fruit trusses and vegetative sinks (three leaves and stem internodes between two trusses). A vegetative unit starts to grow about 3 weeks before the corresponding truss and its growing period is assumed to be equal to that of the corresponding truss. All sinks derive their assimilates for growth from one common assimilate pool (see section on biomass partitioning, above). Daily available biomass is distributed among the total number of sinks per plant, according to their relative sink strength, defined as their potential growth rate, relative to the total sink strength of all sinks together. Sink strength of a truss is proportional to the number of fruits on a truss, the latter being an input to the model.

Partitioning within the vegetative plant part is set at 7:3:1.5 for leaves, stem and roots, respectively. When the available biomass equals or exceeds the total sink strength, each sink organ will grow at its potential rate. Assimilates not used for growth are stored as reserves. The next day these reserves are added to the newly formed assimilates. Photosynthesis is calculated on an hourly basis, but the submodel for partitioning runs on a daily basis. Shorter time steps are not expected to affect partitioning, as temperature is the only climatic factor influencing partitioning and, when temperature fluctuations are not too large, underlying processes such as truss appearance rate and fruit development rate respond almost linearly to temperature (De Koning, 1994).

Model parameters (Table 4.3) and initial settings (e.g. initial organ weights and the developmental stage of vegetative units at first anthesis) are the same as used by Heuvelink (1995a) for validation of the dry matter production module and Heuvelink (1996a) for validation of the dry matter distribution module. The model starts at anthesis of the first truss.

Linking dry mass production and dry mass distribution

Two important interactions (feedback mechanisms) between dry matter production and dry matter distribution in tomato can be distinguished (Fig. 4.18): (i) flower and/or fruit abortion at low source/sink ratio (De Koning, 1994; Bertin, 1995), resulting in fewer fruits on the plant and hence decreased sink strength and increased source/sink ratio; and (ii) partitioning to the vegetative parts determining LAI and hence future light interception and dry matter production. It should be noted that, in TOMSIM, dry matter distribution is fully independent of dry matter production: indirect influence via flower and/or fruit abortion is not modelled.

Leaf area is simulated, based on simulated leaf dry weight and specific leaf area (Fig. 4.18). Leaf dry weight results from leaf growth and removal of leaves. Leaves are removed when the developmental stage of the corresponding truss is 0.9 (at 20°C, the truss growth period is 60 days and thus leaves are removed 6 days before the corresponding truss is harvest-ripe). Specific leaf area depends on many factors (Chapter 3), e.g. light intensity, temperature, CO_2 concentration, concentration of the nutrient solution and sink/source ratio. Quantitative knowledge on the underlying principles of these effects and their interactions is limited and therefore simulation of SLA is difficult. Thornley and Hurd (1974) developed a model assuming a constant structural LAR (LAR for structural dry matter), whereas actual LAR results from leaf area and total plant dry weight (structural and storage). As LWR is fairly constant, this also means a constant structural SLA. Actual SLA then decreases as a result of storage of assimilates in the leaves.

Table 4.3. Parameters used in TOMSIM and their numerical values, if constant (DM = dry matter) (Heuvelink, 1995a).

Parameter	Meaning	Value	Unit
ϵ_0	Leaf photochemical efficiency in absence of oxygen	0.084	mol CO_2/mol photon
$P_{g,max}$	Leaf photosynthesis rate at light-saturation, 20°C and 350 μmol/mol CO_2	29.3	μmol $CO_2/m^2/s$
Γ	Carbon dioxide compensation concentration	Variable	μmol/mol
k	Extinction coefficient for diffuse radiation	0.72[a]	–
σ	Scattering coefficient of individual leaves	0.15	–
f	Regression coefficient in exponential equation relating relative growth rate to maintenance respiration	33	day
$MAINT_{lv}$	Maintenance respiration for leaves at 25°C	0.03	g CH_2O/g DM/day
$MAINT_{st}$	Maintenance respiration for stem at 25°C	0.015	g CH_2O/g DM/day
$MAINT_{rt}$	Maintenance respiration for roots at 25°C	0.01	g CH_2O/g DM/day
$MAINT_{fr}$	Maintenance respiration for fruits at 25°C	0.01	g CH_2O/g DM/day
$Q_{10,c}$	Q_{10}-value for temperature effect on maintenance respiration	2.0	–
WLV_i	Initial leaf weight	12.3	g DM/m^2
WST_i	Initial stem weight	4.4	g DM/m^2
WRT_i	Initial root weight	4.4	g DM/m^2
WSO_i	Initial fruit weight	0.08	g DM/m^2
ASR_{lv}	Assimilate requirements for formation of leaf DM	1.39	g CH_2O/g DM
ASR_{st}	Assimilate requirements for formation of stem DM	1.45	g CH_2O/g DM
ASR_{rt}	Assimilate requirements for formation of root DM	1.39	g CH_2O/g DM
ASR_{fr}	Assimilate requirements for formation of fruit DM	1.37	g CH_2O/g DM
$Q_{10,1}$	Q_{10}-value for temperature effect on leaf dark respiration	2.0	–
$R_{d,20}$	Leaf dark respiration rate at 20°C	1.14	μmol $CO_2/m^2/s$
r_b	Leaf boundary layer resistance to H_2O diffusion	100	s/m
r_s	Leaf stomatal resistance to H_2O diffusion	50	s/m
r_m	Mesophyll resistance to CO_2 transport	Variable	s/m

[a] This value assumes a spherical leaf angle distribution.

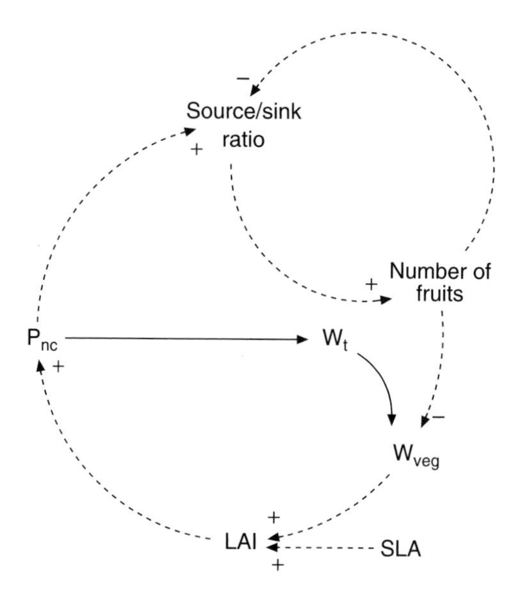

Fig. 4.18. A simplified representation of two important interactions (feedback mechanisms) between dry matter production and dry matter partitioning in an 'indeterminate' tomato crop: (+), positive influence; (–), negative influence. Solid line represents C flow, dashed lines represent information flow. LAI, leaf area index; P_{nc}, crop net assimilation rate; SLA, specific leaf area; W_t, total crop dry mass; W_{veg}, vegetative crop dry mass.

Introducing this hypothesis into an existing tomato crop model greatly improved the model's ability to predict SLA (Gary *et al.*, 1993). However, well-known structural adaptations of leaf anatomy to external conditions are ignored. Taking these adaptations into account, Gary *et al.* (1995) calculated leaf area mainly as a function of temperature and physiological age and allowed structural SLA to vary between a minimum (full satisfaction of growth demand) and a maximum (minimum leaf thickness) value. This may be a promising way to simulate SLA and thus leaf area expansion, although Gary *et al.* (1995) did not validate this part of their model. In TOMSIM, SLA (cm^2/g) is a forcing function (Heuvelink, 1995b), which depends on day of the year (t, day 1 being 1 January):

$$SLA = 266 + 88\ SIN[2\pi(t + 68)/365]$$

Thus only seasonal effects (mainly radiation) (Heuvelink, 1995b) were taken into account. At present no suitable, more explanatory, well-validated and therefore satisfactory prediction of SLA is available.

In the following sections two case studies are presented, illustrating the potential of the photosynthesis-driven tomato crop growth model TOMSIM in the evaluation of greenhouse crop management strategies. Simulations were conducted for 20°C greenhouse air and plant temperature, and outside global radiation of selected months from the 1971–1980 weather records of De Bilt, The Netherlands (Breuer and Van de Braak, 1989), was used. This so-called 'selected year' results in the same average irradiance as observed for the 30-year average global radiation in De Bilt, but it contains a representative variability in radiation. Greenhouse transmissivity for diffuse radiation was set to 63% and light inside the greenhouse was calculated according to the model of Bot (1983). In case study I, CO_2 concentration in the greenhouse was varied between 1000 μmol/mol in winter and 500 μmol/mol in summer; in case study II, CO_2 concentration was fixed at 350 μmol/mol in the greenhouse. Simulations started on 10 January, which was assumed to be the anthesis date of the first truss, and simulations were ended on 7 September (day 250), a reasonable estimate for the date when the top is cut off and plants are left to ripen the remaining trusses in Dutch commercial practice. Plant density was assumed to be 2.5 plants/m^2, with no side shoots retained and all trusses to have seven fruits for the standard growing strategy. Number of fruits per truss is an input to the model; hence, flower and/or fruit abortion is not simulated.

Case study I: optimum fruit number per truss

Fruit yield can be considered as the product of total biomass production and the fraction partitioned to the fruits (harvest index). A larger number of fruits per truss will on the one hand increase partitioning to the fruits (see Fig. 4.15b); on the other hand, as at the same time fewer assimilates are partitioned to the vegetative part, larger fruit numbers will reduce LAI, probably reduce the fraction of intercepted light (see Fig. 4.3) and hence reduce total biomass production. Because of these two counteracting effects, fruit yield shows an optimum response to fruit number per truss (Fig. 4.19). In the range of two to seven fruits per truss, not much influence of fruit pruning on total crop dry mass production was found (Fig. 4.19a), which agrees with experimental data (Heuvelink and Buiskool, 1995). This is the result of a sufficiently high LAI (Fig. 4.19a). When fruit numbers per truss increase further, LAI reduces further and this results in reduced total biomass production, as a result of decreased light interception. Individual fruit size decreased with increasing number of fruits per truss (Fig. 4.19b).

The influence of several factors on optimum fruit number in round tomatoes was quantified by Heuvelink and Bakker (2003), using TOMSIM. Under standard conditions the optimum number of fruits per truss is nine, resulting in a total fruit dry mass of 2.24 kg/m^2 (Table 4.4). Higher fruit numbers per truss reduce mean fruit mass strongly. A lower SLA or advanced leaf removal reduces average LAI and therefore optimum fruit number is reduced (seven instead of nine) (Table 4.4). A 6-day delay in leaf picking

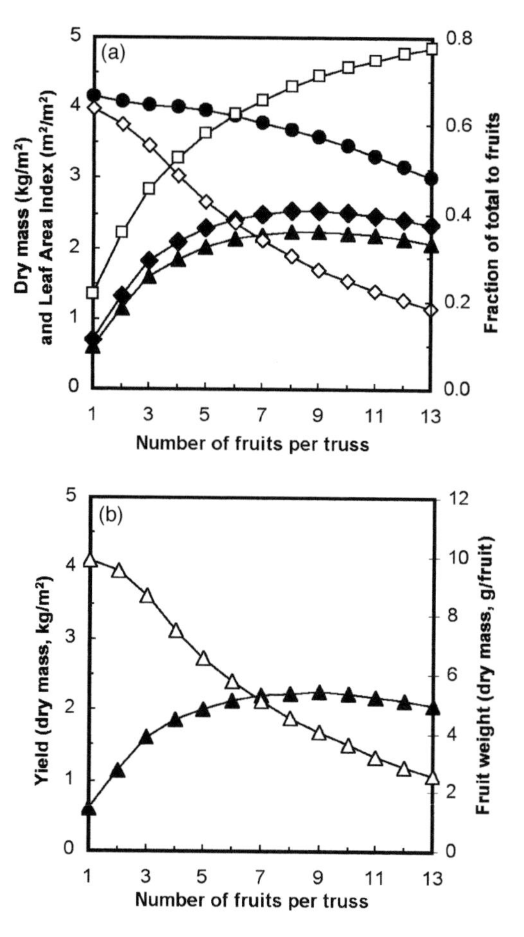

Fig. 4.19. (a) Simulated (TOMSIM) effects of the number of fruits per truss for a round tomato cultivar on leaf area index (◇), the fraction dry mass distributed to the fruits (□) and on fruit dry mass production: total fruit including unripe fruits (♦), harvested ripe fruit (▲), total biomass (●) and (b) average dry weight per fruit harvested (Δ).

increases optimum fruit yield slightly to 2.34 kg/m^2, and this yield is reached at ten fruits per truss. A crop with a higher SLA or delayed leaf removal can attain the same LAI with less dry mass partitioned into the leaves and therefore the number of fruits per truss for maximum fruit yield is higher. Reduced global radiation or CO_2 concentration in the greenhouse reduces total fruit yield and the optimum number of fruits per truss (Table 4.4). The optimum fruit number depends on potential fruit size and hence cultivar. When cultivars would only differ in potential fruit size, optimum fruit number is proportional to the inverse of this potential fruit size (Table 4.4).

Table 4.4. Simulated (TOMSIM) optimum fruit number per truss for a round tomato cultivar and corresponding fruit yield.

Simulation settings	Optimum fruit number per truss	Yield (dry mass, kg/m^2)
Standard conditions (see text)	9	2.24
CO_2 level 350 ppm	6–7	1.56
Global radiation 20% less	8	1.78
SLA 20% lower	7	1.95
Potential fruit size doubled	4–5	2.24
Leaf removal delayed 6 days	10	2.34
Leaf removal advanced 7 days	7	2.05

Optimum fruit number per truss decreases linearly with salinity in the root environment (Fig. 4.20). Salinity was implemented in the TOMSIM model as an effect on leaf expansion only (see case study II for detailed explanation). The slope of the relationship in Fig. 4.20 is –0.8 fruits/dS/m. In these simulations, at salinity levels up to 6 dS/m total fruit yield is not influenced by the number of fruits per truss. However, at 10 dS/m, for example, fruit dry mass yield would be only 0.57 kg/m^2 for seven fruits per truss, whereas this is almost doubled (1.04 kg/m^2) when the optimum fruit number per truss (in this case three) is retained (Fig. 4.20). A reduction in fruit number per truss has been reported

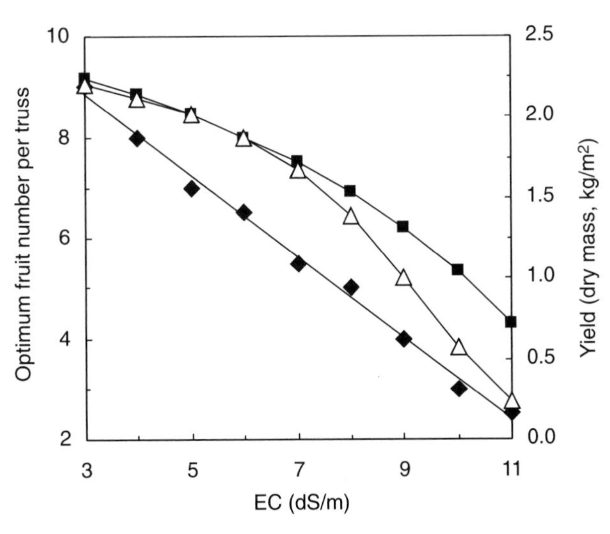

Fig. 4.20. Effect of salt stress on optimum fruit number per truss (♦; regression line $y = -0.8x + 11.3$), and on fruit yield, maintaining optimum fruit number per truss (■) and fruit yield with fixed number of seven fruits per truss (△). EC, electrical conductivity.

based on experiments (Cuartero and Fernadez-Munoz, 1999) and hence would fit in a strategy of the tomato plant to produce the maximum fruit yield under given conditions. Results also imply that under certain conditions fruit pruning may mitigate the negative effect of salinity on fruit yield. Apparently, the increased LAI and hence total biomass production in the pruned plants more than compensates the negative effect of a reduced fraction partitioned into the fruits on total fruit yield in the pruned plants.

The model used in this study is in several ways a simplification of the real tomato crop. Obviously, at 12 fruits per truss, the source/sink ratio is low: an average value of 0.25 was calculated for the present standard simulation. A tomato plant will react to such a low source/sink ratio by flower and/or fruit abortion (Bertin, 1995), thus bringing sink and source more in balance with each other. Here, we simulated with a constant number of fruits per truss during the whole cultivation period. Decreased SLA as a result of high source/sink ratio (Heuvelink and Buiskool, 1995) is not explicitly simulated in the model. It should be noticed, therefore, that an increased number of fruits per truss may have a less strong effect on LAI than predicted (Fig. 4.19), because of a concurrent increase in SLA. Furthermore, conditions are not constant in a greenhouse and hence optimum number of fruits per truss is not constant throughout the year. For example, at low global radiation in spring a lower fruit number per truss will be optimal for yield than that in summer at high global radiation. The optimum number of fruits per truss also varies with cultivars (beefsteak versus round tomato types).

Simulation results are in agreement with practical observations and can be understood on the basis of the underlying processes. More importantly, the model enables the relationships between fruit load, total dry mass production and fruit yield under different conditions to be quantified. It should be noted that the model can evaluate conditions that cannot be attained experimentally, which may be important in theoretical studies.

Case study II: mitigation of salinity effects on yield by improving LAI

Apart from salt-specific toxicities, salinity can reduce crop growth and yield through its impact on plant water relations by: (i) increased fruit dry matter content; (ii) reduced leaf area expansion (leaf size); and (iii) stomatal closure. In a simulation study using TOMSIM, Heuvelink *et al.* (2003) made a comparison of the importance of each of these effects. TOMSIM predicts potential yield, and salinity stress was introduced in the model as follows: (i) increase in fruit dry matter content (5% at EC = 2 dS/m) by 0.2%/dS/m (Li *et al.*, 2001); (ii) decrease in SLA by 8%/dS/m starting from a threshold of 3 dS/m (partly based on Schwarz and Kuchenbuch, 1997) (this effect comes on top of the sinusoidal seasonal pattern for SLA, equation 4.2, and the resulting average LAI, in the simulations with seven fruits per truss, had a slope of about 13% decrease/dS/m); and (iii) increase in stomatal resistance (r_s) by a factor 2 or 4 over the range 1–10 dS/m.

The strongest responses (yield decline/dS/m) were observed when salinity influenced leaf expansion (Table 4.5). Effects of stomatal resistance on yield were rather small, as in a greenhouse (low wind speed) the boundary layer resistance is usually 2–3 times higher than the stomatal resistance; these resistances are in series and influence CO_2 flux and hence photosynthesis. Delayed leaf picking and/or increased plant density had a mitigating effect on yield response to salinity, as light interception reduced less drastically (Fig. 4.21). Since light interception shows a saturation type response to LAI (see Fig. 4.3), reduced leaf area will affect yield only in the lower part of the light interception vs. LAI curve. At EC = 7 dS/m, delayed leaf picking resulted in 10% yield increase (24.2 instead of 22 kg/m^2), whereas at EC = 9 dS/m this was 25% (14.7 instead of 11.8 kg/m^2). Doubling the plant density increased

Table 4.5. Summary of simulated (TOMSIM; nine simulation settings; for each setting six simulations were conducted, i.e. at salinity levels 1, 3, 5, 7, 9 and 11 dS/m) total fruit fresh yields produced by tomato crops from day 10 (first anthesis) until day 250 (decapitation) of the year. DMC indicates whether an effect of EC upon dry matter content is included. Specific leaf area (SLA): No, no effect was included; T3, there is a decrease by salinity (8%/dS/m) starting at EC = 3 dS/m. Stomatal resistance (r_s): either no effect (No) or a multiplying factor linearly increasing to 2 or 4 (F2, F4), respectively, at EC = 10 dS/m. Threshold: EC at which yield has decreased to 90% of maximum. Slope: slope of linear regression line through points on or exceeding the threshold. No yield: EC at which the regression line would result in no yield. Details are published by Heuvelink *et al.* (2003).

Simulation settings			Yield response			
DMC	SLA (dS/m)	r_s	Threshold (%/dS/m)	Slope (dS/m)	No yield	Category[d]
No	No	F2	>11	ND[a]	ND	Resistant
No	No	F4	6.3	1.9	59.7	Resistant
Yes	No	No	3.3	2.8	38.5	Resistant
No	T3	No	6.0	17.8	11.6	Intermediate
Yes	T3	F2	4.4	12.9	13.0	Intermediate/Sensitive
Yes	T3	F4	2.9	12.4	12.1	Intermediate/Sensitive
Yes	T3	No	4.0	13.2	11.6	Intermediate/Sensitive
Yes[b]	T3	No	4.1	12.9	11.9	Intermediate/Sensitive
Yes[c]	T3	No	4.0	11.7	12.6	Intermediate/Sensitive

[a] ND, not determined, as threshold was not reached.
[b] Leaf removal was delayed by about 1 week.
[c] Plant density was doubled.
[d] Qualification of salinity response according to Shannon and Grieve (1999).

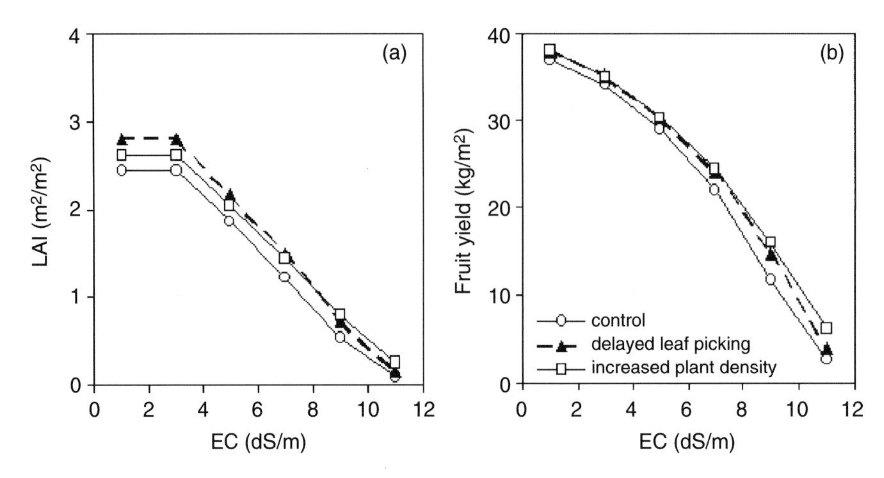

Fig. 4.21. (a) Average leaf area index (LAI) (day 60 to day 250) and (b) total fruit fresh yield, as affected by salinity in three situations. Control: leaves from a vegetative unit were removed when the developmental stage of the corresponding truss (above these leaves) was 0.9 and a plant density at 2.5 plants/m^2; delayed leaf picking: leaf removal at truss stage 1.0 (harvest-ripe) and plant density as control; increased plant density: leaf picking as control, plant density at 5 plants/m^2.

yield by 11% (24.4 instead of 22 kg/m^2) and 36% (16 instead of 11.8 kg/m^2) at EC of 7 or 9 dS/m, respectively (Fig. 4.21).

Simulation results clearly show that salinity effects caused by reduced LAI might be mitigated through actions that increase LAI, e.g. delayed leaf picking or an increased plant density. The effects were quantified by the model (Fig. 4.21), but they still have to be proved by experiments. At present, no literature reports on this mitigation effect are available.

CONCLUSION

Numerous genetic, climatic and cultural factors affect tomato crop growth and yield. As fruit biomass production is mainly determined by canopy photoassimilates, fruit dry matter content can be enhanced by improving either canopy photosynthesis or assimilate distribution within a plant in favour of fruit production. Thus, fruit dry matter content can be manipulated directly or indirectly by light, CO_2, temperature, humidity and water availability during cultivation. Cultural practices, such as leaf and fruit pruning, and plant population can also influence biomass partitioning to the fruits. A better understanding of the rate-limiting steps of the accumulation of dry matter in tomato fruit would identify critical physiological processes that can be improved by plant breeding or by genetic engineering. Plant breeding may provide a long-term solution for yield and quality improvement under different growing conditions, such as low-temperature regime, high salinity or low light levels. Improvement can also be made by a better integration of growing conditions and cultural practices for an optimized balanced plant. In this context, crop modelling has become a major research tool in horticulture in recent decades. A number of descriptive and explanatory models have been developed. These models can be used for yield and quality predictions, climate and crop management, economical and policy analysis and teaching concepts.

For field and greenhouse production of tomatoes, researchers and producers should also consider the environmental and food safety aspects of cultivation. Consumers throughout the world are concerned about genetically modified organisms and new cultural practices such as assimilation lighting. In this vision, sustainable production systems (closed systems) that optimize fruit dry matter production have to be developed and adopted by the tomato industry in the near future.

REFERENCES

Acock, B., Charles-Edwards, D.A. and Hand, D.W. (1976) An analysis of some effects of humidity on photosynthesis by a tomato canopy under winter light conditions and a range of carbon dioxide concentrations. *Journal of Experimental Botany* 27, 933–941.

Acock, B., Charles-Edwards, D.A., Fitter, D.J., Hand, D.W., Ludwig, L.J., Wilson, J.W. and Withers, A.C. (1978) The contribution of leaves from different levels within a tomato crop to canopy net photosynthesis: an experimental examination of two canopy models. *Journal of Experimental Botany* 29, 815–827.

Adams, P. and Holder, R. (1992) Effects of humidity, Ca and salinity on the accumulation of dry matter and Ca by the leaves and fruit of tomato (*Lycopersicon esculentum*). *Journal of Horticultural Science* 61, 137–142.

Bakker, J.C. (1990) Effects of day and night humidity on yield and fruit quality of glasshouse eggplant (*Solanum melongena* L.). *Journal of Horticultural Science* 65, 747–753.

Bakker, J.C., Bot, G.P.A., Challa, H. and Van de Braak, N.J. (1995) *Greenhouse Climate Control: an Integrated Approach.* Wageningen Pers, Wageningen, The Netherlands, 279 pp.

Berry, J. and Björkman, O. (1980) Photosynthetic response and adaptation to temperature in higher plants. *Annual Review of Plant Physiology* 31, 491–543.

Bertin, N. (1995) Competition for assimilates and fruit position affect fruit set in indeterminate greenhouse tomato. *Annals of Botany* 75, 55–65.

Bertin, N. and Gary, C. (1998) Short and long term fluctuations on the leaf mass per area of tomato plants – implications for growth models. *Annals of Botany* 82, 71–81.

Bertin, N. and Heuvelink, E. (1993) Dry matter production in a tomato crop: comparison of two simulation models. *Journal of Horticultural Science* 68, 995–1011.

Bertin, N., Gautier, H. and Roche, C. (2002) Number of cells in tomato fruit depending on fruit position and source-sink balance during plant development. *Plant Growth Regulation* 36, 105–112.

Besford, R.T. (1993) Photosynthetic acclimation in tomato plants grown in high CO_2. *Vegetatio* 104/105, 441–448.

Besford, R.T., Ludwig, L.J. and Withers, A.C. (1990) The greenhouse effect: acclimation of tomato plants growing in high CO_2, photosynthesis and ribulose-1,5-biphosphate carboxylase protein. *Journal of Experimental Botany* 41, 925–931.

Bot, G.P.A. (1983) Greenhouse climate: from physical processes to a dynamic model. PhD dissertation, Wageningen University, Wageningen, The Netherlands.

Breuer, J.J.G. and Van de Braak, N.J. (1989) Reference year for Dutch greenhouses. *Acta Horticulturae* 248, 101–108.

Brouwer, R. (1962) Nutritive influences on the distribution of dry matter in the plant. *Netherlands Journal of Agricultural Science* 10, 399–408.

Bruggink, G.T. and Heuvelink, E. (1987) Influence of light on the growth of young tomato, cucumber and sweet pepper plants in the greenhouse: calculating the effects of differences in light integral. *Scientia Horticulturae* 31, 175–183.

Bunce, J.A. (2000) Acclimation of photosynthesis to temperature in eight cool and warm climate herbaceous C3 species: temperature dependence of parameters of a biochemical photosynthesis model. *Photosynthesis Research* 63, 59–67.

Byrd, G.T., Ort, D.R. and Ogren, W.L. (1995) The effects of chilling in the light on ribulose-1,5-bisphosphate carboxylase/oxygenase activation in tomato (*Lycopersicon esculentum* Mill.). *Plant Physiology* 107, 585–591.

Carrara, S., Pardossi, A., Soldatini, G.F., Tognoni, F. and Guidi, L. (2001) Photosynthetic activity of ripening tomato fruit. *Photosynthetica* 39, 75–78.

Cavero, J., Plant, R.E., Shennan, C., Williams, J.R., Kiniry, J.R. and Benson, V.W. (1998) Application of epic model to nitrogen cycling in irrigated processing tomatoes under different management systems. *Agricultural Systems* 56, 391–414.

Challa, H. (1990) Crop growth models for greenhouse climate control. In: Rabbinge, R., Goudriaan, J., Van Keulen, H., Penning de Vries, F.W.T. and Van Laar, H.H. (eds) *Theoretical Production Ecology: Reflections and Perspectives.* Simulation Monographs 34. Pudoc, Wageningen, The Netherlands, pp. 125–145.

Cockshull, K.E. (1988) The integration of plant physiology with physical changes in the greenhouse climate. *Acta Horticulturae* 229, 113–123.

Cockshull, K.E., Graves, C.J. and Cave, C.R.J. (1992) The influence of shading on yield of glasshouse tomatoes. *Journal of Horticultural Science* 67, 11–24.

Cockshull, K.E., Ho, L.C. and Fenlon, J.S. (2001) The effect of the time of taking side shoots on the regulation of fruit size in glasshouse-grown tomato crops. *Journal of Horticultural Science and Biotechnology* 76, 474–483.

Correia, M.J., Rodrigues, M.L., Osorio, M.L. and Chaves, M.M. (1999) Effects of growth temperature on the response of lupin stomata to drought and abscisic acid. *Australian Journal of Plant Physiology* 26, 549–559.

Cuartero, J. and Fernandez-Munoz, R. (1999) Tomato and salinity. *Scientia Horticulturae* 78, 83–125.

Cure, J.D. and Acock, B. (1986) Crop response to carbon dioxide doubling: a literature survey. *Agricultural and Forest Meteorology* 38, 127–145.

Czarnowski, M. and Starzecki, W. (1990) Carbon dioxide exchange in tomato cluster peduncles and fruits. *Folia Horticulturae* 2, 29–39.

Dai, N., Schaffer, A., Petreikov, M., Shahak, Y., Giller, Y., Ratner, K., Levine, A. and Granot, D. (1999) Overexpression of arabidopsis hexokinase in tomato plants inhibits growth, reduces photosynthesis, and induces rapid senescence. *Plant Cell* 11, 1253–1266.

Dayan, E., Van Keulen, H., Jones, J.W., Zipori, I., Shmuel, D. and Challa, H. (1993) Development, calibration and validation of a greenhouse tomato growth model. I. Description of the model. *Agricultural Systems* 43, 145–163.

De Koning, A.N.M. (1988) The effect of different day/night temperature regimes on growth, development and yield of glasshouse tomatoes. *Journal of Horticultural Science* 63, 465–471.

De Koning, A.N.M. (1989) The effect of temperature on fruit growth and fruit load of tomato. *Acta Horticulturae* 248, 329–336.

De Koning, A.N.M. (1990) Long-term temperature integration of tomato. Growth and development under alternating temperature regimes. *Scientia Horticulturae* 45, 117–127.

De Koning, A.N.M. (1993) Growth of a tomato crop: measurements for model validation. *Acta Horticulturae* 328, 141–146.

De Koning, A.N.M. (1994) Development and dry matter distribution in glasshouse tomato: a quantitative approach. Thesis, University of Wageningen, The Netherlands, 213 pp.

De Koning, J.C.M. (1997) Modelling the effect of supplementary lighting on production and light utilization efficiency of greenhouse crops. *Acta Horticulturae* 418, 65–72.

Dorais, M. and Gosselin, A. (2002) Physiological response of greenhouse vegetable crops to supplemental lighting. *Acta Horticulturae* 580, 59–67.

Dorais, M., Nguyen-Quoc, B., N'tchobo, H., D'Aoust, M.A., Foyer, C., Gosselin, A. and Yelle, S. (1999) What controls sucrose unloading in tomato fruits? *Acta Horticulturae* 487, 107–112.

Dorais, M., Demers, D.A., Micevic, D., Turcotte, G., Hao, X., Papadopoulos, A.P., Ehret, D.L. and Gosselin, A. (2000) Improving tomato fruit quality by increasing salinity: effects on ion uptake, growth and yield. *Acta Horticulturae* 511, 185–196.

Dorais, M., Papadopoulos, A.P. and Gosselin, A. (2001a) Greenhouse tomato fruit quality: the influence of environmental and cultural factors. *Horticultural Reviews* 26, 239–319.

Dorais, M., Papadopoulos, A.P. and Gosselin, A. (2001b) Influence of electrical conductivity management on greenhouse tomato yield and fruit quality. *Agronomie* 21, 367–383.

Dorais, M., Demers, D.A., Papadopoulos, A.P. and Van Ieperen, W. (2004) Greenhouse tomato fruit cuticle cracking. *Horticultural Reviews* 30, 163–184.

Ehret, D.L. and Ho, L.C. (1986a) Effects of osmotic potential in nutrient solution on diurnal growth of tomato fruit. *Journal of Experimental Botany* 37, 1294–1302.

Ehret, D.L. and Ho, L.C. (1986b) The effect of salinity on dry matter partitioning and fruit growth in tomatoes grown in nutrient film culture. *Journal of Horticultural Science* 61, 361–367.

El Ahmadi, A.B. (1977) Genetics and physiology of high temperature fruit-set in tomato. PhD Dissertation, University of California, Davis, California.

Foyer, C.H., Murchie, E., Galtier, N., N'Guyen-Quoc, B. and Yelle, S. (1999) Effects of overexpression of sucrose phosphate synthase on the carbohydrate composition of tomato leaves and fruit. *Acta Horticulturae* 487, 85–92.

Gao, Z., Sagi, M. and Lips, S.H. (1998) Carbohydrate metabolism in leaves and assimilate partitioning in fruits of tomato (*Lycopersicon esculentum* L.) as affected by salinity. *Plant Science* 135, 149–159.

Gary, C., Jones, J.W. and Longuenesse, J.J. (1993) Modelling daily changes in specific leaf area of tomato: the contribution of the leaf assimilate pool. *Acta Horticulturae* 328, 205–210.

Gary, C., Barczi, J.F., Bertin, N. and Tchamitchian, M. (1995) Simulation of individual organ growth and development on a tomato plant: a model and a user-friendly interface. *Acta Horticulturae* 399, 199–205.

Gautier, H., Guichard, S. and Tchamitchian, M. (2001) Modulation of competition between fruits and leaves by flower pruning and water fogging, and consequences on tomato leaf and fruit growth. *Annals of Botany* 88, 645–652.

Gijzen, H. (1995) Interaction between CO_2 uptake and water loss. In: Bakker, J.C., Bot, G.P.A., Challa, H. and van de Braak, N.J. (eds) *Greenhouse Climate Control: an Integrated Approach*. Wageningen Pers, Wageningen, The Netherlands, pp. 51–62.

Goudriaan, J. and Monteith, J.L. (1990) A mathematical function for crop growth based on light interception and leaf area expansion. *Annals of Botany* 66, 695–701.

Grange, R.L. and Hand, D.W. (1987) A review of the effects of atmospheric humidity on the growth of horticultural crops. *Journal of Horticultural Science* 62, 125–134.

Grimstad, S.O. (1995) Low-temperature pulse affects growth and development of young cucumber and tomato plants. *Journal of Horticultural Science* 70, 78–83.

Guan, H. and Janes, H.W. (1991) Light regulation of sink metabolism in tomato fruit. I. Growth and sugar accumulation. *Plant Physiology* 96, 916–921.

Guidi, L., Lorefice, G., Pardossi, A., Malorgio, F., Tognoni, F. and Soldatini, G.F. (1998) Growth and photosynthesis of *Lycopersicon esculentum* (L.) plants as affected by nitrogen deficiency. *Biologia Plantarum* 40, 235–244.

Hao, X., Hale, A.B. and Ormrod, D.P. (1997) The effects of ultraviolet-B radiation and carbon dioxide on growth and photosynthesis of tomato. *Canadian Journal of Botany* 75, 213–219.

Herbers, K. and Sonnewald, U. (1998) Molecular determinants of sink strength. *Current Opinion in Plant Biology* 1, 207–216.

Hetherington, S.E., Smillie, R.M. and Davies, W.J. (1998) Photosynthetic activities of vegetative and fruiting tissues of tomato. *Journal of Experimental Botany* 49, 1173–1181.

Heuvelink, E. (1989) Influence of day and night temperature on the growth of young tomato plants. *Scientia Horticulturae* 38, 11–22.

Heuvelink, E. (1995a) Dry matter production in a tomato crop: measurements and simulation. *Annals of Botany* 75, 369–379.

Heuvelink, E. (1995b) Growth, development and yield of a tomato crop: periodic destructive measurements in a greenhouse. *Scientia Horticulturae* 61, 77–99.

Heuvelink, E. (1995c) Effect of temperature on biomass allocation in tomato (*Lycopersicon esculentum*). *Physiologia Plantarum* 94, 447–452.

Heuvelink, E. (1996a) Dry matter partitioning in tomato: validation of a dynamic simulation model. *Annals of Botany* 77, 71–80.

Heuvelink, E. (1996b) Tomato growth and yield: quantitative analysis and synthesis. Dissertation, Wageningen University, Wageningen, The Netherlands, 326 pp.

Heuvelink, E. (1999) Evaluation of a dynamic simulation model for tomato crop growth and development. *Annals of Botany* 83, 413–422.

Heuvelink, E. and Bakker, M.J. (2003) A photosynthesis-driven tomato model: two case studies. In: Hu, B.G. and Jaeger, M. (eds) *Plant Growth Modeling and Applications*. Tsinghua University Press and Springer, Beijing, pp. 229–235.

Heuvelink, E. and Bertin, N. (1994) Dry-matter partitioning in a tomato crop: comparison of two simulation models. *Journal of Horticultural Science* 69, 835–903.

Heuvelink, E. and Buiskool, R.P.M. (1995) Influence of sink–source interaction on dry matter production in tomato. *Annals of Botany* 75, 381–389.

Heuvelink, E., Bakker, M.J. and Stanghellini, C. (2003) Salinity effects on fruit yield in vegetable crops: a simulation study. *Acta Horticulturae* 609, 133–140.

Heuvelink, E., Bakker, M.J., Elings, A., Kaarsemaker, R. and Marcelis, L.F.M. (2005) Effect of leaf area on tomato yield. *Acta Horticulturae*, GREENSYS 2004, Leuven, Belgium, September 2004 (in press).

Hewitt, J.D. and Marrush, M. (1986) Remobilization of nonstructural carbohydrates from vegetative tissues to fruits in tomato. *Journal of the American Society of Horticulture Science* 111, 142–145.

Ho, L.C. (1996) Tomato. In: Zamski, E. and Schaffer, A.A. (eds) *Photoassimilate Distribution in Plants and Crops: Source–Sink Relationships*. Marcel Dekker Inc., New York, pp. 709–728.

Hoffman, G.J. (1979) Humidity. In: Tibbits, T.W. and Kozlowski, T.T. (eds) *Controlled Environment Guidelines for Plant Research*. Academic Press, London, pp. 141–172.

Holder, R. and Cockshull, K.E. (1988) The effect of humidity and nutrition on the development of calcium deficiency symptoms in tomato leaves. In: Cockshull, K.E. (ed.) *The Effects of High Humidity on Plant Growth in Energy Saving Greenhouses*. Office for Official Publications of the European Communities, Luxembourg, pp. 53–60.

Holder, R. and Cockshull, K.E. (1990) Effects of humidity on the growth and yield of glasshouse tomatoes. *Journal of Horticultural Science* 65, 31–39.

Hunt, R. (1982) *Plant Growth Curves: the Functional Approach to Plant Growth Analysis*. Edward Arnold, London, 248 pp.

Hurd, R.G. (1973) Long-day effects on growth and flower initiation of tomato plants in low light. *Annals of Applied Biology* 73, 211–218.

Hurd, R.G. and Sheard, G.F. (1981) *Fuel Saving in Greenhouses: the Biological Aspects*. Grower Guide 20. Grower Books, London, 55 pp.

Iraqi, D., Gauthier, L., Dorais, M. and Gosselin, A. (1997) Influence du déficit de pression de vapeur (DPV) et de la photopériode sur la croissance, la productivité et la composition minérale de la tomate de serre. *Canadian Journal of Plant Science* 77, 267–272.

Jang, J.C. and Sheen, J. (1994) Sugar sensing in higher plants. *The Plant Cell* 6, 1665–1679.

Jones, T.L., Tucker, D.E. and Ort, D.R. (1998) Chilling delays circadian pattern of sucrose phosphate synthase and nitrate reductase activity in tomato. *Plant Physiology* 118, 149–158.

Kahn, A. and Sagar, G.R. (1969) Alteration of the pattern of distribution of photosynthetic products in the tomato by manipulation of the plant. *Annals of Botany* 33, 753–762.

Kasperbauer, M.J. and Kaul, K. (1996) Light quantity and quality effects on source–sink relationships during plant growth and development. In: Zamski, E. and Schaffer, A.A. (eds) *Photoassimilate Distribution in Plants and Crops: Source–Sink Relationships*. Marcel Dekker Inc., New York, pp. 421–440.

Laing, W.A., Greer, D.H. and Campbell, B.D. (2002) Strong responses of growth and photosynthesis of five C_3 pasture species to elevated CO_2 at low temperatures. *Functional Plant Biology* 29, 1089–1096.

Lentz, W. (1998) Model applications in horticulture: a review. *Scientia Horticulturae* 74, 151–174.

Leonardi, C., Gary, C., Guichard, S. and Bertin, N. (2003) Biomass and nutrient partitioning in tomatoes in relation to air vapour pressure deficit and plant fruit load. *Acta Horticulturea* 614, 561–566.

Li, Y.L. (2000) Analysis of greenhouse tomato production in relation to salinity and shoot environment. Thesis, University of Wageningen, The Netherlands, 96 pp.

Li, Y.L. and Stanghellini, C. (2001) Analysis of the effect of EC and potential transpiration on vegetative growth of tomato. *Scientia Horticulturae* 89, 9–21.

Li, Y.L., Stanghellini, C. and Challa, H. (2001) Effect of electrical conductivity and transpiration on production of greenhouse tomato (*Lycopersicon esculentum* L.). *Scientia Horticulturae* 88, 11–29.

Logendra, S., Putman, J.D. and Janes, H.W. (1990) The influence of light period on carbon partitioning, translocation and growth in tomato. *Scientia Horticulturae* 42, 75–84.

Ludwig, L.J. (1974) Effect of light flux density, CO_2 enrichment and temperature on leaf photosynthesis. In: *Annual Report 1973 Glasshouse Crops Research Institute*. GCRI, Littlehampton, UK, pp. 47–49.

Marcelis, L.F.M. (1996) Sink strength as a determinant of dry matter partitioning in the whole plant. *Journal of Experimental Botany* 47, 1281–1291.

Marcelis, L.F.M., Heuvelink, E. and Goudriaan, J. (1998) Modelling biomass production and yield of horticultural crops: a review. *Scientia Horticulturae* 74, 83–111.

Monforte, A.J. and Tanksley, S.D. (2000) Development of a set of near isogenic and backcross recombinant inbred lines containing most of the *Lycopersicon hirsutum* genome in a *L. esculentum* genetic background: a tool for gene mapping and gene discovery. *Genome* 43, 803–813.

Murchie, E.H., Sarrobert, C., Contard, P., Betsche, T., Foyer, C.H. and Galtier, N. (1999) Overexpression of sucrose-phosphate synthase in tomato plants grown with CO2 enrichment leads to decreased foliar carbohydrate accumulation relative to untransformed controls. *Plant Physiology and Biochemistry* 37, 251–260.

Nederhoff, E.M. (1994) Effects of CO_2 concentration on photosynthesis, transpiration and production of greenhouse fruit vegetable crops. Dissertation, Wageningen Agricultural University, Wageningen, The Netherlands, 213 pp.

Novitskaya, G.V. and Trunova, T.I.D. (2000) Relationship between cold resistance of plants and the lipid content of their chloroplast membranes. *Biochemistry* 371, 50–52.

Osaki, M., Shinano, T., Kaneda, T., Yamada, S. and Nakamura, T. (2001) Ontogenetic changes of photosynthetic and dark respiration rates in relation to nitrogen content in individual leaves of field crops. *Photosynthetica* 39, 205–213.

Pearce, B.D., Grange, R.I. and Hardwick, K. (1993) The growth of young tomato fruit. II. Environmental influences on glasshouse crops grown in rockwool or nutrient film. *Journal of Horticultural Science* 68, 13–23.

Picken, A.J.F. (1984) A review of pollination and fruit set in the tomato (*Lycopersicon esculentum* Mill.). *Journal of Horticultural Science* 59, 1–13.

Picken, A.J.F., Stewart, K. and Klapwijk, D. (1986) Germination and vegetative development. In: Atherton, J.G. and Rudich, J. (eds) *The Tomato Crop: a Scientific Basis for Improvement*. Chapman & Hall, London, pp. 111–166.

Poorter, H. and Garnier, E. (1996) Plant growth analysis: an evaluation of experimental design and computational methods. *Journal of Experimental Botany* 47, 1343–1351 [Erratum: 47, 1969].

Romero-Aranda, R., Soria, T. and Cuartero, J. (2002) Greenhouse mist improves yield of tomato plants grown under saline conditions. *Journal of the American Society for Horticultural Science* 127, 644–648.

Schwarz, D. and Kuchenbuch, R. (1997) Growth analysis of tomato in a closed recirculating system in relation to the EC-value of the nutrient solution. *Acta Horticulturae* 450, 169–176.

Scholberg, J., McNeal, B.L., Jones, J.W., Boote, K.J., Stanley, C.D. and Obreza, T.A. (2000) Field-grown tomato – growth and canopy characteristics of field-grown tomato. *Agronomy Journal* 92, 152–159.

Shannon, M.C. and Grieve, C.M. (1999) Tolerance of vegetable crops to salinity. *Scientia Horticulturae* 78, 5–38.

Slack, G. and Calvert, A. (1977) The effect of truss removal on the yield of early sown tomatoes. *Journal of Horticultural Science* 52, 309–315.

Smillie, R.M., Hetherington, S.E. and Davies, W.J. (1999) Photosynthetic activity of the calyx, green shoulder, pericarp, and locular parenchyma of tomato fruit. *Journal of Experimental Botany* 50, 707–718.

Somerville, C. (1995) Direct tests of the role of membrane lipid composition in low-temperature-induced photoinhibition and chilling sensitivity in plants and cyanobacteria. *Proceedings of the National Academy of Sciences USA* 92, 6215–6218.

Stanghellini, C., Van Meurs, W.T.M., Corver, F., Van Dullemen, E. and Simonse, L. (1998) Combined effect of climate and concentration of the nutrient solution on a greenhouse tomato crop. II. Yield quantity and quality. *Acta Horticulturae* 458, 231–237.

Suto, K. and Ando, T. (1975) Influence of atmospheric humidity and soil moisture contents on the plant water condition as well as on the growth of sweet pepper and tomato plants. *Bulletin of the Vegetable Ornamental Crops Research Station, Ishinden-Ogoso, Japan, Series A* 2, 49–63.

Thompson, D.S., Davies, W.J. and Ho, L.C. (1998) Regulation of tomato fruit growth by epidermal cell wall enzymes. *Plant Cell Environment* 21, 589–599.

Thornley, J.H.M. and Hurd, R.G. (1974) An analysis of the growth of young tomato plants in water culture at different light integrals and CO_2 [carbon dioxide] concentrations. I. A mathematical model. *Annals of Botany* 38, 389–400.

Tremblay, N. and Gosselin, A. (1998) Effect of carbon dioxide enrichment and light. *HorTechnology* 8, 524–528.

Van Oosten, J.J. and Besford, R.T. (1995) Some relationships between the gas exchange, biochemistry and molecular biology of photosynthesis during leaf development of tomato plants after transfer to different carbon dioxide concentrations. *Plant, Cell and Environment* 18, 1253–1266.

Van Oosten, J.J., Wilkins, D. and Besford, R.T. (1994) Regulation of the expression of photosynthetic nuclear genes by high CO_2 is mimicked by carbohydrate: a mechanism for the acclimation of photosynthesis to high CO_2? *Plant, Cell and Environment* 17, 913–923.

Venema, J.H., Posthumus, F. and Van Hasselt, P.R. (1999a) Impact of suboptimal temperature on growth, photosynthesis, leaf pigments and carbohydrates of domestic and high-altitude wild *Lycopersicon* species. *Journal of Plant Physiology* 155, 711–718.

Venema, J.H., Posthumus, F., De Vries, M. and Van Hasselt, P.R. (1999b) Differential response of domestic and wild *Lycopersicon* species to chilling under low light: growth, carbohydrate content, photosynthesis and the xanthophyll cycle. *Physiologia Plantarum* 105, 81–88.

Xiao, S.G., Van der Ploeg, A., Bakker, M. and Heuvelink, E. (2004) Two instead of three leaves between tomato trusses: measured and simulated effects on partitioning and yield. *Acta Horticulturae* 654, 303–308.

Xu, H.L., Gauthier, L., Desjardins, Y. and Gosselin, A. (1997) Photosynthesis in leaves, stem and petioles of greenhouse-grown tomato plants. *Photosynthetica* 33, 113–123.

Yelle, S. (1988) Acclimatation de *Lycopersicon esculentum* Mill. Aux hautes concentrations de l'atmosphère en bioxide de carbone. Thèse de doctorat, Université Laval, Quebec, Canada, 150 pp.

Fruit Ripening and Fruit Quality

M.E. Saltveit

INTRODUCTION

The tomato (*Lycopersicon esculentum*) is a herbaceous perennial that is usually grown as an annual in temperate regions. Tomato cultivars differ considerably in their growth pattern. These differences are especially evident in the relationship between vegetative and fruit growth. In all cultivars, however, developing fruits act as sinks that drain photosynthesized sugars and other nutrients away from nearby leaves. As more fruits are produced, they increasingly monopolize the resources of the plant until vegetative growth nearly stops. Tomato plants stressed early in their fruit development period may be unable to support both additional vegetative growth and continued fruit development. Such a premature cessation of growth results in smaller plants and reduced yields. Fruit quality may also suffer, as insufficient photosynthates and nutrients are available for normal development of all fruit.

The fruiting habit of tomato plants is also variable. Indeterminate pole and cherry tomatoes bear fruit in clusters successively along each stem. Fruit set occurs over a protracted period and it may take a long time before the fruit load becomes enough of a sink to limit vegetative growth. Plants often therefore grow large, and multiple harvests are required to pick the fruits as they mature and ripen over a long period of time. Most of these indeterminate tomato cultivars produce fruit destined for the fresh market. In contrast, both processing and fresh-market bush-tomato cultivars are determinate plants that grow to a certain size and then initiate flowering and fruit set over a relatively short period of time. The simultaneous development of large numbers of fruit inhibits further vegetative growth and the plants remain small and compact, i.e. a bush. High-density plantings further limit vegetative growth and favour a concentrated fruit set. The uniform maturity and ripening of a large proportion of fruit on the plant allows a single pick and mechanical harvesting to be economical.

RIPENESS CLASSIFICATION

The traditional six ripeness stages for fresh-market tomatoes are based almost entirely on the external colour change of the fruit from green to red (i.e. destruction of chlorophyll and synthesis of lycopene) (Colour Plate 1). Fruit greenness (i.e. chlorophyll content) varies among cultivars and growing location. Heavily shaded fruit can be almost white, while fruit exposed to sunlight are usually dark green. Greenness also varies over the surface of the fruit, often being greatest at the stem end and least at the blossom end.

When fruit reach about 80% of their final size and acquire the ability to continue to develop and ripen normally after harvest, they are said to be mature-green. Mature-green (stage 1) fruit show no external red coloration, but the blossom end may show some whitening (Table 5.1). Depending on the cultivar, other changes in external appearance (such as loss of a surface bloom or hairs, corking around the peduncle and waxiness of the epidermis) may appear at this stage. Immature and early-stage mature-green fruit can be

Table 5.1. Ripeness classification of fresh-market tomatoes based on changes in external and internal colour and tissue softening (source: Sargent and Moretti, 2002).

Stage	Description
0. Immature	The fruit is not sufficiently developed to ripen to an acceptable level of horticultural quality. Many immature fruit will eventually ripen, but to an inferior quality.
1. Mature-green (MG)	The fruit will ripen to an acceptable level of horticultural quality. The entire surface of the fruit is either green or white, no red colour visible. Stages within the MG classification include: MG1 – firm locular tissue, knife cuts seeds MG2 – softened locular tissue, seeds not cut with knife MG3 – some gel in the locule, no red colour MG4 – locular tissue predominantly gel, some red colour in columella
2. Breaker	There is a definite break in colour from green to tannish-yellow, pink or red on the blossom end of the fruit
3. Turning	More than 10% but less than 30% of the surface of the fruit shows a definite colour change to tannish-yellow, pink, red, or a combination of colours
4. Pink	More than 30% but less than 60% of the surface of the fruit shows pink or red colour
5. Light-red	More than 60% but less than 90% of the surface of the fruit shows red colour
6. Red-ripe	More than 90% of the surface of the fruit shows red colour

induced to 'ripen' but they will be of inferior quality. Under proper conditions of temperature and humidity, tomatoes progress through the six well-defined stages described in Table 5.1 to the red-ripe stage.

There are three distinct phases of pigment changes during ripening. Chlorophyll is the predominant pigment during growth and development up to the mature-green stage. During this time the chlorophyll/carotenoid ratio is about 10:1. As the fruit ripens from mature-green to the breaker stage, there is a destruction of chlorophyll and an increase in carotenoids. The chlorophyll/carotenoid ratio is about 1:1 at this stage. From breaker to red-ripe there is a surge in lycopene synthesis as chlorophyll content falls to zero (Fig. 5.1).

MEASUREMENTS OF FRUIT RIPENESS

Ripening is accompanied by changes in colour and firmness and the development of characteristic aromas and flavours (Grierson and Kader, 1986; Dorais *et al.*, 2001; Kader, 2002). When mature-green fruit are harvested, it is important to separate them into groups on the basis of the time they will take to ripen. Fruit of advanced maturity can be shipped to local markets, fruit that are less mature to more distant markets, and immature fruit discarded.

As fruit ripen, the external colour changes in characteristic ways. While fruit of most cultivars change from a uniform green to red as they ripen, a few

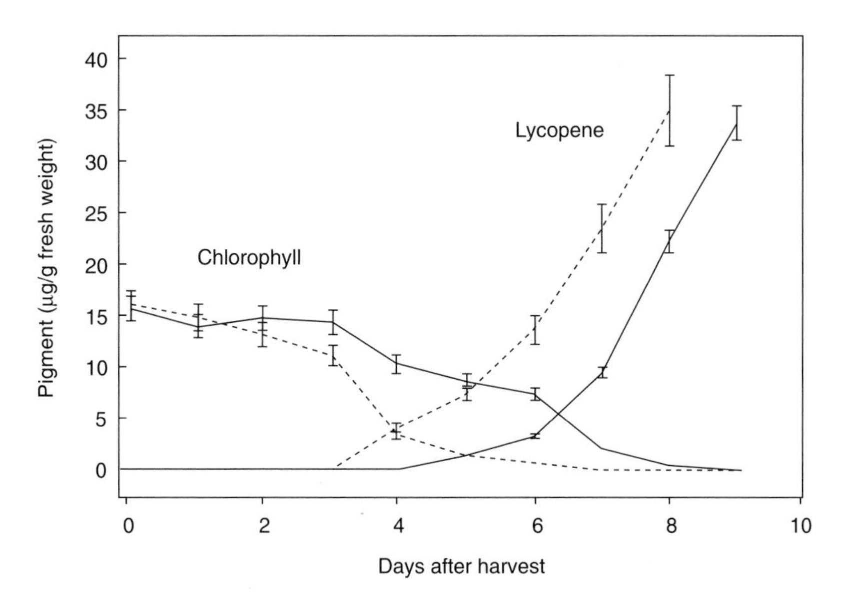

Fig. 5.1. Changes in chlorophyll and lycopene content in ripening tomato fruit exposed to ethylene (100 μl/l) (dashed line) for 5 days or to air (solid line).

cultivars have fruit that turn yellow (reduced or incomplete synthesis of lycopene) or are variegated in colour. In general, fruit colour becomes lighter, less green and more red, and less yellow and more blue (see section on pigments, below) as fruit progress from mature-green to red-ripe.

External changes in colour can be measured subjectively by comparing visual changes in the intensity and distribution of pigments with colour charts or pictures. Instruments have been developed that objectively measure external colour changes while the fruit move down a packing line (Abbott *et al.*, 1997). However, fruit that show colour may be too advanced in ripeness and too soft to tolerate shipment to distant markets. Much progress has been made in the last few years in developing commercial instruments for sorting by colour. Electronic colour sorters are used in some tomato-packing operations in Europe and the USA. Portable instruments are available that measure surface colour (e.g. Minolta Chromameter). Measurements by such instruments are often highly correlated with levels of pigments quantified by extraction and spectrophotometric analysis. Since lycopene first appears in the columella of ripening fruit, instruments that can detect colour changes inside the fruit could identify earlier stages of ripeness than instruments measuring external colour (Colour Plate 1).

Many instruments have been developed to measure firmness non-destructively by measuring the compression of the fruit under a constant load (Abbot *et al.*, 1997). Compression tests are usually used to test fruit firmness. Because of the heterogeneity of the tomato fruit tissues (e.g. pericarp walls and locular regions), a number of readings around the fruit are needed to achieve an accurate assessment of firmness (Fig. 5.2). For example, a 500 g weight can compress a medium-sized ripe fruit (6 cm diameter, 115 g) by about 1.5 mm in 20 seconds. Puncture tests that work well with fruit having thick cortex or mesocarp tissue (e.g. apples, cucumbers, pears, peaches, melons) also give meaningful readings for tomatoes. There is a very good relationship between ripeness stage and penetration force, measured with or without the tomato skin; measurements with the skin are consistently higher than those without the skin.

Other physical changes that take place in the ripening fruit (e.g. changes in acoustical properties, infrared reflectance, nuclear magnetic resonance (NMR) images, vibrational harmonics) are being studied as possible indicators of fruit maturity. Some are successful in the laboratory, but are not commercially viable because they are too delicate, expensive or slow.

RESPIRATORY AND ETHYLENE CLIMACTERIC

Harvested tomatoes exhibit a climacteric rise in respiration (i.e. carbon dioxide production and oxygen consumption) and ethylene production that coincides with the onset and progression of ripening (Hobson and Davies,

Fig. 5.2. Tomato firmness being measured with a compression tester.

1971; Biale and Young, 1981; Grierson and Kader, 1986). Respiration involves the oxidation of carbohydrates (primarily sugars) and organic acids to carbon dioxide and water, with the release of energy and the production of intermediate carbon compounds. The climacteric rise in respiration during ripening is thought to be necessary to supply the increasing need of the tissue for energy and the intermediate compounds that are required for the synthetic reactions upon which ripening depends. However, some non-climacteric fruit also undergo profound changes during ripening without the apparent need for a significant increase in the rate of respiration.

The division between climacteric (e.g. apples, bananas and tomatoes) and non-climacteric fruit (e.g. lemons, oranges and strawberries) is not absolute, as the magnitude of the respiratory rise varies from $< 20\%$ to $> 100\%$ among species and cultivars. Although the range of respiration rates for mature-green and turning tomato fruit are listed as similar by Hardenburg *et al.* (1986) (Table 5.2), in reality the rate of carbon dioxide production by a tomato fruit at the turning stage of ripeness is about twice what it was for the same fruit at the mature-green stage of ripeness (Grierson and Kader, 1986; Saltveit, 1993) (Fig. 5.3). The seeming inconsistency in these two statements is caused by the fact that there is as much variability in the rate of respiration among cultivars, growing locations and individual fruits as there is within one fruit as it progresses through the climacteric during ripening. The climacteric is easily measured from one ripening fruit, but it is difficult to resolve when a number of ripening fruits are grouped together. Both biotic (e.g. insects, diseases) and abiotic (e.g. injury, drought, temperature extremes)

Table 5.2. Respiration rate of mature-green and turning fresh-market tomato fruit at different temperatures (source: Hardenberg *et al.*, 1986).

	Respiration rate (mg CO_2/kg/h)	
Temperature	Mature-green	Turning
10°C	12–18	13–16
15°C	16–28	24–29
20°C	28–41	24–44
25°C	35–51	30–52

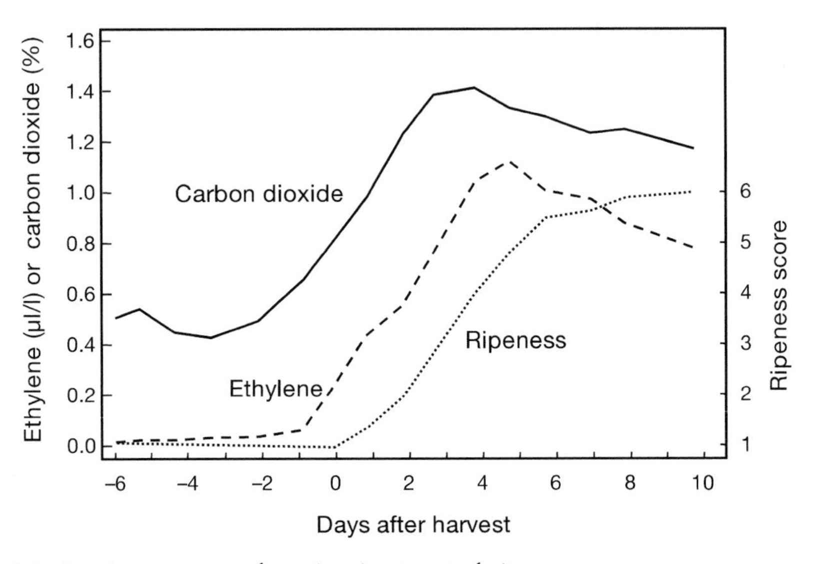

Fig. 5.3. Respiratory pattern for a ripening tomato fruit.

stresses can significantly affect the onset and magnitude of the respiratory climacteric and ripening. In general, modest levels of stress tend to promote ripening, while severe levels of stress may retard ripening.

The moderate rise in respiration during ripening is in comparison with the tenfold to 100-fold increase in the production of the plant hormone ethylene that accompanies the ripening of tomatoes. The climacteric rise in ethylene production occurs in both attached and detached fruit, but the respiratory rise is diminished in fruit ripening on the plant (Saltveit, 1993). Ethylene biosynthesis in pre-climacteric immature and mature-green fruit and ripening climacteric fruit is by the conventional methionine to *S*-adenosylmethionine (SAM) to 1-aminocyclopropane-1-carboxylic acid (ACC) to ethylene pathway (Abeles *et al.*, 1992).

The ethylene climacteric is thought to initiate and coordinate uniform ripening throughout the fruit. Exposure of pre-climacteric mature-green fruit to hormonal levels of ethylene (or to an ethylene analogue such as propylene) hastens the onset of the climacteric and ripening. In contrast, inhibitors of ethylene synthesis (e.g. atmospheres with around 3% oxygen, treatment with aminoethoxyvinylglycine, AVG) or ethylene perception (e.g. atmospheres with around 5% carbon dioxide, treatment with silver or 1-methylcyclopropene (1-MCP)) retard ripening (Leshuk and Saltveit, 1990; Abeles *et al.*, 1992; Saltveit, 2001, 2003). Once the ethylene concentration within the fruit surpasses a 'threshold' level, it will promote its own biosynthesis (i.e. positive feedback) and autocatalytic ethylene production will cause a rapid increase in production and accumulation within the tissues (Abeles *et al.*, 1992).

The atmosphere within a tomato fruit is effectively isolated from the surrounding atmosphere by an impermeable skin and cuticle; about 95% of gas exchange occurs through the stem scar. Therefore, once ethylene has started its positive feedback climacteric rise, few external treatments can modulate its synthesis and that portion of ripening affected by ethylene. Reduced temperatures and lowered oxygen atmospheres will slow overall metabolism but ripening will continue, albeit at a slower pace. However, certain inhibitors of ethylene action (e.g. 1-MCP, ethanol vapours) appear to stop ethylene-enhanced fruit ripening at almost any stage of ripeness (Saltveit and Sharaf, 1992). Ripening continues once the inhibitory effect of these compounds has dissipated.

QUALITY CHARACTERISTICS

Tomatoes are harvested at different stages of ripeness for different purposes. Processing tomatoes are mechanically harvested red-ripe and immediately transported to a processing plant. Fruit destined for the fresh market are hand harvested at the mature-green, partially ripe or fully ripe stage. Mature-green fruit are picked because they are firm enough and have sufficient shelf-life to survive the stress of being shipped considerable distances and are ripened to acceptable levels of quality at distant markets. Ethylene to stimulate and coordinate the natural ripening of harvested fruit must be used with caution, to ensure that forcing immature fruit to ripen does not produce inferior quality. Quality characteristics of fresh-market fruit are similar to those of processing tomatoes, but characteristics that are readily apparent to the consumer (colour, size, shape, firmness and aroma) dominate the others.

Fruit quality is strongly affected by temperature (Dorais *et al.*, 2001). Temperature directly influences metabolism and, indirectly, cellular structure and other components that determine fruit quality such as colour, texture, size and organoleptic properties. An air temperature of 23°C improved the taste of tomatoes, increased fruit dry matter and K:Ca ratio, and reduced the proportion of softer and mealy fruit as compared with fruit grown at 17°C

(Janse and Schols, 1992). Growth at 17°C produced softer and less juicy and aromatic fruits. Such fruit have a less resistant cuticle, despite a higher content of reducing sugars and a lower content of titratable acids (Janse and Schols, 1992). Elevated temperatures favour increased numbers of hollow fruit in the winter and miscoloured fruit in the summer, and the number of misshapen and soft fruits (pollination defect).

Suboptimal temperatures favour large, irregularly shaped fruit with separated carpels and exposed placenta. Low night temperatures (≤ 12°C) slow fruit ripening and increase the tomato plant's susceptibility to pointed-fruit malformation, which is associated with deformed locules (i.e. placenta retarded or absent, and an absence of seeds and jelly) (Tomer *et al.*, 1998). Temperature below 10°C for a substantial part of the night during flower development cause the formation of fasciated fruit (Rylski, 1979). Tomato flower and fruit development were severely affected when grown at day/night temperatures of 17°C/7°C compared with 26°C/18°C (Lozano *et al.*, 1998). Fruit display homeotic and meristic transformations and alterations in the fusion pattern of the organs, which could be related to expression of MADS-box genes. As a consequence, abnormal fruits of low economic value are produced. Nevertheless, Gent and Ma (1998) reported that tomato fruit ripened earlier, fruit size increased and the incidence of irregular fruit was reduced when there was a large day/night temperature differential early in fruit development (e.g. a differential of 14°C compared with 5°C).

The gross composition of tomato fruit changes only slightly during ripening from mature-green to red-ripe (Table 5.3). High quality red-ripe tomatoes contain around 93% water and 5–8% dry matter. The carbohydrate

Table 5.3. Composition of mature-green and ripe tomato fruit (source: Jones, 1999).

Component[a]	Mature-green	Ripe
Water (%)	93.0	93.5
Calories	24	22
Protein (g)	1.2	1.1
Fat (g)	0.2	0.2
Carbohydrates		
Total (g)	5.1	4.7
Fibre (g)	0.5	0.5
Ash (g)	0.5	0.5
Calcium (mg)	13	13
Phosphorus (mg)	27	27
Iron (mg)	0.5	0.5
Sodium (mg)	3.0	3.0
Potassium (mg)	244	244

[a]Per 100 g.

concentration (mainly as equal amounts of glucose and fructose) increases progressively through maturation and ripening, and can account for 50% of the dry matter. Ripe fruit are a good source of vitamins A and C and potassium. A steroid glycoside, tomatine, decreases during ripening from 0.08% in mature-green fruit to 0.00% in ripe fruit. Tomatine is toxic to mammals and has a bitter taste.

Pigments

Fruit colour is probably the most important attribute that determines overall quality. A number of changes occur when tomatoes progress from the mature-green to red-ripe stage of ripeness (Table 5.4). The most obvious external changes are associated with the loss of chlorophyll and the accumulation of lycopene (see Table 5.1 and Colour Plate 1). Lycopene content first becomes apparent at the blossom end of the fruit and progresses towards the stem end. Therefore the fruit can be partially green and partially red during ripening. Once ripe, however, high quality fruit have uniform red distributed over their entire surface. Incomplete coloration is a serious disorder. Lycopene and β-carotene increase from 0.0 to around 48 and 4.3 μg/g, respectively, during ripening (Table 5.4). Carotenoid pigments are synthesized in the chloroplast and chromoplast of the plant cell by an extension of the normal isoprenoid pathway. Eight isoprene units condense through a number of reactions to form lycopene, which then undergoes additional reactions to form the carotenes.

The ability of the fruit to synthesize lycopene and β-carotene is almost the same for harvested mature-green fruit as it is for fruit ripening on the plant. Exposure of harvested mature-green fruit to ethylene stimulates normal ripening with the synthesis and accumulation of lycopene and β-carotene, often to a greater extent than for harvested fruit left to ripen without ethylene stimulation. Temperature has a considerable effect on pigment synthesis. The optimal temperature is 16–21°C. Temperatures above 30°C significantly reduce lycopene and carotenoid synthesis. In some cultivars, yellow fruit will be produced because lycopene synthesis is reduced more than carotene

Table 5.4. Composition of tomato fruit at different stages of ripeness. Fruit were harvested at the mature-green, breaker or red-ripe stage of ripeness (source: Cantwell, 2000).

Stage of ripeness	Soluble solids (%)	Reducing sugars (%)	pH	Titratable acidity (%)	β-carotene (μg/g)	Lycopene (μg/g)	Ascorbic acid (mg/100 g)
Mature-green	2.37	0.81	4.20	0.28	0.0	0.0	12.5
Breaker	2.42	0.85	4.17	0.39	0.40	0.52	18.0
Red-ripe	5.15	1.62	4.12	0.43	4.33	48.3	22.5

synthesis, while in other fruit chlorophyll loss will be delayed and the fruit will remain partially green.

Breeding for better fruit colour is an active area of research, not only because the fruit will have a better visual appearance to the consumer and produce a superior processed product, but also because of the health benefits of the carotenoids (see section on vitamins, below). Carotenoid biosynthesis is under direct nuclear control. The existing diversity of fruit pigmentation is controlled by the participation of a great number of non-allelic genes distributed on almost all chromosomes. Many mutations have been identified (e.g. *rin*, *nor*, Nr) in which carotenoid biosynthesis is disrupted, or other steps in fruit ripening have been affected (e.g. chlorophyll degradation, chloroplast to chromoplast conversion, fruit softening). Some of these mutations have been introduced into existing commercial lines to produce slow-ripening fruit.

Fruit colour can be determined either subjectively or objectively. Reference to standard charts showing colour images of fruit at different stages of ripeness (Colour Plate 1) can be used to enhance the accuracy and precision of subjective evaluations of fruit maturity. Objective measurements can be made by spectroscopically measuring the concentration of individual pigments extracted in an organic solvent. However, such procedures are more suited to laboratory investigations than to determining fruit ripeness in a commercial setting. Pigment concentration, visual appearance and ripeness stage can all be correlated with readings from instruments measuring fruit colour. There are a number of instruments commercially available to measure tomato fruit colour objectively. The Hunterlab Color and Color Difference Meter, which is a tristimulus colorimeter, was one of the first electronic instruments to measure colour on the X, Y and Z scale of the Commission International de l'Eclavinge (CIE) system. Currently, such instruments as the Minolta Color Meter give readings in any number of systems but the most common is the L*, a*, b* colour system, where L* is the lightness (0 = black, 100 = white), a* is red (positive values) to green (negative values), and b* is yellow (positive values) to blue (negative values) (Table 5.5). The chroma (a measure of colour intensity) does not show a consistent change with ripening, while the hue angle shows a significant decrease as chlorophyll content decreases and lycopene content increases. These readings can be correlated with the amount of specific pigments determined spectrophotometrically.

A simple procedure to measure chlorophyll and lycopene content is to blend 10 g of pericarp tissue with 10 ml of acetone and centrifuge the mixture to clarify the solution (e.g. 5000 × g for 5 min). Lycopene concentrations can be determined in pericarp tissue varying from green to red by measuring the absorbance of the acetone tissue extract at 503 nm. This wavelength is best for lycopene because the influence of carotenoids (473 nm) is negligible (Beerh and Siddappa, 1959). Non-destructive Minolta Chromameter measurements can also be taken of the pericarp tissue before homogenization and the readings can be correlated with measured lycopene concentration. The best-fit

Table 5.5. Colour changes during the ripening of tomato fruit (source: Cantwell, 2000).

Stage of development	Ripening stage	L*	a*	b*	Chroma	Hue angle
Mature-green	1	62.7	−16.0	34.4	37.9	115.0
Breaker	2	55.8	−3.5	33.0	33.2	83.9
Pink	4	49.6	16.6	30.9	35.0	61.8
Light red	5	46.2	24.3	27.0	36.3	48.0
Red-ripe	6	41.8	26.4	23.1	35.1	41.3
Over-ripe	6+	39.6	27.5	20.7	34.4	37.0

L* indicates lightness (high value) to darkness; a* changes from green (negative value) to red; b* changes from blue (negative value) to yellow. Chroma and hue are calculated from a* and b* values and indicate intensity and colour, respectively. Chroma is given by the equation $(a^{*2} + b^{*2})^{0.5}$ while hue is given by the equation $\tan^{-1}(b^*/a^*)$. The lower the hue value, the redder the tomato fruit.

exponential equation (i.e. highest r^2) can vary with cultivar and method of tissue preparation (Beaulieu and Saltveit, 1997) and so individual curves should be periodically established for each cultivar under examination.

Size and shape

Tomatoes are usually sized by diameter because of their generally spherical shape. Fruit that will not pass through a round opening when they are oriented with the largest transverse diameter across the opening are designated extra small fruit when the diameter is 48–54 mm, small when the diameter is 54–58 mm, medium at 58–64 mm, large at 64–73 mm, extra large at 73–88 mm and maximum large when the hole is > 88 mm (Sargent and Moretti, 2002).

There are characteristic shapes for many tomato cultivars (e.g. spherical, pear, oblong). High quality fruit should exhibit the characteristic shape of that cultivar. Fruit are considered misshapen not only when grossly deformed (e.g. catfacing) but also when their shape deviates from the norm (e.g. a normally round cultivar producing pear-shaped fruit).

Surface appearance

In some cultivars, a surface bloom is lost at more advanced stages of ripeness, as is the development of corking around the stem scar. Scuffing the surface during sorting, grading and packing may lead to the development of russeting or blisters on the fruit surface during storage under high relative humidity.

Firmness

Firmness of pericarp tissue is a key component for both processing and fresh-market cultivars, especially when the latter are shipped long distances. Softening during the ripening of mature-green fruit is promoted by exposure to elevated levels of ethylene and increasing temperatures up to 30°C, and is inhibited by temperatures > 35°C. Exposing tomato fruits to high temperatures (35–40°C) for a short period (1–2 days) delays ripening for a few days without affecting quality. A great deal of research and effort has been spent to develop cultivars and postharvest procedures to maintain firmness of ripe tomatoes. The exact molecular changes that produce fruit softening are unknown, but it is known that a number of cell wall hydrolytic enzymes contribute to tissue softening and intercellular adhesion (Fisher and Bennett, 1991). The major classes include polygalacturonases, pectinases, pectinmethylesterases and carboxymethylcellulases. Pericarp firmness is a quantitative trait controlled by nuclear genes. Fruit with genetically engineered lower levels of polygalacturonase are slightly firmer as they ripen, but other parameters of ripening are unaffected. Postharvest practices have been developed to control softening through the use of calcium dips, and through their effect on ripening (e.g. controlled atmospheres).

Composition and flavour

Soluble solids (SS) and titratable acids (TA) are important components of flavour. They exert their effect not only through the amount present (about 50% of the dry weight is SS and 12% TA), but also through their ratio. Fruit high in both acids and sugars have excellent flavour, while tart fruit have low sugar content and bland fruit have low acidity (Table 5.6). Sugars accumulate through their importation from photosynthesizing leaves and after harvest from the hydrolysis of stored starch (see Table 5.4). Storage of harvested fruit at elevated temperatures hastens not only ripening but also the respiratory loss of stored carbohydrates. The major sugars are about equal amounts of

Table 5.6. Relationship between the level of sugars and titratable acids on the taste and flavour of fresh tomato fruit (source: Kader *et al.*, 1978).

	Sugars (soluble solids)	
Titratable acidity	High	Low
High	Good	Tart
Low	Bland	Tasteless

glucose and fructose (each ~ 22% dry weight) with a small amount of sucrose (1%). Alcohol-insoluble solids (e.g. proteins, pectins, hemicelluloses and celluloses) comprise about 25% of the fruit's dry weight (DW). Minerals (mainly K, Ca, Mg and P) make up 8% of the dry weight. Soluble solids are usually measured with a refractometer calibrated in °Brix. Juice is expressed from excised tissue with a garlic press or from homogenized tissue and a few drops are put on the optical prism of the refractometer. While readings are predominantly influenced by the amount of sugars in the sample, the other water-soluble constituents (e.g. organic acids and soluble pectins) in the sample may also contribute significantly to the reading.

There is a large variation among tomato genotypes for pH and titratable acidity. A ripe tomato is acidic and its pH values range from 4.1 to 4.8. Storage and handling procedures should maintain the pH of red-ripe fruit at or below 4.7 to prevent growth of microorganisms such as *Clostridium botulinum*. Storage at elevated temperatures facilitates the metabolism of organic acids, with an increase in pH. Fungal infections (e.g. with *Fusarium solani*) also tend to increase the pH of infected and adjacent tissue. Titratable acids are composed primarily of the organic acids citric acid (~ 9% DW) and malic acid (~ 4% DW). While organic acids only constitute 0.4% of the fresh fruit, they are important attributes of taste. Variation in acid content has a much greater impact on flavour than the limited variation in sugar content that exists among cultivars.

Volatiles

The major components of flavour in a ripe tomato are flavour volatiles, sugars and acids. Over 400 volatile aroma compounds have been identified in tomato fruit, but fewer than 17 have a significant impact on characteristic tomato aroma (Buttery, 1993). Overall acceptance of ripe fruit by an experienced, trained taste panel was highly correlated with the flavour volatile *cis*-3-hexenol, soluble solids/titratable acidity, sucrose equivalents/titratable acidity, and titratable acidity (Baldwin *et al.*, 1998). Tomato-like flavour was correlated with geranylacetone, 2+3-methylbutanol and 6-methyl-5-hepten-2-one. Sweetness was correlated with sucrose equivalents, pH, *cis*-3-hexenal, *trans*-2-hexenal, hexanal, *cis*-3-hexenol, geranylacetone, 2+3-methylbutanol, *trans*-2 heptenal, 6-methyl-5-hepten-2-one and 1-nitro-2-phenylethane. Sourness was correlated with soluble solids, pH, acetaldehyde, acetone, 2-isobutylthiazole, geranlyacetone, beta-ionone, ethanol, hexanal and *cis*-3-hexenal. Levels of aroma compounds affected perception of sweetness and sourness and measurements of soluble solids show a closer relationship to sourness, astringency and bitterness than to sweetness.

Soluble solids can be easily measured with a refractometer. Measurement of titratable acidity is more difficult, while measurement of flavour volatiles is very difficult. Technical refinements have made collection of volatile samples easier, but the definitive identification of the various components requires an array of volatile standards, and a gas chromatograph equipped with a flame ionization detector. Unknown compounds in the sample separated on the gas chromatographic column can be identified by diverting some of the effluent from the column through an attached mass spectrometer. However, the threshold concentration detected by humans for various aroma compounds ranges over many orders of magnitude, so the presence of a relatively large amount of one compound does not mean that it contributes significantly to the overall aroma. Sensory observations of compounds singularly and in combination with others must be used to confirm both its specific odour and its contribution to aroma.

Fruit harvested mature-green and improperly ripened, and ripe fruit stored at chilling temperatures, do not produce the characteristic volatiles associated with high quality tomatoes. Identification of the important flavour volatiles is necessary so that ripening and storage procedures can be devised that will optimize their production and maximize fruit quality. Genetic engineering for better flavour will also require identification of the few volatiles that contribute the most to flavour, so that their biosynthetic pathways can be selectively modified.

Vitamins

Ripe tomatoes are good sources of many vitamins, including vitamins A and C (Table 5.7). The large per capita consumption of tomato products (e.g. fresh fruit, sauces, condiments) makes tomatoes an excellent source of these vitamins. Plants do not contain vitamin A, but humans can convert the carotenoids they

Table 5.7. Vitamin content of ripe tomato fruit (source: Davies and Hobson, 1981).

Vitamin	Range (per 100 g)
A (β-carotene)	540–760 μg
B1 (thiamin)	50–60 μg
B2 (riboflavin)	20–50 μg
B3 (pantothenic acid)	50–75 μg
B6 complex	80–110 μg
Nicotinic acid (niacin)	500–700 μg
Folic acid	6.4–20 μg
Biotin	1.2–4.0 μg
C	15–23 μg
E (α-tocopherol)	40–1200 μg

do contain into vitamin A. The biological activities of vitamin A and previtamin A (i.e. specific carotenoids) are not equivalent. Among the carotenoid pigments in tomatoes, β-carotene has the highest vitamin A activity, with α-carotene and γ-carotene having 55% and 45% of its activity, respectively. In humans, 6 mg of β-carotene is equivalent to 1 mg of vitamin A. The international unit (IU) is used to compare the biological activity of various sources of vitamin A. One IU of vitamin A is supplied by 0.6 μg of β-carotene.

The vitamin A content of a tomato is determined by its carotene content. In general, cultivars with better colour (deeper red, uniform colour throughout the fruit) have higher vitamin A activity, though this is not always the case. Cultivars containing high β-carotene levels have been developed, but they often have an orange hue that detracts from consumer acceptance. Marketing the health benefits of these lines may increase their acceptance.

Lycopene and the remaining carotenoid pigments have no vitamin A activity, but lycopene is beneficial to human health. It is a long-chain unsaturated carotenoid that imparts the red colour to ripe tomato fruit. It is the most powerful antioxidant in the carotenoid family and, along with vitamins C and E, helps to detoxify free radicals. Diets rich in carotenoid-containing foods are associated with a reduction in some diseases, including lung, bladder, cervical and skin cancers and disorders of the digestive tract (Stahl and Sies, 1996; Clinton, 1998). Benefits occur when about 50 mg of lycopene are consumed daily. This amount is supplied by about seven medium-sized ripe tomatoes, two cups of spaghetti sauce, or two glasses of tomato juice. Tomato cultivars with high lycopene content have been developed.

The vitamin C content of tomatoes increases as they ripen (see Table 5.4). Mature-green and breaker fruit that were ripened with ethylene lost less vitamin C by the time they reached the red-ripe stage than did fruit allowed to ripen without added ethylene. However, both were lower than fruit ripened on the plant.

There is a large variation in vitamin C content among tomato species and cultivars. Vitamin C ranges from 8 to 120 mg per 100 g and so there are ample genetic resources to increase the content of vitamin C. In fact, there has been a fairly steady increase in vitamin C content of newly released tomato cultivars for many decades. Vitamin C exists in two water-soluble, biologically active forms: ascorbic acid and dehydroascorbic acid. Measurements of vitamin C activity should include measurements of both forms. Ascorbic acid is converted to dehydroascorbic acid by a number of reactions, while dehydroascorbic acid is converted to ascorbic acid by a coupled reaction with glutathione reductase via dehydroascorbic acid reductase. Although ascorbic acid content shows significant losses during fruit ripening and storage, the content of vitamin C shows remarkable stability when expressed as the sum of ascorbic acid and dehydroascorbic acid.

Physiological disorders

A number of disorders affect the quality of fresh-market tomatoes. These disorders result from a combination of environmental, production and handling procedures, or are genetic in origin.

A high vapour pressure deficit (i.e. low humidity) increases the number of fruit affected by blossom-end rot due to a high foliage transpiration rate that limits the supply of xylem sap (and, therefore, of calcium) to the fruit. However, fruit produced under low humidity are firmer, juicier and less mealy, and have more soluble sugars and fewer physiological disorders such as cracking and gold specks than fruit produced under low vapour pressure deficit. A low vapour pressure deficit (0.1–0.3 kPa or 0.6–2 g/kg) influences fruit colour (marblings) and increases the incidence of gold specks. Under these conditions, a low rate of plant transpiration decreases the transport of nutrient elements and the increase of the root pressure favours cracking of the fruit cuticle (Dorais *et al.*, 2003). Fruit are generally smaller, softer and misshapen.

Anther scarring

This disorder is characterized by the presence of a long scar along the blossom end of the fruit. It may have resulted from some injury to the flower at an early stage of fruit development.

Blistering

Abrasions of the skin and cuticle of mature-green fruit during postharvest handling may foster the development of outgrowths of pericarp cells during subsequent holding (Fig. 5.4). High relative humidity, prolonged holding and immature fruit contribute to the development of this disorder.

Blotchy ripening

This disorder is uneven ripening with some areas (usually near the calyx end of the fruit) retaining chlorophyll and not accumulating sufficient lycopene to produce normal red fruit. Early and mid-season crops are particularly susceptible. Improper plant nutrition, insect feeding and environmental stresses (e.g. chilling or solar radiation) may contribute to the occurrence of this disorder. Blotchy areas contain less nitrogenous compounds, organic acids, starch, sugars and dry matter. Low concentrations of inorganic nitrogen and potassium in the soil are related to increased incidence of blotchy ripening. When possible, the air temperature should be maintained below 30°C to decrease the incidence of this disorder.

Blossom-end rot

The blossom end of green fruit develops a water-soaked area near the blossom scar. The area dries, turns brown and then black as the tissue dies (Fig. 5.5). The name is unfortunate, since this is a physiological disorder and any rot

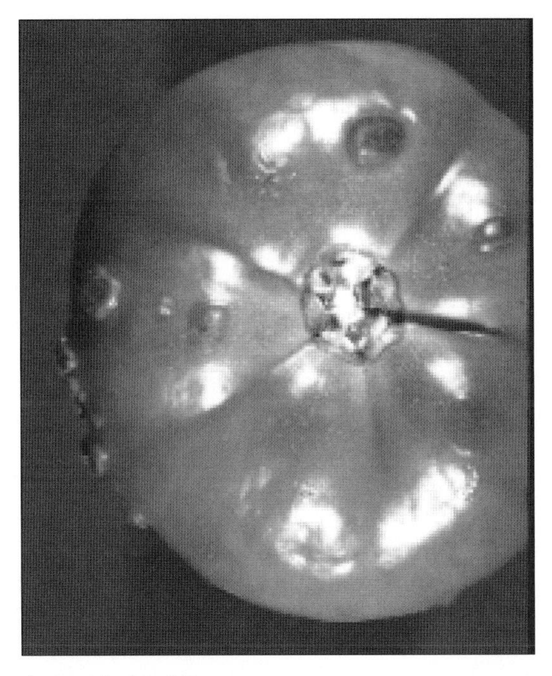

Fig. 5.4. Tomato fruit with skin blisters.

Fig. 5.5. Tomato fruit with blossom-end rot.

that develops is a secondary growth on the dead tissue. Low calcium levels in the blossom-end tissue predispose susceptible cultivars to this disorder. Calcium is relatively immobile in plants and the more rapid expansion of cells near the blossom end, in comparison with the stem end of the fruit, can dilute the calcium imported during growth of the young fruit to levels conducive to the development of blossom-end rot (Ho, 1999). Fluctuating soil moisture, high nitrogen fertilization and root pruning during cultivation reduce the uptake of calcium. Calcium uptake and translocation can be enhanced in greenhouse-grown tomatoes by lowering the relative humidity during the day and increasing it during the night. Increased transpiration during the day increases mineral uptake by the roots, while increased root pressure during the night causes the calcium-containing xylem sap to move into developing fruit. The incident of blossom-end rot increases significantly when the concentration of calcium in the fruit falls below 0.08% (DW), while the disorder seldom occurs at fruit levels above 0.12%.

Cracking

If the ripening fruit expands too rapidly, concentric cracks can develop around the stem end of the fruit, or radial cracks can develop that extend from the stem to the blossom end (Fig. 5.6). A rapid influx of water and solutes and reduced strength and elasticity of the tomato skin and pericarp wall contribute to the occurrence and severity of cracking. Not only is this disorder unsightly, but breaks in the epidermis also increase water loss and the entry of pathogens. Cracking can be minimized by the uniform application of water to avoid periods of water stress, adequate calcium nutrition, and the selection of crack-resistant cultivars. Cracking and splitting are inherited tendencies and cultivars differ greatly in susceptibility. The development of numerous very fine cracks is another disorder, known as russeting (see below).

Green shoulder

The shoulders of ripening fruit near the calyx remain green, while the rest of the fruit turns red (Fig. 5.7). While generally undesirable, this condition is actually preferred by consumers in some countries. Incorporating the ‘uniform ripening’ gene can eliminate this disorder in susceptible cultivars. Predisposing factors include exposure to excessive heat and an inadequate supply of potash and phosphate fertilization.

Misshapen fruit

Incomplete pollination and differential growth of various parts of the fruit can produce misshapen fruit. Examples include catfacing, puffiness and non-characteristic shapes. Poor pollination can cause portions of the fruit to grow and expand to produce conspicuous protuberances.

Fig. 5.6. Tomato fruit with concentric or radial cracks.

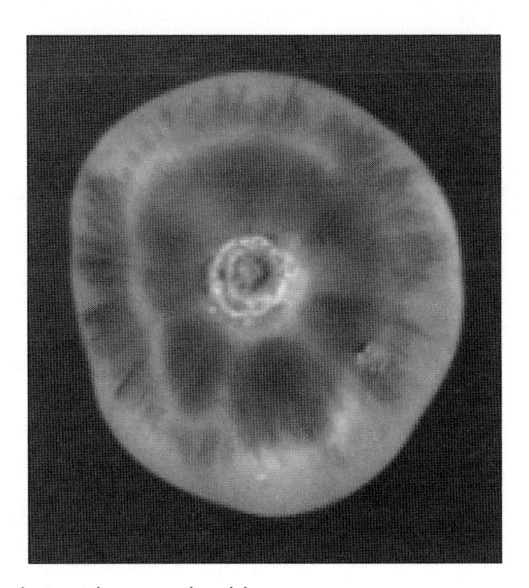

Fig. 5.7. Tomato fruit with green shoulders.

CATFACING In this type of misshapen fruit, the differential growth of various locules produces a convoluted shape in contrast to the smooth shape of most fruit (Fig. 5.8).

PUFFINESS This is also known as boxiness, or hollowness, in which one or more seed cavities (i.e. locules) of the fruit are empty of some or all tissue (Fig. 5.9).

Fig. 5.8. Tomato fruit with catfacing.

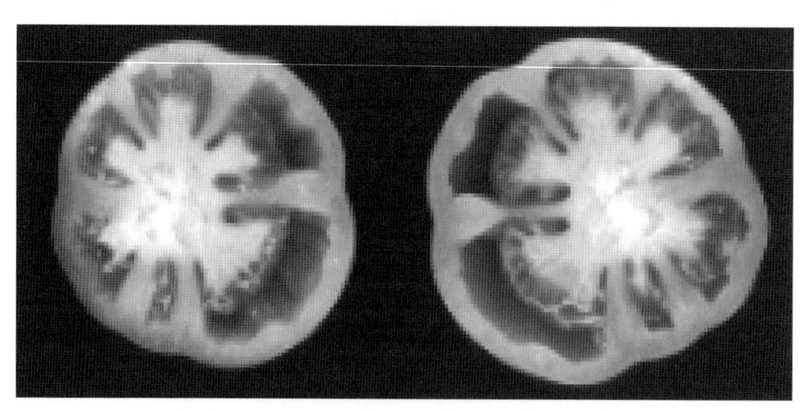

Fig. 5.9. Tomato fruit with puffiness. Note the absence of tissue in some of the locules.

Externally, the fruit surface is not uniformly curved but is flattened above the affected locule. Puffy fruit are less dense than good fruit and so they can be separated by flotation in water. Genotype and growing conditions that cause improper pollination, fertilization or seed development contribute to the occurrence of this disorder. Early-season greenhouse crops appear to be more prone.

Russeting

Unlike cracking, where the cracks extend several millimetres into the pericarp, russeting is the development of numerous fine cracks in the tomato skin. Microbial infection is not a significant problem, but water loss is increased and visual appearance decreased. This disorder has caused significant losses in greenhouse crops in The Netherlands.

Sheet pitting

Holding harvested mature-green tomato fruit at chilling temperatures (e.g. 2.5°C) for 7 days or longer produces sheet pitting (Fig. 5.10). This disorder is produced when epidermal cells below the cuticle lose water and collapse to form depressed areas.

Sunscald

Direct sunlight may increase the temperature of exposed fruit tissue by 10°C or more above ambient air temperatures. A few hours at 30°C will prevent normal pigment synthesis and tissue softening during ripening. Fruit temperatures above 40°C are lethal and the exposed tissue will die, turn white, dry out and form a flat parchment-like covering over the affected area. Green fruit are more sensitive to solar injury than ripe fruit. Plant architecture that shades the fruit during growth and bin covers that shade the harvested fruit during transport to packing facilities are the most effective ways to reduce sunscald.

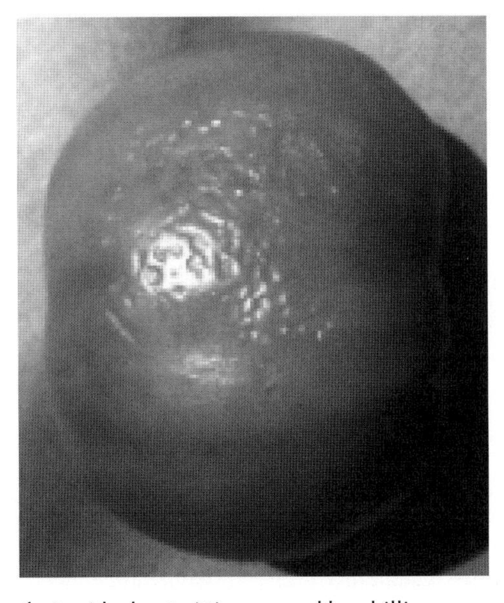

Fig. 5.10. Tomato fruit with sheet pitting caused by chilling.

GENETIC IMPROVEMENTS

Desired genetic improvements include higher soluble solids (this is especially important in processing cultivars), concentrated fruit set (to take advantage of mechanical harvesting), firmer fruit, uniform ripening, slower ripening and prolonged storability. Although the genetic base of the cultivated tomato is very narrow, very high levels of genetic variability exist in wild *Lycopersicon* spp. (Stevens, 1994). Increased resistance to chilling, disease, drought and insects, along with increased soluble solids and shelf-life, are just some of the many traits that could be improved through the incorporation of genes from wild species. Traditional plant breeders have also made successful crosses with *Solanum* species, which may serve as an additional source of useful genes.

Table 5.8. Ripening and quality characteristics for ripened tomato fruit. Fruit were harvested at the breaker stage and ripened to red-ripe at 20°C (source: Cantwell, 2000).

	Conventional cultivar		Long shelf-life cultivar	
Characteristic	Average	Range	Average	Range
Ripening physiology				
Days from breaker to red-ripe	6.3	5.3–7.3	7.2	6.0–9.3
Max. respiration (μl CO_2/g/h)	20.9	16.4–27.5	19.6	14.9–29.7
Max. ethylene production (nl C_2H_4/g/h)	4.8	2.3–9.4	2.8	1.5–9.6
Firmness[a]				
Compression (mm)	1.58	1.11–2.40	1.30	0.98–1.95
Colour[b]				
L* value	43.1	36.6–46.6	43.3	36.8–46.9
a* value	24.0	21.2–26.7	22.5	19.3–26.5
b* value	19.2	16.3–23.4	19.8	16.3–22.7
Chroma	30.8	34.7–46.9	30.1	27.4–35.3
Hue	38.5	34.7–46.9	41.4	36.0–46.2
Composition				
Soluble solids (%)	4.47	4.07–5.11	4.46	1.00–5.10
Acidity (pH)	4.36	4.13–4.65	4.39	4.18–4.57
Titratable acid (%)	0.33	0.28–0.45	0.36	0.28–0.46

[a] Firmness was measured as the deformation in mm of a fruit under a 0.5 kg weight for 20 seconds.
[b] L* indicates lightness (high value) to darkness; a* changes from green (negative value) to red; b* changes from blue (negative value) to yellow. Chroma and hue are calculated from a* and b* values and indicate intensity and colour, respectively. Chroma is given by the equation $(a^{*2} + b^{*2})^{0.5}$ while hue is given by the equation $\tan^{-1}(b^*/a^*)$. The lower the hue value, the redder the tomato fruit.

Incorporating specific genes from wild species into commercial cultivars has produced lines with greater disease resistance and higher soluble solids. Much additional work is needed to exploit fully these sources of useful genes. Changes are occurring rapidly in the biotechnogical modification of tomatoes, and up-to-date information can be found in recent reviews (e.g. Powell and Bennett, 2002) and current research reports.

A large number of ripening mutants are also available and some of these mutant genes have been incorporated into commercial lines. Single-gene ripening mutants such as *rin* (ripening inhibitor), Nr (never ripe) and *nor* (non-ripening) have helped in studies of the physiology of ripening. In the homozygous state *rin* and *nor* block ripening, while in the heterozygous state lines with *rin* and *nor* have enhanced storability.

Incorporation of such genes and directed genetic engineering have produced cultivars with longer shelf-life than conventional cultivars (Tables 5.8 and 5.9). The climacteric peak in both carbon dioxide and ethylene production is reduced in these long shelf-life cultivars and this results in the fruit taking a longer time to proceed from breaker to red-ripe (Table 5.8). The fruit are firmer than conventional cultivars, but neither colour nor composition is significantly altered. In other cultivars, the days from breaker to red-ripe are significantly increased, but there is only a noticeable effect on the maximum rate of respiration in the slowest ripening cultivar (Table 5.9). The maximum rate of ethylene production is variable, but usually lowest in the slowest ripening cultivars and colour is poorly developed, but softening is comparable to normal cultivars. These different responses show that the complex and interrelated components of ripening and quality cannot yet be manipulated at will.

Table 5.9. Quality characteristics of cultivars of normal and long shelf-life tomato fruit. Fruit were harvested at the breaker stage and stored at 20°C until 6 days past ripe (source: Cantwell, 2000).

Cultivar	Weight (g)	Days from breaker to table-ripe	Maximum respiration (μl CO_2/g/h)	Maximum ethylene (nl/g/h)	Colour (a*)[a]	Firmness[b] (deformation in mm)
Normal cultivars						
Shady Lady	193	6.5	20.6	5.4	26.7	1.67
Sunbelt	189	6.8	20.1	4.6	27.3	1.85
XPH 12109	168	7.3	17.4	2.3	26.6	1.28
Long shelf-life cultivars						
T 1011	170	11.5	21.0	1.6	24.8	0.99
FMX 209	172	>14	17.6	2.3	20.5	1.36
Longevity	113	>18	12.5	3.2	8.8	1.58

[a] As defined in footnote b of Table 5.8.

[a] Firmness was measured as the deformation in mm of a fruit under a 0.5 kg weight for 20 seconds.

For ethylene, both the biosynthetic pathway and the signalling pathway have been elucidated in great detail. However, the molecular cause of fruit softening and the identification of ethylene-dependent and independent ripening events still await clarification. Continued dissection of the genetic and molecular basis of ripening with modern tools of molecular biology promises to allow purposeful modification of the quality and ripening characteristics of harvested tomato fruit.

REFERENCES

Abbott, J.A., Lu, R.F., Upchurch, B.D. and Stroshine, R.L. (1997) Technologies for nondestructive quality evaluation of fruits and vegetables. *Horticultural Reviews* 20, 1–120.

Abeles, F.B., Morgan, P.W. and Saltveit, M.E. (1992) *Ethylene in Plant Biology*, 2nd edn. Academic Press, New York, 414 pp.

Baldwin, E.A., Scott, J.W., Einstein, M.A., Malundo, T.M.M., Carr, B.T., Shewfelt, R.L. and Tandon, K.S. (1998) Relationship between sensory and instrumental analysis of tomato flavour. *Journal of the American Society for Horticultural Science* 123, 906–915.

Beaulieu, J.C. and Saltveit, M.E. (1997) Inhibition or promotion of tomato fruit ripening by acetaldehyde and ethanol is concentration-dependent and varies with initial fruit maturity. *Journal of the American Society for Horticultural Science* 122, 392–398.

Beerh, O.P. and Siddappa, G.S. (1959) A rapid spectrophotometric method for the detection and estimation of adulterants in tomato ketchup. *Food Technology* 13, 414–418.

Biale, J.B. and Young, R.E. (1981) Respiration and ripening in fruits – retrospect and prospect. In: Friend, J. and Rhodes, M.J.C. (eds) *Recent Advances in the Biochemistry of Fruit and Vegetables*. Academic Press, New York, pp. 1–39.

Buttery, R.G. (1993) Quantitative and sensory aspects of flavor of tomato and other vegetables and fruits. In: Acree, T.E. and Teranishi, R. (eds) *Flavor Science: Sensible Principles and Techniques*. American Chemical Society, Washington, DC, pp. 259–286.

Cantwell, M. (2000) Optimum procedures for ripening tomatoes. In: *Management of Fruit Ripening*. Postharvest Horticultural Series No. 9, University of California, Davis, California, pp. 80–88.

Clinton, S.K. (1998) Lycopene: chemistry, biology, and implications for human health and disease. *Nutritional Review* 56, 35–51.

Davies, J.N. and Hobson, G.E. (1981) The constituents of tomato fruit – the influence of environment, nutrition and genotype. *Critical Review of Food Science Nutrition* 15, 205–280.

Dorais, M., Papadopoulos, A.P. and Gosselin, A. (2001) Greenhouse tomato fruit quality. *Horticultural Reviews* 26, 236–319.

Dorais, M., Demers, D.A., Van Ieperen, W. and Papadopoulos, A.P. (2003) Greenhouse tomato fruit cuticle cracking. *Horticultural Reviews* 30, 163–184.

Fisher, R.L. and Bennett, A.B. (1991) Role of cell wall hydrolases in fruit ripening. *Annual Review of Plant Physiology and Plant Molecular Biology* 42, 675–703.

Gent, M.P.N. and Ma, Y.Z. (1998) Diurnal temperature variation of the root and shoot affects yield of greenhouse tomato. *HortScience* 33, 47–51.

Grierson, D. and Kader, A.A. (1986) Fruit ripening and quality. In: Atherton, J.G. and Rudich, J. (eds) *The Tomato Crop*. Chapman & Hall, London, pp. 241–280.

Hardenburg, R.E., Watada, A.E. and Wang, C.Y. (1986) *The Commercial Storage of Fruits, Vegetables and Florist and Nursery Stocks*. Agriculture Handbook No. 66, USDA, ARS, Washington, DC, 130 pp.

Ho, L.C. (1999) The physiological basis for improving tomato fruit quality. *Acta Horticulturae* 487, 33–40.

Hobson, G.E. and Davies, J.N. (1971) The tomato. In: Hulme, A.C. (ed.) *The Biochemistry of Fruits and Their Products*, Vol. 2. Academic Press, New York, pp. 437–482.

Janse, J. and Schols, M. (1992) Specific temperature and transpiration effects on the flavour of tomatoes. In: *Annual Report*. Glasshouse Crops Research Station, Naaldwijk, The Netherlands, pp. 39–40.

Jones, J.B. (1999) *Tomato Plant Culture: in the Field, Greenhouse, and Home Garden*. CRC Press, Boca Raton, Florida.

Kader, A.A. (ed.) (2002) *Postharvest Technology of Horticultural Crops*, 3rd edn. DANR Special Publication No. 3311, University of California, Davis, California.

Kader, A.A., Morris, L.L., Stevens, M.A. and Albright-Holton, M. (1978) Composition and flavour quality of fresh market tomatoes as influenced by some postharvest handling procedures. *Journal of the American Society for Horticultural Science* 103, 6–13.

Leshuk, J.A. and Saltveit, M.E. (1990) Controlled atmospheres and modified atmospheres for the preservation of vegetables. In: Calderon, M. (ed.) *Food Preservation by Modified Atmospheres*. CRC Press, Boca Baton, Florida, pp. 315–352.

Lozano, R., Angosto, T., Gomez, P., Payan, C., Capel, J., Huijser, P., Salinas, J. and Martinez-Zapater, J.M. (1998) Tomato flower abnormalities induced by low temperature are associated with changes of expression of MADS-Box genes. *Plant Physiology* 117, 91–100.

Powell, A.L.T. and Bennett, A.B. (2002) Tomato. In: Valpuesta, V. (ed.) *Fruit and Vegetable Biotechnology*. Woodhead Publishing, Cambridge, UK, pp. 185–221.

Rylski, I. (1979) Effect of temperatures and growth regulators on fruit malformation in tomato. *Scientia Horticulturae* 10, 27–35.

Saltveit, M.E. (1993) Internal carbon dioxide and ethylene levels in ripening tomato fruit attached to or detached from the plant. *Physiologia Plantarum* 89, 204–210.

Saltveit, M.E. (2001) A summary of CA requirements and recommendations for vegetables. In: *Optimal Controlled Atmospheres for Horticultural Perishables, Postharvest Horticulture Series* no. 22A. Postharvest Technology Center, University of California, Davis, California, pp. 71–94. Also available on CD at: http://postharvest.ucdavis.edu as *Postharvest Horticulture Series* No. 22.

Saltveit, M.E. (2003) A summary of CA requirements and recommendations for vegetables. *Acta Horticulturae* 600, 723–727.

Saltveit, M.E. and Sharaf, A.R. (1992) Ethanol inhibits ripening of tomato fruit harvested at various degrees of ripeness without affecting subsequent quality. *Journal of the American Society for Horticultural Science* 117, 793–798.

Sargent, S.A. and Moretti, C.L. (2002) Tomato. In: Gross, K.C., Wang, C.Y. and Saltveit, M.E. (eds) *Agricultural Handbook 66 – The Commercial Storage of Fruits, Vegetables and Florist and Nursery Crops.* Available at: http://www.ba.ars.usda.gov/hb66/contents.html

Stahl, W. and Sies, H. (1996) Lycopene: a biochemically important carotenoid for humans? *Archives of Biochemistry & Biophysics* 336, 1–9.

Stevens, M.A. (1994) Processing tomato breeding in the 90's: a union of traditional and molecular techniques. *Acta Horticulturae* 376, 23–34.

Tomer, E., Moshkovits, H., Rosenfeld, K., Shaked, R., Cohen, M., Aloni, B. and Pressman, E. (1998) Varietal differences in the susceptibility to pointed fruit malformation in tomatoes: histological studies of the ovaries. *Scientia Horticulturae* 77, 145–154.

Irrigation and Fertilization

M.M. Peet

INTRODUCTION

The principles of water and fertilizer management discussed in this chapter apply equally to greenhouse and field production. Because of greater standardization and degree of control, more precise information is available on irrigation and fertilization requirements in soilless culture than is available for field crops, but information developed in soilless culture (such as essential elements, pH and salinity) should also be applicable to field-grown tomato crops when the specific soil characteristics are taken into account. To some extent, fertilization practices for greenhouse soilless culture and field production are converging, as drip irrigation with constant or periodic feeding of fertilizers (fertigation) becomes increasingly common for field tomato crops, especially in plasticulture production (the use of plastic mulch and drip irrigation).

Irrigation and fertilization guidelines for crop production are given in this chapter and additional information is available in Chapters 8 and 9. This chapter highlights the water and fertilization requirements of the tomato plant at various stages of development, interactions between nutrients, and interactions between watering and fertilization regimes as they relate to various physiological disorders, including blossom-end rot, goldspot, oedema, fruit cracking and russeting.

WATER QUALITY

pH

Root-zone pH affects nutrient availability. In acid soils, calcium, phosphorus, magnesium and molybdenum are the nutrients most likely to be deficient. The optimum pH value for most crops is about 6.0, but field tomatoes are considered to be moderately tolerant of soil acidity. Tomatoes in soilless culture appear to be able to tolerate even lower pH values.

Salinity

High quality irrigation water (Class 1 in Table 6.1) is a prerequisite for both soil and soilless culture. Soil salinity is measured by the electrical conductivity (EC), in units of mmho/cm, dS/m, or mS/cm (where S is siemens). These units are equivalent and are approximately equal to 640 mg salt/l. High salinity reduces plant uptake of both water and nutrients. Water with increased salts (EC), sodium, chloride and sulphates (Class 2) can be used in substrate and soil culture systems where the root zone can be leached (salts removed as water moves down). For example, in rockwool culture, frequent irrigation flushes the root zone and prevents salt accumulation.

In rockwool systems, salinity may be increased above recommended levels to improve fruit quality. This can be done in several ways, the simplest of which is to increase the concentration of all nutrients in the fertilizer solution. However, this is relatively expensive and can increase discharge of N and P to the environment. An alternative to raising concentrations of all nutrients is to add sodium chloride to the fertilizer water and, at the same time, carefully balance the nutrient solution and hold the water supply constant. Tomato crops can be grown in nutrient solution containing 100 ppm chloride without too much difficulty, but even for tomatoes it is better to use water sources with < 50 ppm sodium and 70 ppm chloride (OMAFRA, 2001). If possible, another water source should be substituted for Class 2 water. It is also important to subtract secondary and trace elements present in the irrigation water from the amount to be added as part of the fertilization programme. Class 2 water should not be used in nutrient film technique (NFT) or other types of recirculating systems (OMAFRA, 2001) without pre-treatment. Class 3 water should not be used in any type of soilless system.

In the soil, salinity rises rapidly as water is depleted. Hot, dry conditions coupled with high salinity can cause severe wilting and permanent damage (Fig. 6.1). High salinity can also result in toxic concentrations of ions in

Table 6.1. Minimum standards for greenhouse crop irrigation water (OMAFRA, 2001).

Water class	EC[a] (dS/m)	Sodium (ppm)	Chloride (ppm)	Sulphate (ppm)
1	< 0.5	< 30	< 50	< 100
2	0.5–1.0	30–60	50–100	100–200
3	1.0–1.5	60–90	100–150	200–300

[a] Electrical conductivity or soluble salts level.

Fig. 6.1. Tomato plant subjected to heat and salinity stress. Leaves also show symptoms of calcium deficiency.

plants. Soil-grown tomatoes are moderately sensitive to soil salinity, compared with other vegetables, with a maximum threshold for yield loss of 2.5 dS/m and a 10% yield decrease/dS/m above this threshold (Maas, 1984). Where salinity is high,but not extreme, fruit quality and soluble solids increase with increasing soil salinity, just as in soilless systems.

IRRIGATION REGIMES

Starting seeds

Seeds are started by pre-soaking the growing medium, such as rockwool or oasis plugs, with clear water or a dilute complete nutrient solution (EC 1.0–1.4 dS/m). After seed sowing, vermiculite is placed over the seeds and the growing medium is gently watered. The trays are covered with a clear polyethylene sheet to prevent drying out, and maintained at a temperature of 25°C. The cover is removed when 60–75% of the seedlings have germinated, usually after 60–96

h. Once the true leaves have emerged, flats should be watered with a complete nutrient solution, starting with a fertilizer EC of 2.0 dS/m and gradually increasing the concentration up to 3 dS/m. Flats should not be overwatered at this stage, as excess water in the root zone reduces temperature and aeration, both of which then reduce root growth. Maintaining excessively dry flats will also decrease growth and leaf size and increase medium EC, which can in turn reduce seedling survival (OMAFRA, 2001).

After transplanting

Amount of water available

A greenhouse irrigation system should be able to provide at least 8 l/m^2/day. A mature tomato crop uses 2–3 l of water per plant per day when light levels are high (OMAFRA, 2003). Figure 6.2 shows a typical water application schedule for greenhouse vegetables in Ontario, Canada, where approximately 1000 l/m^2 was applied over the course of the production year. The plant uses most of this water (90%) in transpiration and only uses 10% for growth. Although this represents a large volume of water, water use efficiency is higher in greenhouses than in field production. To produce 1 kg of fresh tomatoes requires 60 l of irrigation water in Israeli field production, 40 l in unheated plastic houses in Spain, 30 l in unheated glasshouses in Israel, 22 l in climate-controlled glasshouses in The Netherlands and only 15 l when the drain water is reused in those glasshouses (Stanghellini *et al.*, 2003).

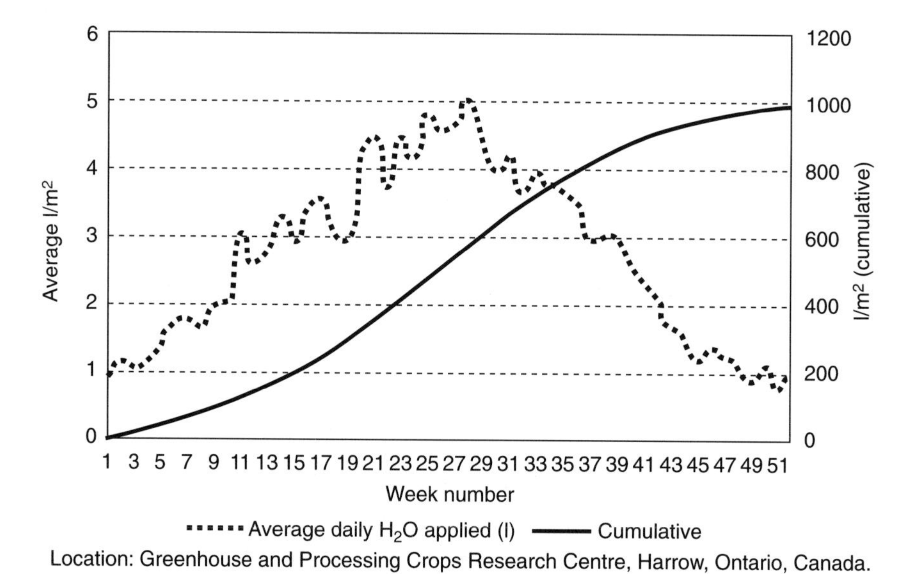

Fig. 6.2. Water application schedule for greenhouse vegetables (OMAFRA, 2003).

Irrigation based on solar radiation

The amount of transpiration depends on radiation, vapour pressure deficit (VPD) and other conditions in the greenhouse, such as air movement and the location of the heating pipes. Figure 6.3 shows the close relationship between transpiration in a tomato crop in The Netherlands and seasonal radiation patterns. Within a single day, changes in transpiration also closely follow changes in outside radiation (Fig. 6.4). When radiation is higher, leaf temperatures increase, which in turn increases VPD and transpiration (Fig. 6.5).

Because of this close relationship between incoming solar radiation and water loss through transpiration, computerized irrigation in commercial greenhouses is frequently based on received solar radiation (measured in joules), although factors such as air temperature, stage of crop growth, plant populations and growing medium may also be considered. In rockwool systems, 100 ml of water are typically provided per watering cycle, but the amount provided may be adjusted based on the amount of light received by a certain time of day. Morning irrigation often starts after 150–200 J have accumulated. A rule of thumb for irrigation rates is three times the radiation level per day. For example, for 1000 J, 3000 ml/m^2 should be provided. At high irradiances (> 2000 J), a higher ratio of irrigation to irradiance is often used.

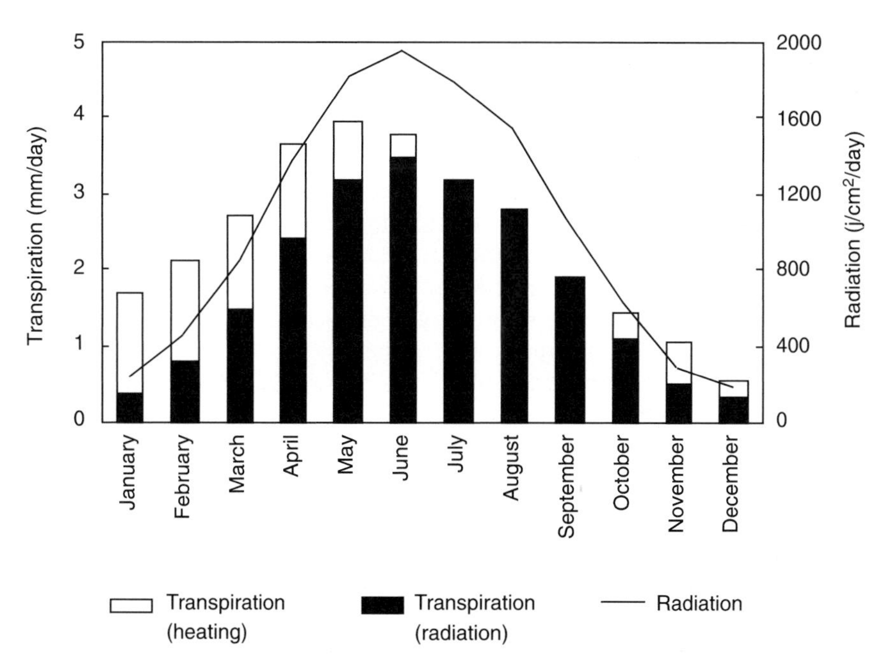

Fig. 6.3. Transpiration of a greenhouse tomato crop throughout the year in The Netherlands, showing the importance of solar radiation and heating for transpiration. (Data from KWIN, 2001.)

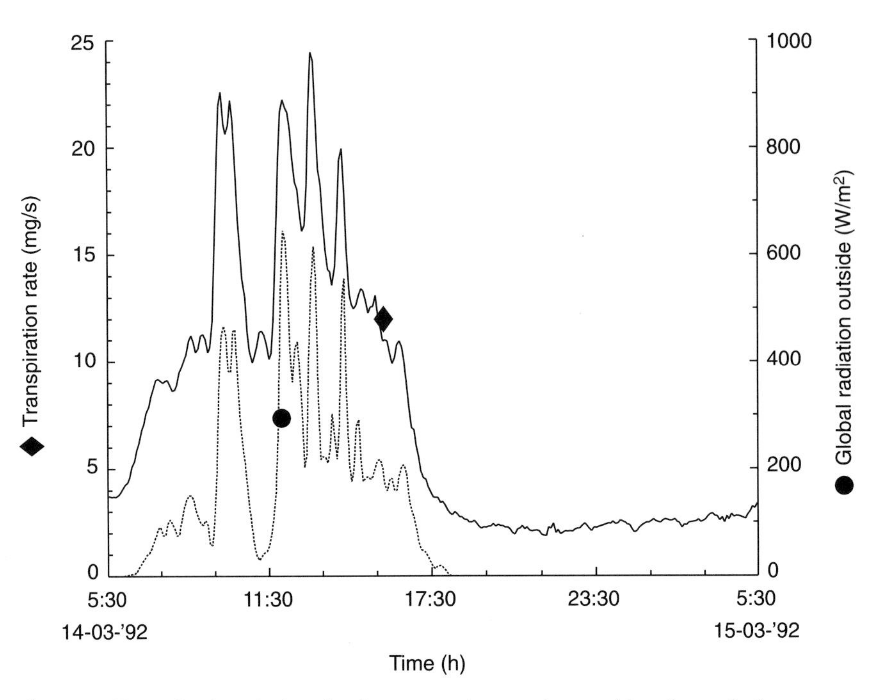

Fig. 6.4. Transpiration during daytime strongly correlates with solar radiation (greenhouse tomato crop in The Netherlands). (Reprinted, with permission, from *Journal of Experimental Botany* 45, 56, Van Ieperen and Madery, 1994.)

Another concern in greenhouse irrigation is the concentration of the fertilizer solution. When air temperature increases, the water uptake rate increases more than nutrient uptake. Therefore, the concentration of the remaining nutrient solution in the root zone increases and salinity rises. If this effect is not controlled, both water and nutrient uptake may be reduced, resulting in wilting and slower growth. Generally this tendency is overcome by feeding a more concentrated solution in the winter, when both low root zone temperatures and low light may reduce nutrient uptake, and feeding a more dilute solution in the summer when uptake rates are high.

The frequency of watering is also a tool in crop management. With frequent irrigation, plants experience less stress because the water status is more constant. However, if emitters are not pressure compensated, frequent irrigation can accentuate the effects of small differences in emitter performance. With frequent irrigation, there is also less opportunity to flush the root zone, potentially resulting in a build-up of EC. In managing plant growth in soilless growing systems, stress resulting from either periodic water deficits or high salinity levels moves the plant in the generative direction, while higher water availability levels (less stress) moves the plant towards more vegetative growth.

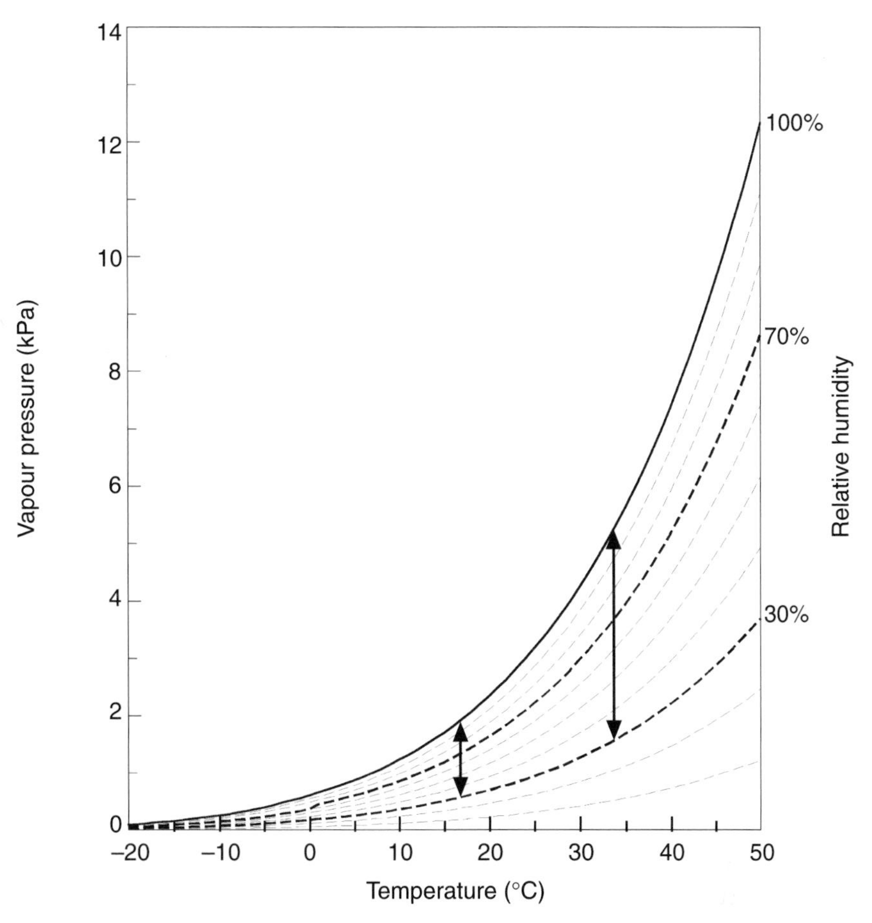

Fig. 6.5. Water vapour pressure as a function of temperature. Vapour pressure deficit and hence driving force for transpiration, increases with temperature at equal relative humidity (RH), as shown by the two double-headed arrows.

Irrigation based on drainage

Irrigation management in soilless media may be based on the amount of water draining out (also referred to as overdrain or percentage leach). Some overwatering is necessary, especially in greenhouse drip irrigation systems, to compensate for lack of uniformity in the irrigation system, plant-to-plant differences and differences between locations. For example, outer rows may require more water. Typically, in rockwool production in the greenhouse, 30–50% overwatering daily is recommended in sunny weather and 10–20% in cloudy weather, for an average of 20–30%. To adjust irrigation based on drainage, samples should be taken at the same time daily. Usually irrigation cycles are started 1–2 h after sunrise and ended 1–2 h before sunset. This helps to reduce the incidence of russeting and fruit cracking in the summer, as well as disease incidence (OMAFRA, 2001). During winter months irrigation may be needed at night, because heating systems are running and relative humidity may be low. In some cases, night-time irrigation is also needed in the summer when daytime light and temperature are both high.

Most of the overdrainage should occur during the peak light period (1200–1600 h), with only minimal overdrainage (1–2%) in early morning. If there is no overdrainage after the third cycle, however, the bags are too dry and the irrigation programme should be extended later into the evening or started earlier in the day. If slabs are too dry, root hairs will be damaged, EC may become excessive, calcium uptake may be reduced (resulting in blossom-end rot) and plant growth will become harder. In this case, plants have thin stems and small, dark leaves. If there is leachate after the first cycle, the slabs are too wet, and start-up should be delayed or irrigation ended earlier (OMAFRA, 2001). Excessively wet slabs will decrease aeration, increase the potential for root disease and result in softer vegetative growth and poor rooting.

By matching the amount of water provided to plant demands, fertilizer runoff and water use can be minimized. In addition to adverse environmental impacts, provision of water in excess of plant requirements can reduce the amount of air available in the root zone. In either soil or soilless systems, low oxygen levels (< 3 gm/l) reduce not only nutrient uptake but growth and yield as well (Adams, 1999). In the field, this is most frequently a problem with poorly draining soils or those with subsurface hardpans. In soilless culture, low oxygen levels generally occur in hot weather because the available oxygen in the solution decreases as the root zone temperature rises. Root zone temperatures above 23°C are said to be deleterious for greenhouse tomato production. Low root zone oxygen can occur in either the field or the greenhouse but is most common in peat-based greenhouse substrates, because they have a higher water-holding capacity, i.e. drain less freely, than rockwool and perlite. In peat, overwatering can lead to iron deficiency, which must be corrected with a reduction in amount of watering and an addition of iron to the feed (Adams, 1999).

Effect of humidity on transpiration

Humidity also affects water uptake. This can be expressed as absolute humidity (g/m^3), specific humidity (g water/kg air) or relative humidity (RH) (ratio between the mass of water vapour in the air and the mass it can hold at the saturation point). The amount of water vapour that a given volume of air can hold is dependent on temperature, almost doubling for every 10°C rise in temperature. Transpiration rate increases as the difference between the fully saturated atmosphere inside the leaf (100% RH) and the water vapour content outside the leaf increases. Mathematically, this difference is described as the vapour deficit, or vapour pressure deficit (VPD), and is measured by comparisons of wet and dry bulb thermometers. Since water vapour exerts a pressure, VPD is expressed in units of pressure: millibars (mbar) or kilopascals (kPa).

The effects of relative humidity and temperature on VPD are shown in Table 6.2 and Fig. 6.5. For the same relative humidity, the VPD is higher at a higher temperature, which increases transpiration. Nutrient uptake and photosynthesis are optimal at 4–8 mbar (the area of Table 6.2 shown in bold). Transpiration is reduced when VPD is too low, and leaves may appear thicker and larger. Stems are also thick, but root systems may be weak and plants are more susceptible to disease (OMAFRA, 2001). Transpiration is particularly important for calcium uptake, since calcium moves only in the xylem. Lack of calcium can result in blossom-end rot (BER), particularly when coupled with other stress factors. This issue is discussed further in the section on physiological disorders, below. At low humidity (high VPD), transpiration may be excessive, stressing the plant. Indications of excessive transpiration in greenhouse tomatoes include small, thin leaves and stems, but a strong root

Table 6.2. Vapour pressure deficit in millibars (mbar) in relation to temperature and humidity (adapted from OMAFRA, 2001).

Temperature (°C)[a]	Relative humidity									
	50%	55%	60%	65%	70%	75%	80%	85%	90%	95%
15	8.5	**7.7**	**6.8**	**6.0**	**5.1**	**4.3**	3.4	2.6	1.7	0.8
16	9.1	8.2	**7.3**	**6.4**	**5.5**	**4.6**	3.6	2.7	1.8	0.9
17	9.7	8.7	**7.8**	**6.8**	**5.8**	**4.9**	3.9	2.9	1.9	1.0
18	10.3	9.3	8.3	**7.2**	**6.2**	**5.2**	**4.1**	3.1	2.1	1.0
19	11.0	9.9	8.8	**7.7**	**6.6**	**5.5**	**4.4**	3.3	2.2	1.1
20	11.7	10.5	9.4	8.2	**7.0**	**5.9**	**4.7**	3.5	2.3	1.2
21	12.4	11.1	9.9	8.7	**7.5**	**6.2**	**5.0**	3.7	2.5	1.2
22	13.2	11.9	10.6	9.3	**7.9**	**6.6**	**5.3**	**4.0**	2.6	1.3
23	14.1	12.6	11.2	9.8	8.4	**7.0**	**5.6**	**4.2**	2.8	1.4
24	14.9	13.4	11.9	10.4	9.0	**7.5**	**6.0**	**4.5**	3.0	1.5
25	15.8	14.3	12.7	11.1	9.5	**7.9**	**6.3**	**4.8**	3.2	1.6
26	16.8	15.1	13.4	11.8	10.1	8.4	**6.7**	**5.0**	3.4	1.7
27	17.8	16.0	14.2	12.5	10.7	8.9	**7.1**	**5.4**	3.6	1.8
28	18.9	17.0	15.1	13.2	11.3	9.5	**7.6**	**5.7**	3.8	1.9
29	20.0	18.0	16.0	14.0	12.0	10.0	8.0	**6.0**	**4.0**	2.0
30	21.2	19.1	17.0	14.8	12.7	10.6	8.5	**6.4**	**4.2**	2.1

[a] Refers to plant tissue temperatures, not air temperatures.
Note: **Bold area** indicates optimal range for most greenhouse crops of 4–7 mbar.
1 mbar = 0.1 kPa
0.1 kPa = 0.7 g/m^3 = 3% RH

system (OMAFRA, 2001). VPD cannot be completely controlled in greenhouses, but increasing temperature, venting and air movement will generally increase VPD, while increasing irrigation water, misting and fogging will generally decrease VPD.

FERTILIZATION

Principles

In applying fertilizer to a plant grown either in soil or in a soilless medium, the goal is to match the nutrient uptake of the crop as closely as possible to the amount provided as fertilizer. This will prevent both salt build-up and nutrient deficiency. In addition, it will prevent excessive fertilizer runoff. Historically, growers have added more N and P than the plants could use immediately to avoid depletion in the root zone. This was done by preplant incorporation of fertilizers in field crops or by maintaining high levels of nutrient solution in soilless culture (> 200 ppm N, > 30 ppm P). In both field and greenhouse production, growers are now under pressure to reduce their inputs of nitrogen and phosphorus, in particular, as these are the most important contaminants in groundwater and surface water systems.

Environmental considerations

Soil testing is a valuable tool to reduce excess fertilizer usage. As shown in Tables 6.3 and 6.4, less P and K are added when soil levels are already high. Nitrogen addition is usually standardized on the assumption that most is lost from the soil over the winter. In addition to soil testing, new monitoring and nutrient delivery capabilities enable growers to implement 'just in time' and 'just enough' fertilization practices. In field production of tomatoes using drip irrigation, petiole sap testing guidelines (Table 6.5) and tissue analysis (Table 6.6) and/or weekly fertigation guidelines (Table 6.7) allow fertilizer additions to be made when they are needed by the crop, rather than attempting to supply all nutrients preplant or in one or two large additions during the growing season (Tables 6.3 and 6.4).

In soilless culture, runoff can be reduced by replacing the open systems or 'run to waste' systems used in rockwool and perlite culture, which require 20–30% excess water application, with recirculating systems, such as those used in NFT. In recirculating systems, EC, pH and nutrient levels are constantly monitored and additions are made to the solution as necessary. The solution cannot be reused indefinitely, even with nutrient replacement and pH control, because not all ions are taken up in equal proportion. Those not taken up accumulate over time in the nutrient solution. This is why EC

Table 6.3. Recommended nutrients based on soil tests for transplants in bare ground in New York State (adapted from *Integrated Crop and Pest Management Recommendations for Commercial Vegetable Production*, 1999, Cornell Cooperative Extension, Ithaca, New York, p. 273). The pH should first be corrected to 6.0–6.5. Zinc may be needed if the organic matter is low and the pH is > 7. If these conditions exist, apply zinc sulphate in the transplant water. For transplants, add 50 g zinc sulphate per 100 l transplant water for a 0.2% solution (Reiners *et al.*, 1999).

	P_2O_5 (kg/ha)			K_2O (kg/ha)			
	Soil phosphorus level			Soil potassium level			
N (kg/ha)	Low	Medium	High	Low	Medium	High	Comments
112.5	168.75	112.5	56.25	202.5	135.0	67.5	Total recommended
56.2	168.75	112.5	56.25	202.5	135.0	67.5	Broadcast and disk in[a]
56.2	0	0	0	0	0	0	Apply when first clusters set fruit[b]

[a] If equipment is available, apply half of the phosphorus and potassium in bands 10 cm deep and 10 cm from the row at planting.
[b] Nitrogen can be applied as a split application. Apply half at fruit set and the rest when fruit are 2.5 cm in diameter.

Table 6.4. Recommended nutrients based on soil tests for transplants in plastic mulch in New York State. The pH should first be corrected to 6.0–6.5. Zinc may be needed if the organic matter is low and the pH is >7. If these conditions exist, apply zinc sulphate in the transplant water. For transplants, add 50 g zinc sulphate per 100 l transplant water for a 0.2% solution (Reiners *et al.*, 1999).

	P_2O_5 (kg/ha)			K_2O (kg/ha)			
	Soil phosphorus level			Soil potassium level			
N (kg/ha)	Low	Medium	High	Low	Medium	High	Comments
112.5	168.75	112.5	67.5	202.5	135.0	67.5	Total recommended
45.0	101.25	45.0	0	135.0	67.5	0	Broadcast and disk in[a]
22.5	22.5	22.5	22.5	22.5	22.5	22.5	Apply 1 week after transplanting
22.5	22.5	22.5	22.5	22.5	22.5	22.5	Apply when first fruits are 2.5 cm in diameter
22.5	22.5	22.5	22.5	22.5	22.5	22.5	Apply when first fruits turn colour

[a] If equipment is available, apply half of the phosphorus and potassium in bands 10 cm deep and 10 cm from the row at planting.

Table 6.5. Testing guidelines for tomato leaf petiole fresh sap nitrate-nitrogen and potassium (adapted from Sanders, 2004–2005).

	Fresh petiole sap concentration (ppm)	
Development stage/time	NO_3-N	K
First buds	1000–2000	3500–4000
First open flowers	600–800	3500–4000
Fruits 25 mm diameter	400–600	3000–3500
Fruits 50 mm diameter	400–600	3000–3500
First harvest	300–400	2500–3000
Second harvest	200–400	2000–2500

Table 6.6. Adequate ranges and toxicity values for nutrient content of the most recently matured whole leaf, including petiole, in tomato (adapted from Hochmuth *et al.*, 1991).

Time of sampling	Status	%				ppm							
		N	P	L	Ca	Mg	S	Fe	Mn	Zn	B	Cu	Mo
5-leaf stage	Adequate	3.0	0.30	3.0	1.0	0.30	0.30	40	30	25	20	5	0.2
	range	5.0	0.60	5.0	2.0	0.50	0.80	100	100	40	40	15	0.6
First flower	Adequate	2.8	0.20	2.5	1.0	0.30	0.30	40	30	25	20	5	0.2
	range	4.0	0.40	4.0	2.0	0.50	0.80	100	100	40	40	15	0.6
	Toxic (>)	–	–	–	–	–	–	–	1500	300	250	–	–
Early fruit set	Adequate	2.5	0.20	2.5	1.0	0.25	0.30	40	30	20	20	5	0.2
	range	4.0	0.40	4.0	2.0	0.50	0.60	100	100	40	40	10	0.6
	Toxic (>)	–	–	–	–	–	–	–	–	–	250	–	–
First ripe fruit	Adequate	2.0	0.20	2.0	1.0	0.25	0.30	40	30	20	20	5	0.2
	range	3.5	0.40	4.0	2.0	0.50	0.60	100	100	40	40	10	0.6
During harvest period	Adequate	2.0	0.20	1.5	1.0	0.25	0.30	40	30	20	20	5	0.2
	range	3.0	0.40	2.5	2.0	0.50	0.60	100	100	40	40	10	0.6

alone cannot be used to determine when the solution is depleted. Disinfection, using ultraviolet light, ozone or, more recently, heat treatment and slow sand filtration, allows for longer periods of recirculation, but the solution is eventually discarded and a fresh one is made up because of concerns about disease transmission, solution imbalances and root exudates. Although heat treatment and UV are the most popular disinfection methods in The Netherlands, slow sand filtration is of particular interest because of the potential for pathogen control by resident microflora rather than completely sterilizing the system (van Os, 2000).

Table 6.7. Suggested fertigation schedule for field-grown tomato plants in the south-eastern USA on low-potassium soils (Sanders, 2003–2004). Before applying plastic mulch, the soil pH should be adjusted to 6.5 and enough fertilizer applied either to meet soil test recommendations or to supply 56 kg each of N, P_2O_5 and K_2O/ha. On low-potassium soils, it may be necessary to apply an additional 84, hence 140 kg K_2O/ha. All fertilizers should be thoroughly incorporated. The first soluble fertilizer application should be applied through the drip irrigation system within a week after field-transplanting the tomatoes and continued through the last harvest. On low to low-medium boron soil, 0.23 kg actual boron should also be included in the fertigation programme.

Days after planting	Daily nitrogen (kg/ha)	Daily K_2O (kg/ha)	Cumulative nitrogen (kg/ha)	Cumulative K_2O (kg/ha)
Preplant			56	140
0–14	0.56	0.56	64	148
15–28	0.78	1.575	75	170
29–42	1.13	2.25	90	202
43–56	1.69	3.38	114	249
57–77	2.81	5.6	173	367
78–98	3.38	6.75	237	495

Crop requirements

Additional fertilization recommendations are available in Chapters 8 and 9. The fertigation recommendations for south-eastern US soils (Table 6.7) and New York preplant recommendations for transplants in bare ground (Table 6.3) and in plastic mulch (Table 6.4) are given as guidelines. For both bare ground and plasticulture production in the field, it is suggested that soil pH values initially be raised to 6.5–6.8 and that the soil be tested before preplant fertilizers are added. N, P, K, Mg and Ca are usually added preplant, based on soil test results, even if the crop is to be fertigated later. Preplant applications can be broadcast and disked in or side-dressed in bands 10 cm deep and 10 cm from the row at planting. The application can also be split between the two methods of incorporation.

For greenhouse crops, tomato seedlings are transplanted into 7.5 or 10 cm rockwool blocks soaked with a complete nutrient solution at a pH of 5.5 and an EC of 1.5–2.0 dS/m. The bag must be completely saturated to activate the wetting agent, ensuring that the whole slab is available to the plant roots (OMAFRA, 2001). Within 24–48 h after planting, a slit is cut in the bag from the bottom up to allow for drainage. The plants are watered for 30 min and then no further nutrient solution is applied for 2–3 days, depending on the weather. This ensures rapid root penetration into the slab as well as allowing the slab to dry out. After 2–3 days, a 15 min watering period is followed by 2–3 days without water. At each subsequent watering, the EC should gradually be raised to 3.0–3.5 dS/m, based on available light, rate of growth, plant vigour, available moisture and temperature regime (OMAFRA, 2001). A fertigation schedule for the remaining crop period is shown in Table 6.8. Demand, uptake and removal of the major nutrients are shown in Table 6.9. In greenhouse production both yields and crop removal levels are higher.

Nutrient deficiencies and excess

For colour plates illustrating nutrient deficiencies and toxicities, see Roorda van Eysinga and Smilde (1981).

Nitrogen (N)

Under hot, bright growing conditions, nitrogen must be adequate for plants to grow rapidly, but high nitrogen levels encourage vegetative growth, which

Table 6.8. Fertigation schedule for greenhouse tomatoes grown in rockwool slabs (OMAFRA, 2001).

	Nutrient (kg/ha)														
	N	NH_4	P	K	Ca	Mg	Fe	Mn	Zn	B	Cu	Mo	S	Cl	HCO_3
Slab saturation	200	10	50	353	247	75	0.8	0.55	0.33	0.5	0.05	0.05	120	18	25
4–6 weeks after planting	180	10	50	400	190	75	0.8	0.55	0.33	0.5	0.05	0.05	120	18	25
Normal feed	190	22	50	400	190	65	0.8	0.55	0.33	0.5	0.05	0.05	120	18	25
Heavy fruit load	210	22	50	420	190	75	0.8	0.55	0.33	0.5	0.05	0.05	120	18	25

Table 6.9. Nutrient demand, uptake and removal by tomato crops grown in the field and in greenhouses (adapted from Halliday and Trenkel, 1992; Jones, 1999; OMAFRA, 2001).

Nutrient (kg/ha)	Outdoor crop (yield 40–50 t/ha)	Greenhouse crop (yield 100 t/ha)
N	100–150	200–600
P_2O_5	20–40	100–200
K_2O	150–300	600–1000
Ca	–	45*
MgO	20–30	290*

*Values from OMAFRA, 2001.

can be detrimental to reproductive growth under low light. Typically, nitrogen levels are kept relatively low until fruit set to encourage reproductive growth, then raised. In soilless culture, the ratio of nitrogen to other nutrients is closely controlled to encourage either reproductive or vegetative growth, depending on the grower's perception of crop needs.

Tomato plants have different absorption rates of nitrogen at various stages of growth. The nitrogen absorption by tomato plants during the formation of the first two to three flower trusses is about the same as that for potassium. As the fruit load increases, so does the potassium uptake, resulting in a K:N uptake of 2:1. An excessive amount of nitrogen under low light conditions early in the production cycle results in an overly vegetative plant prone to disease and poor flower development, fruit set and size (OMAFRA, 2001). Lower nitrogen feed during this phase helps to control plant growth and make the plant more generative. Plant growth can also be slowed by increasing the K:N ratio in the fertilizer, increasing EC by irrigating less frequently and in lower amounts, and reducing relative humidity. For more details on this approach to controlling plant growth, see Chapter 9.

DEFICIENCY Nitrogen deficiency is sometimes hard to detect without a well-fertilized control for comparison. Growth may be reduced overall, so the plants are stunted, but leaves may look healthy, except for being a paler green than normal. Symptoms appear first on the lower leaves and continue to be more pronounced there, as nitrogen is a mobile element, moving from older to younger tissue. Symptoms at the top of the plant include flowers that are pale rather than deep yellow and a main stem that is thin at the top. The whole plant has a spindly appearance. Rather than foliage being lush or succulent, the leaves are small, erect and 'hard'. Over time, the whole plant can turn yellow, flowers drop and fruit remain small. Plants may mature early, but fruit yield and quality decline. Nitrogen deficiency can appear in waterlogged soils (heavy clay) or sandy soils (after heavy leaching) or can be induced by heavy applications of straw or other organic material with a high C:N ratio.

TOXICITY Dark green leaves, sometimes thickened and brittle, indicate excess nitrogen. At the top of the plant, stems remain thick and new leaves may curl into a ball, a condition sometimes referred to as 'bullishness'. Clusters and flowers are large, but fruit set may be poor. Although leaf growth is initially promoted, it is eventually restricted under excessive nitrogen. Plants may be more susceptible to diseases and insects (Jones, 1999).

Nitrogen form is also an important consideration in guarding against nitrogen toxicity. Tomatoes are much more sensitive to nitrogen in the ammonium form than in the nitrate form, especially under low light. The early symptoms of ammonium toxicity are small chlorotic spots on the leaves, which later turn necrotic (brown and dead). Spot size may increase, covering the entire inter-veinal area, giving the leaf a scorched appearance, and leaf margins may roll up. Lesions may also occur on the stems. As the plant matures, the vascular tissue at the base of the plants begins to deteriorate, with wilting occurring during periods of high atmospheric demand, followed by plant death (Jones, 1999).

Phosphorus (P)

Although phosphorus is used in much smaller quantities than nitrogen and potassium, it must be provided continuously. Initially, phosphorus is important for early root growth, especially under cool soil conditions when P uptake decreases, but later it is necessary for vegetative growth and fruit set. Phosphorus is stored well in soil but is easily leached in peat media, and less available at high pH.

DEFICIENCY Symptoms of phosphorus deficiency occur first on the lower leaves and stems, as phosphorus, like nitrogen, is phloem-mobile. Plants appear stunted and leaves are unusually dark green. A characteristic red or purple colour on the undersides of the leaves (including the veins) and stem appears later, starting on the older leaves. Leaves are small and curve slightly downwards. In severe cases, leaves first develop chlorosis, followed by necrosis. Plants become slender, with thin stems, and cluster development is poor. Roots become brown and develop few lateral branches. Low rooting zone and air temperatures can also reduce P uptake, and cause the purple pigmentation typical of P deficiency.

TOXICITY Phosphorus toxicity is uncommon but can occasionally be seen as slow plant growth, with visual symptoms related to Zn deficiency. These symptoms can be severe, as large sections of the leaves turn light brown, giving a burned appearance. This is particularly noticeable under anaerobic rooting conditions (Jones, 1999).

Potassium (K)

Potassium is required for best fruit quality and to regulate growth. As a major nutrient with a positive charge, potassium balances the negative charges of organic acids produced within the cell, and those of anions such as sulphates, chlorides and nitrates. Potassium levels are particularly important at transplanting, to control subsequent plant growth and to prevent ripening disorders. The ratio of nitrogen to potassium is also important in controlling growth, as discussed below.

DEFICIENCY Yield increases with potassium additions are greatest when nitrogen is not a limiting factor. Potassium deficiency is first expressed as dark green foliage, which later turns purplish brown. Marginal chlorosis and necrosis appear first on the lowest leaves, then progress up the plant. Like nitrogen and phosphorus, potassium is phloem-mobile, and young leaves are the last affected. Chlorosis almost always occurs first at the margins of older leaves, which often curve downwards. Later chlorosis moves into the inter-veinal areas towards the centre of the leaf, and necrosis of leaf margins follows. In advanced stages, the small veins lose their colour, older leaves become severely scorched and drop, young leaves turn yellow and remain small, plant growth is restricted, and fruit ripening is uneven. Postharvest quality is also poor (Jones, 1999).

Inter-veinal chlorosis is also symptomatic of low magnesium, and the two deficiencies are sometimes confused. Compared with magnesium deficiency, potassium deficiency is more likely to occur along the leaf margins and is also more likely to develop into necrotic spots. Problems with fruit quality, such as blotchy ripening, boxy fruit and even to some extent greenback, are associated with low levels of potassium and in most cases can be counteracted with high-potassium feeds (OMAFRA, 2001). Uptake of potassium can also be reduced under anaerobic rooting conditions and low root zone temperatures.

TOXICITY Potassium toxicity per se is rare. However, very high rates of potassium may induce Ca or Mg deficiency or salinity damage. Calcium deficiency can lead to BER (see below). Reductions in yield occur at very high levels of K, when the K:N ratio in the liquid feed is too high, or when both N and K are too high. In this case, yield reductions are attributed to increased salinity in the growth medium. High levels of potassium improve fruit shape, decrease fruit size, and reduce the proportion of hollow fruit (a disorder associated with early growth under poor light conditions) even under conditions where total yields are somewhat reduced.

Calcium (Ca)

Calcium is considered a secondary nutrient and is not required in as large quantities as nitrogen and potassium. In contrast to nitrogen, phosphorus and potassium, calcium is immobile in the phloem, i.e. it does not move from older to younger leaves. Calcium moves mostly through the xylem, along with water, and its uptake is reduced by low temperature, drought, high salts in the growth medium, or high atmospheric humidity. If, for any reason, the supply of calcium to meristems of young shoots (or roots) is interrupted even for a short period, localized deficiency and dieback may occur. Since little Ca is translocated from leaves to fruits, leaf analysis is of limited use as diagnostic tool for preventing calcium shortages in the fruit that result in BER.

DEFICIENCY Localized calcium deficiency shows up as chlorosis and dieback of the leaves and tip (see Fig. 6.1) and fruit blossom-end rot, a physiological disorder discussed in greater detail below. Typical signs in the top of a calcium-deficient plant include: restriction of growth; loss of turgor; small dark green leaves later turning yellow or orange or purplish; short internodes; leaves curling downwards and inwards; scorched leaf margins; and dieback of the growing tip. Roots are poorly developed and brown, with few root hairs and dieback of the tips. Clusters are weak with poor fruit set and fruit ripening. In most cases, the calcium deficiency is not in the soil but is induced by cultural or environmental factors. The most likely cause is water stress on the plant resulting from inadequate or uneven watering, frequent and large variations in relative humidity, or a high level of salts.

TOXICITY Calcium, magnesium and potassium compete for the same sites of absorption by the plant and so increasing one may decrease another.

Magnesium (Mg)

Like calcium, magnesium is considered a secondary nutrient, but an average tomato crop takes up 290 kg/ha (OMAFRA, 2001). Magnesium deficiency is common but rarely results in yield reduction. Deficiencies are usually related to competition with other cations (especially K^+), low soil pH, or poor conditions in the root zone. Adverse root conditions could include soil compaction, waterlogging, water stress, or poor aeration of hydroponic solutions. Magnesium deficiency also sometimes appears at times of heavy fruit load. Under any of these conditions, the plant might not be able to take up sufficient magnesium. Since magnesium is mobile, reserves may be moved from old leaves to new. Magnesium problems might also develop in soilless culture if the magnesium in the nutrient solution drops below the minimum recommended level or becomes unbalanced relative to the other cations (i.e. K^+, Ca^{2+}, NH_4^+, H^+).

DEFICIENCY Initially leaf margins on the lower leaves turn yellow. These yellow areas quickly develop in the inter-veinal areas of the lower leaves while the large veins remain green. Magnesium deficiency first appears in the lower or middle leaves, while manganese and iron deficiencies first appear in young leaves. The yellowing later spreads to the top of the plant while affected areas of the lower leaves form necrotic spots between the veins. In severe situations, the lower leaves dry out completely and the whole plant turns yellow, withers and dies. Fruit production is the least affected and is affected only when magnesium deficiency is serious enough to limit the photosynthetic productivity of the plant. However, it may be associated with reduced fruit quality (OMAFRA, 2001) and BER in the fruit (Jones, 1999). Magnesium deficiency is promoted by high pH, high K, or low N concentration in the growth medium, as well and low root zone temperature (OMAFRA, 2001).

EXCESS High magnesium levels may cause potassium and calcium deficiency because of competition in uptake, resulting in slow growth.

K:N ratio

Nitrogen is the element that, more than any other, controls the growth rate of the plant. Up to a certain point, the more N supplied, the more rapidly the plant grows. The ratio between potassium and nitrogen is also important: the higher the ratio, the slower is the growth. The optimal ratio of potassium to nitrogen varies with growth stage (Adams, 1999). When the first truss is in flower, the K:N ratio should be 1.2:1, which is the same K:N requirement as in most plants during the vegetative stage. This ratio increases as the fruit load on the plant increases, since about 70% of the potassium absorbed moves into the fruit. By the time the ninth cluster flowers open, the ratio should be 2.5:1. Low potassium during times of high fruit load reduces tomato quality, especially flavour. Other nutrients are generally supplied in a constant ratio to each other.

Nitrogen forms and nutrient interactions

The form of nitrogen provided is important (Adams, 1999). Too much ammonium-nitrogen (NH_4-H) will reduce the calcium content of the crops substantially, may reduce growth, and can result in BER. Ammonium-nitrogen is particularly likely to harm the plant early in the season when conversion of NH_4-N to NO_3-N is slow. Later, up to 10% of the nitrogen requirement can be supplied in the ammonium form, but a level of more than 20% will result in BER (Adams, 1999) in soilless systems.

Excessive fertilization can also create an imbalance. For example, high potassium levels will reduce calcium and magnesium uptake. In general, nitrogen and phosphorus have antagonistic effects and induce or accentuate potassium deficiency. Other conditions that reduce calcium uptake include the presence of high concentrations of Na and Mg. Calcium (and to a smaller extent magnesium) antagonize potassium uptake. Ammonium greatly decreases the rate of potassium uptake. Potassium deficiency tends to induce or accentuate iron deficiency.

Controlling growth with irrigation and fertilization

Immediately after transplanting, tomato plants usually show rapid vegetative growth. At some point, the plant must start putting its energy into fruit production, i.e. the generative stage of plant growth. A number of practices, including specific irrigation and fertilization methods, can be used to steer plant growth between generative and vegetative growth. This concept is discussed in more detail in Chapter 9, but recommendations related to irrigation and fertilization are summarized in Table 6.10.

Table 6.10. Summary of irrigation and fertilization practices that can be used to steer plants between vegetative and generative growth.

	Effect on plant growth	
Practice/environmental condition	Vegetative	Generative
Humidity level	High (low VPD)	Low (high VPD)
Solution and slab EC	Lower	Higher
Length of each irrigation event	Shorter	Longer
Frequency of irrigation events	More frequent	Less frequent
Timing of irrigation start in the morning	Early	Later
Timing of last irrigation in the afternoon	Continue until later in afternoon or evening	Stop sooner in afternoon or evening

Foliar sprays

Foliar sprays should be avoided, because of the potential for damage to the foliage from spraying in hot or bright weather and because leaves do not absorb nutrients well, but they are sometimes used under emergency conditions while the situation in the root zone is being corrected. Foliar sprays of urea dissolved in water at 2.5 g/l can be applied to correct a severe nitrogen deficiency. For phosphorus deficiency, foliar sprays with potassium or ammonium phosphate are possible, but are not recommended as they can cause serious leaf damage. For potassium deficiency, the crop may be sprayed with a solution of 20 g potassium sulphate/l. To correct calcium deficiency quickly, plants can be sprayed with a solution of 2–7 g calcium nitrate/l or a 0.3% calcium chloride solution. Such sprays are rarely beneficial in correcting BER, since, as discussed above, calcium absorbed through the leaves cannot move into the fruit. Hence, only the Ca that lands on the fruit surface is utilized (Adams, 1999). Magnesium deficiencies can also be corrected with sprays of Epsom salts.

Organic methods

N requirements are harder to predict when working with organic materials. Organic materials tend to tie up nitrogen as they decompose due to microbial action and may require more nitrogen, at least early in the season, compared with inert materials. Later in the season, organic soils, substrates and compost additions release some of this N to the plant. Most organic farmers follow the philosophy of 'feeding the soil, rather than feeding the plant' and try to add nutrients in forms that will be available over the long term, rather than immediately.

PHYSIOLOGICAL DISORDERS RELATED TO IRRIGATION AND FERTILIZATION

Blossom-end rot (BER)

At the anatomical level, the earliest symptoms of BER are areas of white or brown locular tissue. Symptoms next appear in the fruit placenta in the case of internal BER, or in the blossom-end pericarp in the case of external BER (Adams and Ho, 1992). Externally, the disorder begins as a small, water-soaked spot at or near the blossom scar of green tomatoes. As the spot enlarges, the affected tissue dries out and turns light to dark brown, gradually developing into a well-defined sunken leathery spot (Fig. 5.5). Internal BER, consisting of black necrotic tissues in the parenchyma around the young seeds and the distal placental tissues (Adams and Ho, 1992), often develops in the same fruit.

Causes

Interactions between daily irradiance, air temperature, water availability, salinity, nutrient ratios in the rhizosphere, root temperature, air humidity and xylem tissue development in the fruit all contribute to the incidence of BER (Dorais and Papadopoulos, 2001). Although BER shows up in distal fruit tissue, and a gradient in fruit calcium concentration has been shown (Adams and Ho, 1992), Dorais and Papadopoulos (2001) cited evidence that calcium content and cation distribution within the fruit is not directly related to BER. Rather, they suggested that lack of coordination between accelerated cell enlargement (due to high import of assimilates) and inadequate supply of calcium (due to poor development of xylem within the fruit) is generally linked to susceptibility to BER.

This disorder generally appears at the beginning of the rapid phase of fruit growth, when calcium becomes deficient in the distal part of the fruit. Osmotic stress, linked to high salinity in the nutrient solution and water stress, frequently causes fruit calcium deficiency, because it leads to reduced calcium absorption and distribution to the distal part of the fruit (Guichard *et al.*, 2001). Since high salinity increases such fruit quality parameters as fruit dry matter content, sugar content, acidity and shelf-life, it is sometimes raised in soilless culture, despite lower fruit production and the risk of BER. Nederhoff (1999) found that increasing EC (8 dS/m) at night but applying normal levels (2 dS/m) during the day showed some potential for improving tomato fruit quality with minimum loss of production. However, BER was not reduced in this treatment, although Van Ieperen (1996) had previously noted a reduction of BER when EC was low during the day and high at night.

Control

BER is now relatively well understood, but control is still not always achievable in practice. The following general guidelines should be helpful:

- As with all physiological disorders, cultivars differ in susceptibility and so alternative cultivars should be considered whenever there is a problem.
- The root zone must be conducive to uptake, i.e. not too saline, flooded, dry or otherwise restricted.
- Water and calcium must go to the fruit, as opposed to the leaves, which means avoiding daytime high temperatures and high VPD. Misting or fogging inside the greenhouse should reduce BER incidence.
- The root zone calcium supply must be adequate and concentrations of competing cations should not be excessive, because high levels of K and Mg in the nutrient solution replace Ca.

When the ratio of Ca is low relative to either K or Mg, plants are susceptible to BER because of the increased concentration of organic acids in the fruit. These high concentrations reduce uptake and availability of Ca. Recent work summarized by Dorais and Papadopoulos (2001) suggested that, when salinity is used to increase fruit quality, BER can be prevented by avoiding high leaf K:Ca ratios and by maintaining ion activity ratios (mol/l) between K and the total of Ca and Mg of 0.1, and an ion activity ratio of 0.3 between Mg and Ca in the root zone. Dorais and Papadopoulos (2001) also suggest avoiding high (> 26°C) root temperatures and low oxygen concentration, avoiding excessive canopy transpiration (by de-leafing, shading, roof sprinkling and greenhouse fogging), keeping a proper fruit:leaf ratio, and spraying young expanding fruit with 0.5–0.65% calcium chloride solution. Bertin *et al.* (2000) showed that small fruit loads, which lead to an increase in fruit size, increase BER. This suggests that maintaining a high fruit load would decrease BER.

Goldspot, gold speck

Gold specks or flecks are often observed around the calyx and shoulders of mature fruit, particularly in summer. In green fruit, the specks are white and less abundant. These specks decrease the attractiveness of the fruit and significantly shorten its shelf-life (Janse, 1988). Cells with the characteristic gold appearance were identified by Den Outer and van Veenendaal (1988) as containing a granular mass of tiny calcium salt crystals, probably calcium oxalate.

Causes

These specks are considered to be symptoms of excess calcium in the fruit. De Kreij *et al.* (1992) found that, under conditions of high air humidity and high Ca:K ratios, more calcium was transported into the fruit and the incidence of gold speck increased. Increasing the P level also increased Ca uptake rate and increased speckling. As summarized in Ho *et al.* (1999), the level of goldspot was decreased by increasing NO_3 or reducing Cl, NH_4, K or EC in the feed, presumably because these reduced the uptake of calcium.

As temperatures increase during the growing season, the incidence of goldspot also increases, particularly when average greenhouse temperatures are higher than usual. Since the amount of calcium in the fruit does not increase, high temperatures may increase the proportions of calcium deposited in the fruit as oxalate (Ho *et al.*, 1999).

Control

The disorder can be reduced by avoiding susceptible cultivars (Ilker *et al.*, 1977). Cultivars resistant to BER tend to be more susceptible to goldspot (Ho *et al.*, 1999). Sonneveld and Voogt (1990) found that raising the EC of the nutrient solution reduced gold speck incidence, as did increasing the K:Ca ratio and increasing Mg. Presumably in all three cases the mechanism was prevention of excess Ca uptake. Ho *et al.* (1999) recommended: (i) lowering Ca in the hydroponic solution from 200 mg/l to 120 mg/l; (ii) lowering fruit temperature; (iii) applying low but sufficient N (180 mg/l); (iv) applying sufficient but not too high K (400 mg/l); and (v) avoiding P depletion (> 5 mg/l). Following these guidelines, goldspot can be reduced, while avoiding BER.

Fruit cracking and russeting

Cracks (Fig. 5.6) may occur in circles around the stem scar (concentric cracking) or may radiate from the stem scar (radial cracking). 'Russeting' is a related disorder of the tomato skin in which minute hairline cracks, invisible to the naked eye, cover up to 25% of the fruit surface (Bakker, 1988). It is also called rain check, crazing, swell cracking, shrink cracking, hair cracking or cuticle blotch (Emmons and Scott, 1997). The fruit has a rough feel and when examined closely the surface appears crazed rather than smooth. It is sometimes described as a poor skin 'finish' and significantly reduces the shelf-life of harvested fruit (Hayman, 1987). Cracks appear as fruit approach maturation, 6–7 weeks after fruit set, and incidence of the disorder is highest early and late in the production season (Bakker, 1988). Many articles do not distinguish between the two forms of cracking and conditions conducive to one form of cracking are also usually conducive to the other forms.

Causes of fruit cracking

Fruit cracking is a complicated disorder. In areas where rainfall is common during ripening, losses can be very heavy since cracked fruit cannot be sold in the fresh market. Environmental and cultural factors associated with the disorder were summarized by Peet (1992) as follows. Fruit cracking occurs when there is a rapid net influx of solutes and especially water into the fruit at the same time that ripening or other factors reduce the strength and elasticity of the tomato skin. Increases in fruit temperature raise gas and hydrostatic pressures of the pulp on the skin, resulting in immediate visible cracking in ripe fruit. In the greenhouse, excess watering has been shown to increase the incidence of radial cracking, and there are also a few reports in field tomato crops of increased cracking at higher levels of soil moisture (Peet and Willits, 1995).

Although it has been difficult to breed for cracking resistance per se (Stevens and Rick, 1986), commercial cultivars bred for firm fruit and tough skin in order to decrease handling and shipping losses in field tomato production in North America are often quite resistant to fruit cracking. This is probably because these qualities are components of resistance to cracking. In addition, much of the crop is harvested at the green-mature stage, when it is less susceptible to cracking. With the increasing production of vine-ripened tomatoes in high-rainfall areas, cracking may become more of a problem for commercial field production.

Causes of russeting

Dorais *et al.* (2004) summarized the causes of russeting or cuticle cracking as an interplay of genetic, environmental and cultural practices. The percentage of harvested fruit affected in greenhouse conditions can range from 10% to 95% of the total. Guichard *et al.* (2001) concluded that the occurrence of netting and cracking corresponds to the loss of elasticity in the epidermis of the fruit coupled with high rates of fruit expansion. Emmons and Scott (1997) did not find an increase in russeting in field-grown tomatoes in Florida by pruning either leaf or fruit, but did find that the amount of russeting correlated with the amount of rain during the entire 2-week period before harvest.

Control of fruit cracking

Cultural practices that result in uniform and relatively slow fruit growth, such as constant, preferably relatively low soil moisture, offer some protection against fruit cracking (Peet, 1992). In field crops, cracking is usually attributed to fluctuations in the water supply. The classic occurrence is when a long period of drought is followed by heavy rain. Cultural practices that reduce diurnal fruit temperature changes may also reduce cracking. In the field, these practices include maintaining vegetative cover. Greenhouse growers should maintain minimal day and night temperature differences and increase temperatures gradually from night-time to daytime levels. For both field and greenhouse tomato growers, harvesting before the pink stage of ripeness and selection of crack-resistant cultivars probably offer the best protection against cracking.

Control of russeting

To control russeting in soilless systems, the following practices are suggested: (i) selecting resistant cultivars; (ii) avoiding conditions that favour large fruit growth spurts (abundant but infrequent irrigation, high day/night temperature differences or excessive humidity fluctuations during the day); and (iii) high leaf/fruit ratios (Guichard *et al.*, 2001). Dorais *et al.* (2004) suggested that russeting can be reduced by: (i) keeping a high fruit load on the plant; (ii) spraying with a boron/calcium mist; and (iii) avoiding high temperatures, large day/night temperature differentials, high relative humidity, direct exposure of the fruit to solar radiation and large variation in fruit water status. They suggested that improvements are likely to come from cultivar selection and better matching of water supply with plant demand. In the field, Emmons and Scott (1997) suggested: (i) staking plants; (ii) avoiding direct exposure of the fruit to the sun by protecting foliage from disease and damage during harvest; (iii) using resistant cultivars; and (iv) harvesting before rains, if possible.

Oedema

This disorder is often mistaken for bacterial or fungal diseases. In its early stages the blister-like swellings on the leaf resemble an undifferentiated callus-type growth (Fig. 6.6). The granulated appearance of the fresh blisters is caused by the splitting of the epidermis, presumably under pressure from within (Grimbly, 1986). This exposes the swollen, water-filled parenchyma cells, which eventually erupt. Rupture of these cells over a period of time causes twisting and distortion of the leaves, and produces a necrotic area when the cells dry out.

Cause

This disorder is seen in a number of crops besides tomatoes, including cabbage and sweet potatoes. In all cases, it is caused by water provided to the leaves exceeding that used in transpiration for a period of several days. Sagi and Rylski (1978) showed that, under high humidity and excess water, the symptoms increased as light intensity decreased, presumably representing the diminishing ability of the plant to transpire. They also found much greater susceptibility to the disorder in an Israeli field cultivar than in a cultivar selected for greenhouse production in Northern Europe.

Control

In greenhouses or growth chambers, the disorder can be prevented or remedied. Decreasing watering and promoting transpiration by such measures as increased ventilation, higher temperatures and higher light should be effective. In the field, the disorder only appears when there are long

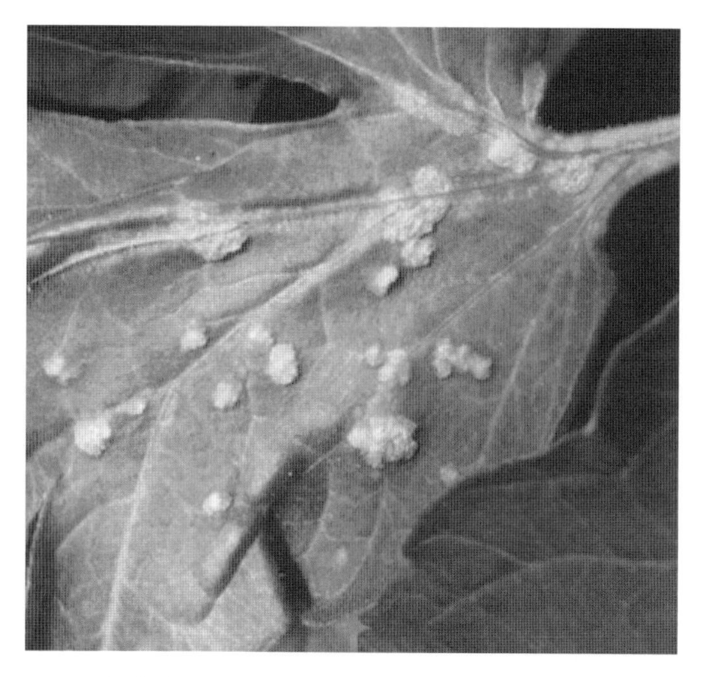

Fig. 6.6. Tomato leaf with oedema.

periods of excess water and low transpiration. There is little that can be done in the field, assuming that irrigation has already been stopped, except for trying different cultivars.

CONCLUDING REMARKS

Providing plants with adequate, but not excessive, amounts of water and fertilizer can be done using modelling techniques (such as soil tension, evaporative demand and irradiance) and monitoring techniques (such as measuring runoff and tissue analysis). Balancing fertilizer cations to prevent physiological disorders, such as BER, is increasingly well understood, but not always achievable in practice because of the complex interplay of nutrition and environmental influences. Similarly, the role of cation balance and EC in determining fruit quality is now better understood, but additional research is required on the question of how to avoid trade-offs between fruit quality and yield. The challenge for the future will be developing economically viable systems that reduce runoff of nitrogen and phosphorus, while maintaining adequate yields and fruit quality.

REFERENCES

Adams, P. (1999) Plant nutrition demystified. *Acta Horticulturae* 481, 341–344.

Adams, P. and Ho, L.C. (1992) The susceptibility of modern tomato cultivars to blossom-end rot in relation to salinity. *Journal of Horticultural Science* 67, 827–839.

Bakker, J.C. (1988) Russeting (cuticle cracking) in glasshouse tomatoes in relation to fruit growth. *Journal of Horticultural Science* 63, 459–463.

Bertin, N., Guichard, S., Leonard, C., *et al.* (2000) Seasonal evolution of the quality of fresh glasshouse tomatoes under Mediterranean conditions as affected by air vapour pressure deficit and plant fruit load. *Annals of Botany* 85, 741–750.

De Kreij, C., Janse, J., Van Goor, B.J. and Van Doesburg, J.D.J. (1992) The incidence of calcium oxalate crystals in fruit walls of tomato (*Lycopersicon esculentum* Mill.) as affected by humidity, phosphate and calcium supply. *Journal of Horticultural Science* 67, 45–50.

Den Outer, R.W. and van Veenendaal, W.L.H. (1988) Gold speckles and crystals in tomato fruits (*Lycopersicon esculentum* Mill.). *Journal of Horticultural Science* 63, 645–649.

Dorais, M. and Papadopoulos, A.P. (2001) Greenhouse tomato fruit quality. *Horticultural Reviews* 26, 239–319.

Dorais, M., Demers, D.A., Papadopoulos, A.P. and Van Ieperen, W. (2004) Greenhouse tomato fruit cuticle cracking. *Horticultural Reviews* 30, 163–184.

Emmons, C.LW. and Scott, J.W. (1997) Environmental and physiological effects on cuticle cracking in tomato. *Journal of the American Society for Horticultural Science* 122, 797–801.

Grimbly, P. (1986) Disorders. In: Atherton, J.G. and Rudich, J. (eds) *The Tomato Crop.* Chapman & Hall, London, pp. 369–390.

Guichard, S., Bertin, N., Leonardi, C. and Gary, C. (2001) Tomato fruit quality in relation to water and carbon fluxes. *Agronomie* 21, 385–392.

Halliday, D.J. and Trenkel, M.E. (eds) (1992) *IFA World Fertilizer Use Manual.* International Fertilizer Industry Association, Paris, France, pp. 289–290, 331–337.

Ho, L.C., Hand, D.J. and Fussell, M. (1999) Improvement of tomato fruit quality by calcium nutrition. *Acta Horticulturae* 481, 463–468.

Hochmuth, G. (1994) *Plant Petiole Sap-testing Guide for Vegetable Crops.* Florida Cooperative Extension Service Circular No. 1144. IFAS, Gainesville, Florida.

Hochmuth, G., Maynard, D., Vavrina, C. and Hanlon, E. (1991) *Plant Tissue Analysis and Interpretation for Vegetable Crops in Florida.* Florida Cooperative Extension Service Special Series SSVEC-42. IFAS, Gainesville, Florida.

Ilker, R., Kader, A.A. and Morris, L.L. (1977) Anatomical changes associated with the development of gold fleck and fruit pox symptoms on tomato fruit. *Phytopathology* 67, 1227–1231.

Janse, J. (1988) Goudspikkels bij tomaat: een oplosbaar probleem. *Groenten en Fruit* 43, 30–31.

Jones, J.B. Jr (1999) *Tomato Plant Culture in the Field, Greenhouse, and Home Garden.* CRC Press, Boca Raton, Florida, 199 pp.

KWIN (2001) *Kwantitatieve Informatie voor de Glastuinbouw.* Proefstation voor Bloemisterij en Glasgroente, Naaldwijk/Aalsmeer, The Netherlands 38, 20–21.

Maas, E.V. (1984) Crop tolerance. *California Agriculture* 38, 20–21.

Nederhoff, E. (1999) Effects of different day/night conductivities on blossom-end rot, quality and production of greenhouse tomatoes. Proceedings International Symposium on Growing Media and Hydroponics (ed. Papadopoulos, A.P.). *Acta Horticulturae* 481, 495–501.

OMAFRA (2001) *Growing Greenhouse Vegetables*. Publication 371. Ontario Ministry of Agriculture, Food and Rural Affairs, Toronto, Canada, 116 pp.

OMAFRA (2003) *Growing Greenhouse Vegetables, 2003 Supplement*. Publication 371S. Ontario Ministry of Agriculture, Food and Rural Affairs, Toronto, Canada, 8 pp.

Peet, M.M. (1992) Fruit cracking in tomato. *HortTechnology* 2, 216–223.

Peet, M.M. and Willits, D.H. (1995) Role of excess water in tomato fruit cracking. *HortScience* 30, 65–68.

Reiners, S., Petzoldt, C.H., Hoffmann, M.P. and Schoenfeld, C.C. (1999) *Integrated Crop and Pest Management Recommendations for Commercial Vegetable Production*. Cornell Cooperative Extension, Ithaca, New York.

Roorda van Eysinga, J.P.N.L. and Smilde, K.W. (1981) *Nutritional Disorders in Glasshouse Tomatoes, Cucumbers and Lettuce*. Centre for Agricultural Publishing and Documentation, Wageningen, The Netherlands, 130 pp.

Sagi, A. and Rylski, I. (1978) Differences in susceptibility to oedema in two tomato cultivars growing under various light intensities. *Phytoparasitica* 6, 151–153.

Sanders, D.C. (2004–2005) *Vegetable Crop Guidelines for the Southeastern US*. North Carolina Vegetable Growers Association, Raleigh, North Carolina, 73 pp.

Sonneveld, C. and Voogt, W. (1990) Response of tomatoes (*Lycopersicon esculentum*) to an unequal distribution of nutrients in the root environment. *Plant and Soil* 124, 251–256.

Stanghellini, C., Kempkes, F.L.K. and Knies, P. (2003) Enhancing environmental quality in agricultural systems. *Acta Horticulturae* 609, 277–283.

Stevens, M.A. and Rick, C.M. (1986) Genetics and breeding. In: Atherton, J.G. and Rudich, J. (eds) *The Tomato Crop*. Chapman & Hall, London, pp. 35–110.

Van Ieperen, W. (1996) Effects of different day and night salinity levels on vegetative growth, yield and quality of tomato. *Journal of Horticultural Science* 71, 99–111.

Van Ieperen, W. and Madery, H. (1994) A new method to measure plant water uptake and transpiration simultaneously. *Journal of Experimental Botany* 45, 51–60.

Van Os, E.A. (2000) New developments in recirculation systems and disinfection methods for greenhouse crops. *Proceedings 15th Workshop on Agricultural Structures and ACESYS (Automation, Culture, Environment & Systems) IV Conference*, pp. 81–91.

Crop Protection

A.A. Csizinszky, D.J. Schuster, J.B. Jones and J.C. van Lenteren

INTRODUCTION

Reliable productivity and fruit quality are dependent upon adequate control of: (i) bacterial fungal and viral plant pathogens; (ii) insects that damage foliage and fruit by sucking, chewing or transmitting disease; (iii) nematodes feeding on roots; and (iv) weeds competing with the tomato plants for moisture and nutrients. During the growing season, the combination of agronomical, biological and chemical techniques known as integrated pest management (IPM) should be used to protect the crop from plant pathogens, insects, nematodes and weeds.

The first step in the IPM programme is the selection of disease-resistant or tolerant cultivars that give adequate yields. Then the crop should be planted in a well-prepared soil with optimum moisture and nutrients when climatic conditions, especially temperatures, are favourable for rapid plant growth and development. During the season, plants are surveyed for the presence or signs of plant pests and the land is kept free of weeds. In greenhouses, humidity control is of the utmost importance, especially in relation to diseases. The development of pathogens or insects is favoured by humidity levels that are low (oidium, thrips, spider mites) or high (botrytis, mildew, leaf mould). High humidity often leads to the condensation of moisture on leaves, spore germination and diseases. On the other hand, the development of tomato powdery mildew is greatest at 80% relative humidity (RH) and progressively less with increasing humidity to a minimum level at 95% RH (Fig. 7.1).

Applications of chemical control agents are based, in the case of insects, on population thresholds or, in the case of pathogens, the combination of first symptoms of the disease and expected climatic conditions that favour further development and spread of the pathogenic organisms. Biocontrol agents (e.g. insect parasites), pheromones, growth retardants or environmentally friendly products, where available, should be used first to control plant pests. All pest-control chemicals must be applied according to their pesticide labels. Misuse of chemicals can lead to contamination of the workers, higher permitted

Fig. 7.1. Heavy sporulation of powdery mildew (*Oidium lycopersicum*) on a leaf of a susceptible tomato cultivar. (Photograph by Dr W.H. Lindhout.)

pesticide residues on the fruit and pollution of the environment. Where several chemicals are available for control of a pest, alternating the use of different chemicals reduces the chance that the target pest will develop resistance to one particular chemical.

WEED CONTROL

Weeds compete for nutrients, water and light with the crop; they also serve as hosts to plant pests and interfere with harvest operations. Some weed species may release allelopathic chemicals into the soil that are detrimental to the growth and development of other plants.

Weeds are classified as annuals, biennials or perennials and may belong to the grasses (*Monocotyledonae*) or to the broadleaves (*Dicotyledonae*). They reproduce sexually (resulting in seeds) or asexually (vegetatively). Since there are many kinds of weed, with much variation in growth habit and method of propagation, they cannot be controlled and managed by a single method. Germinating seeds, annual plants and seedlings of biennial and perennial weeds are easier to control than perennials that reproduce asexually. The critical weed-free period for transplanted tomatoes is during the first 28–35 days after transplanting and for direct-seeded tomatoes 49–63 days after seeding. During these periods one or two weedings will be sufficient to control weeds, thus preventing yield reduction.

An important first step in planning a weed control programme is to know the types of weeds and their density in the field where the tomatoes will be planted, in order to prepare a weed map. The survey of weeds for constructing the map should be made two or three times each season, because weeds may germinate and grow at various times of the year. The survey should include a history of the previous crops grown in the field and the herbicides applied for those crops. In every case the residual effects of herbicides that were used for the previous crops have to be considered when planting tomatoes. Germination tests of tomato seeds in soil suspected to have herbicide residues will detect adverse affect on tomatoes.

Weeds that belong to the same family (*Solanaceae*) as the tomato (e.g. nightshade) are difficult to control in tomato fields. Crop rotation programmes, in which cereals, maize, garlic, onion or safflower are planted in rotation with tomatoes, can reduce nightshade and other broadleaf weed populations. Herbicides that give a good control of nightshades, but cannot be used selectively for tomatoes, can be used for the crops preceding tomatoes, thereby reducing nightshade infestation in the following tomato crop.

The soil between the crops has to be cultivated to kill the emerging weeds. Rotavating or disking the soil will also move seeds and reproductive organs of asexually propagated perennial weeds closer to the surface to facilitate germination and thus reduce the weed population for the tomato crop. During the season, weeding by manual methods (pulling and hoeing) is tedious and expensive and has been virtually eliminated in many parts of the world. Mechanical cultivation is used with the non-mulched tomato production system. Cultivation is most effective when the weeds are at the two to four true-leaf stage, or 10–15 cm tall. At that stage, they present little competition for the crop and can be removed with little damage to the tomatoes. Both hand hoeing and mechanical cultivation should be only deep enough to control the weeds. Deep cultivation could lead to crop root damage and to rapid soil drying.

Weed control by herbicides alone in a non-mulched tomato production system is difficult, because few herbicides are registered for tomatoes (Table 7.1). Herbicides are classified by their physiological and chemicophysical properties, chemical structure and soil relations. Herbicides are also classified as selective (controlling weeds without injuring the crop) or non-selective (controlling all vegetation). Another classification refers to the time of application: preplant, before seeding or transplanting the crop; and postplant, after the crop has been seeded or transplanted.

Regardless of the type of herbicide used, care must be taken to use the chemicals at the permitted rate and timing to avoid crop damage. Selection

Table 7.1. Weed control chemicals (active ingredient and product name) for tomatoes.

Herbicide	Time of application	Weeds controlled	Conditions for most effective control
Chloramben (Amiben)	Before weeds emerge after transplanting	Grasses and broadleaves	Not recommended on sandy soils and soils low in organic matter and under arid conditions
DCPA (Dacthal)	4–6 weeks after transplanting, or on direct seeded plants 10–15 cm high	Germinating annuals, grasses	Apply to weed-free soil or to moist soil between the plant rows
Glyphosate (Roundup)	Prior to planting or sowing	Annual and perennial weeds	Perennials should be treated in the autumn before the spring crop planted
MCDS (Enquik)	Postemergence desiccant, apply between rows only	Broadleaves; poor control of grasses	
Metam-sodium (Vapam)	3–4 weeks before planting as a soil drench, or injected into the soil, 10 cm deep	Grasses, broadleaves and soil-borne fungi	Compact soil with roller or cover with plastic immediately after application
Methyl bromide[a] + chloropicrin (Dowfume MC33; Terr-O-Gas 67)	Preplant under plastic sheet	All weeds and soil-borne pests	Good soil moisture important for pest control
Metribuzin (Sencor)	Postemergence, post-transplanting after stand establishment	Annual grasses and broadleaves	Apply when direct-seeded plants reach 5–6 true-leaf stage and on small emerged weeds
Napropamid (Devrinol)	Apply prior to weed emergence and incorporate to 3–5 cm deep into the soil. For direct-seeded or transplanted tomatoes	Germinating annuals, grasses and broadleaves	Soil should be well worked and free of weeds Wet soil to 3 cm depth within 24 h after application

Paraquat (Gramoxone)	Prior to seeding or transplanting to control emerged weeds; to control weeds in row middles during the season	Grasses and broadleaves	Direct spray when weeds are 3–15 cm tall; use non-ionic spreader Apply at low pressure during season to control drift
Pebulate (Tillam)	Incorporate granules with rotary tiller into soil before planting	Germinating annual grasses and broadleaves. Poor control of nightshade and mustard	Apply to well-prepared seedbed, free of clods Use only in mineral soils
Sethoxydim (Poast)	Postemergence on actively growing grasses	Annual and perennial grasses	Add paraffin oil-based adjuvant for good results
Trifluarin (Treflan)	Prior to seeding or transplanting and for direct-seeded crop at time of thinning	Germinating annuals	Incorporate not more than 10 cm deep prior to transplanting or seeding Incorporate into soil after thinning with rolling cultivator

[a] Banned in developed countries since 2005.

of proper nozzle type and calibration of the sprayer to deliver the proper amount of herbicide are very important. Foliage-applied herbicides at high pressure will form small droplets that dry up rapidly and the herbicide cannot penetrate the foliage; addition of a surfactant or adjuvant to the spray tank for foliage-applied herbicides will result in good leaf coverage of the weeds.

Soil-applied herbicides are either sprayed on the surface or are incorporated. To be effective, surface-applied herbicides (e.g. napropamid) require moisture, rainfall or sprinkler irrigation shortly after application. Incorporated herbicides (e.g. pebulate and trifluarin) are widely used in areas with low rainfall. To have good weed control, the incorporated herbicide has to be at a depth where the germinating weed seeds lie. For example, pebulate must be incorporated 7–10 cm deep for control of nutsedge (*Cyperus* spp.) but only 2–5 cm deep to control nightshades (*Solanum* spp.).

The combination of preplant soil-applied herbicides with full-bed plastic mulch production systems is used under the subtropical conditions in Florida. Black, or white-on-black, plastic film covers over the soil will kill weeds (except nutsedge) by excluding light and by acting as a barrier to weed growth. The combination of plastic mulch with wide-spectrum preplant-applied pesticides controls not only weeds but all soil-borne pests. Weeds growing between the plastic-mulched beds are controlled by herbicides or by mechanical and hand cultivation, or a combination of herbicides and cultivation. In the case of herbicide weed control, the sprayer has to be shielded to prevent damage to the crop. Weed control with organic mulches (composted plant residues applied in a layer over the plant bed) and living mulches (leguminous plants or cereals grown in the field, then cut down before the tomatoes are seeded or transplanted) is still in the experimental stage and not yet accepted by growers.

Biological weed control (the use of another organism that is a natural enemy of the weed) is also in the experimental rather than application stage. Under hot sunny climatic conditions, soil solarization, a non-pesticidal method of soil disinfestation, has been investigated as an alternative to chemical weed control. The effectiveness of solarization on weed control depends on seasonal conditions (including rainfall and cloud cover) and on soil depth. Therefore, soil cultivation after solarization should be minimized, to prevent bringing viable weed seeds to the soil surface where they may germinate and reinfest the treated soil.

In summary, weeds are controlled by a combination of methods in the tomato field, depending on the economic conditions and technological development of the region. Due to environmental concerns, fewer herbicides will be available for weed control in the future than are used at present. Alternative non-chemical methods for weed control are in the experimental stage and will have to be perfected before they can be used to protect the tomato crop from weeds.

NEMATODES

Nematodes are microscopic roundworms that live in aqueous media or in films of moisture. Plant parasitic nematodes live in soil and attack the roots of the plants. Crop losses caused by the nematodes occur as a consequence of reduced uptake of water and nutrients by the affected plants. Nematodes may also carry plant pathogenic organisms, or the plant pathogens may invade the weakened plants and cause disease. The aboveground symptoms of nematode infestation are chlorosis, stunted growth, wilting, early senescence and smaller and fewer fruits. Nematodes are not evenly distributed in the field and the symptoms of infection on the plants appear at random in patches, which expand as the infection spreads.

There are many different genera and species of nematode that attack tomatoes throughout the world. Nematode development is most rapid in soils at 21–26°C and the parasites are not active at < 16°C. Consequently, nematodes are very important plant pests in tropical and subtropical regions. The most widespread nematodes that attack tomatoes and can devastate the crop are the root-knot nematodes, *Meloidogyne* spp., the presence of which on the roots can be detected by pulling the plant from the soil and inspecting the roots for the presence of swellings (galls or knots). The galls cannot be rubbed off from the roots if caused by root-knot nematodes. Other plant parasitic nematodes that have been reported to attack tomatoes are: the sting nematode, *Belonolaimus longicaudatus*; the reniform nematode, *Rotylenchus reniformis*; root lesion nematodes, *Pratylenchus* spp.; false root-knot nematodes, *Nacobbus* spp.; potato cyst nematodes, *Globodera* spp.; stunt nematodes, *Tylenchorhynchus* spp.; dagger nematodes, *Xiphinema* spp.; and stubby root nematodes, *Trichodorus* spp. Nematodes have a wide range of hosts among cultivated plants and weeds and they are spread by infested transplants, farm machinery, animals, farmyard manure and surface runoff water. Yield losses are greater when the plants are infested by the nematodes at the seedling stage.

Because nematodes are virtually impossible to control after the crop is seeded or transplanted, control methods have to be planned and carried out prior to planting. Selection of a nematode-resistant cultivar is an important first step in the control programme. There are several fresh-market and processing tomato cultivars that are resistant to the root-knot nematodes *M. incognita*, *M. arenaria* and *M. javanica*. The resistance of tomatoes to nematode damage is gradually reduced above 26°C and the plants become fully susceptible above 33°C. Nevertheless, nematode populations are reduced in the soil after a nematode-resistant tomato crop.

Clean cultivation between crops is an effective method to control nematodes. Ploughing or disking should be done as soon as possible after the harvest, to prevent population growth of nematodes on the old crop and to expose the nematodes to the drying action of sun and wind. Cultivation of the field, at least two to three times between crops, will also destroy weeds that serve as hosts to the nematodes.

Crop rotation of tomatoes with nematode-resistant cover crops will also reduce nematode populations. Some cover crops are resistant to only one nematode genus, but host for another genus of nematodes, in which case the cover crop should be chosen against the nematode that is the most difficult to control or causes the greatest damage to the tomato crop. For example, a cover crop of sorghum (*Sorghum bicolor*) decreases the densities of root-knot and reniform nematodes but increases the densities of sting nematodes. A crop of sorghum–sudangrass (*S. bicolor* × *S. sudanense*) also reduces the root-knot nematode population but increases the number of stubby-root nematodes.

Soil solarization will reduce nematode populations. This is a hydrothermal process in which the sun's energy is used to heat the moist soil, covered with a clear plastic film, for 6–12 weeks. The most successful use of soil solarization is in heavier loamy to clay soils in arid environments with 6–12 weeks of intense sunshine. On light sandy soils in subtropical environments, solarization results depend on rainfall and cloud cover.

Efforts have also been made to control plant parasitic nematodes with biological agents: bacterial parasites, fungi and mites. At present, there are no reliable biological methods that would control nematodes. Organic amendments (composted plant materials and municipal sludge) applied to a sandy soil with low water-holding capacity and < 2% organic matter have increased tomato yields, in spite of a high root-knot nematode population. It is thought that the yield increase with this organic amendment was due to the increased availability of water and nutrients for the tomato plants, rather than nematode control by the compost. Alternate flooding and drying of the land in 2–3-week cycles between crops is also practised to reduce nematode populations.

Under intensive cultivation, especially where the land has to be used repeatedly for tomato production, nematodes are controlled by preplant-applied nematicides. The nematicides are used in an IPM system, in combination with non-chemical methods, to protect the crop. They must be applied according to label and the recommended waiting period has to be followed between applying the nematicide and planting the crop, to avoid damage to the tomatoes. Well-prepared soil, free of plant residues and with good moisture, is necessary for the chemicals to be effective against the nematodes and the soil temperature should be > 15°C.

The most effective chemical for nematode control is the soil fumigant, methylbromide-chloropicrin (Terr-O-Gas 67; MC-33). According to the Montreal Protocol, the production and use of methylbromide is banned in developed countries as of 2005 and in developing countries from 2015. Critical use exemptions will be allowed when no suitable, widely accepted alternatives are available. These exemptions will be reviewed annually. In preparation for the phase-out of methylbromide, various chemicals – fumigants and non-fumigants, alone or in combination with non-chemical methods – have been tested in large-scale field trials. So far, 1,3-dichloropropene + 17% chloropicrin (Telone C-17), alone or in combination with solarization, has been found to be the best alternative replacement for methylbromide against nematodes. Other preplant soil fumigants, such as dichloropropene-dichloropropane (D-D; Vidden D), ethylene dibromide (Dowfume) and ethylene dibromide-chloropicrin (Terr-O-Cide 15; Terr-O-Cide 30), are less effective for nematode control. Among the non-fumigant nematicides, fensulfothion (Dasanit), a granular organophosphate, is used for nematode control. Dasanit has to be incorporated 10–15 cm deep into the soil by cross-working with a disc harrow or other tillage equipment. Oxamyl (Vydate L) is applied as a post-transplant soil drench around the plants if the soil is infected by nematodes. Soil treatment with Vydate L should be initiated as soon as the nematode infestation is discovered to avoid yield reduction.

INSECT AND MITE PESTS

Insects and mites are important arthropod pests of tomatoes. The type of damage inflicted depends upon the type of mouthparts of the damaging life stage. Many insects have chewing mouthparts, or modifications of them, and their damage appears as holes in leaves, stems or fruit, leaf mines, leaf rolls, etc. Some insects other than mites also have piercing–sucking mouthparts or modifications of them. Damage by these arthropods usually appears as speckling, spotting or distortion of leaves, stems or fruit that at times can be confused with pathogenic or nutritional causes. In addition, some insects with piercing–sucking mouthparts are important pests as vectors of many viruses that are responsible for some very serious diseases. Thus, recognition of damage can be a first step in identifying the cause and, ultimately, developing a management strategy. For this reason, pests will be presented and discussed here according to the type of damage inflicted.

Because of the large complex of arthropod pests that damage tomato plants and because of the ability of many arthropods to develop resistance to insecticides, most successful and long-term management programmes will integrate many tactics. Tomatoes fit IPM programmes well, because they can tolerate up to 30% defoliation in the lower canopy before bloom and up to 50% defoliation in the lower canopy after bloom (Keularts *et al.*, 1985). The manipulation of cultural practices probably is the oldest tactic for managing pests and often involves the use of good common sense. Some practices are adapted for specific pests but most are used for their negative impact on many pest species. Pests can be excluded from plant production facilities using fine-mesh screening to cover all entrances and air intakes. Row covers can be used to exclude pests from field plantings. Good sanitation within and adjacent to fields and the timely destruction of crops following the last harvest can reduce reservoirs of insects as well as the plant diseases for which they are vectors. Avoiding the use of certain plants as cover crops during fallow periods and separating new plantings in time and space from old plantings of the same or different crops can also reduce pest reservoirs. Delaying planting can create or extend crop-free periods as well as avoid early migrations of overwintering or oversummering pests. The use of ultraviolet light-reflective and coloured plastic films to cover the soil under plants can be used to interfere with the normal settling of certain insects. The film should cover as much of the soil as possible, because it is only effective as long as the insects can see it. Pesticide residue on the plastic can also reduce the effectiveness of the mulch.

Growing resistant varieties can contribute to the management of insects as well as of the disease organisms that they transmit. Unfortunately, resistance to arthropod pests is not yet a reality in commercial tomato varieties.

Most arthropod pests are attacked by numerous parasitoids and predators. These natural enemies can be conserved by eliminating or reducing the use of broad-spectrum pesticides and by using selective or biorational pesticides. Natural control can be augmented in some cases by the release of natural enemies reared specifically for this purpose. Certain weedy or crop areas can also serve as reservoirs or refuges of natural enemies.

The application of pesticides is the method used most often to control arthropod pests. Twice-weekly scouting to ascertain the densities of pests and natural enemies is essential to the wise selection and timing of applications. Eliminating unnecessary pesticidal applications, reducing the number of applications and using selective pesticides reduces the probability of the development of pesticide resistance. Natural enemies will also be conserved and can thus contribute to reducing pest densities.

Chewing damage: seedlings

Crickets

Field crickets, such as *Acheta* (= *Gryllus*) *assimilis* Fabricius, may sometimes damage tomato seedlings. They are dark brown to black insects, 2–3 cm long, with long antennae and with the hind pair of legs enlarged for jumping. They migrate into the perimeters of fields from weedy, trashy or storage areas and chew seedlings off above the soil line. Aboveground plant parts may occasionally be attacked. Mole crickets, such as *Scapteriscus vicinus* Scudder, *S. borellii* Giglio-Tos, *Gryllotalpa gryllotalpa* L., *G. hexadactyla* Perty and others, are similar to field crickets but are longer (4 cm) and are lighter brown, with a claw-like front pair of legs for digging. Mole crickets also damage seedlings, chewing off plants at or just below the soil surface. They construct characteristic tunnels just below the soil surface into which excised seedlings may be pulled and wholly consumed, leaving no trace. If stems are not completely severed, plants may wilt.

If field crickets are present on immediate field perimeters, weeds should be mown and debris and trash should be cleaned and removed. Fields should be deep ploughed to destroy field cricket eggs and be kept weed free. The soil should be kept moist so that crickets remain close to the surface. Granular or bait formulations containing pesticides or products containing the entomopathogenic nematode *Steinernema scapterisci* Nguyen & Smart may be applied. Broad-spectrum soil fumigants will control mole crickets in the treated soil zone, but crickets can move back into the treated soil as the fumigant dissipates.

Cutworms

There are many species of caterpillar pest that are generally called cutworms. Some of the most prominent species include: the variegated cutworm, *Peridroma saucia* (Hübner); the black cutworm, *Agrotis ipsilon* (Hufnagel); and the granulate cutworm, *Agrotis subterranea* (Fabricius). Larvae attack seedlings as they emerge or as transplants are set. Plants are cut at or just above the soil surface. Because larvae hide during the day, they are often overlooked. Larvae grow to about 4 cm long and vary in appearance from a dirty grey to a yellowish or dark brown with variable longitudinal or diagonal lines. Cutworm larvae appear smooth or 'greasy' to the naked eye and curl up tightly when disturbed.

To control cutworms, land preparation should begin early enough to remove alternative host plants at least 10–14 days before planting. Bait formulations containing the bacterium *Bacillus thuringiensis* Berliner, or granular or bait formulations containing insecticide, should be applied if cutworm larvae are present. Broad-spectrum soil fumigants will control cutworm larvae in the treated soil zone.

Chewing damage: foliage

Flea beetles

Flea beetles are 3–4 mm long and shiny, with enlarged hind legs adapted for jumping. The dark brown to black beetles, primarily in the genera *Epitrix* and *Phyllotreta*, fly into tomato fields from overwintering sites. They chew small pits or holes in leaves. Damage on large plants is not usually of economic significance but damage to seedlings can result in death or stunted growth. Early plantings are usually the most affected. Larvae occur in the soil but generally are not damaging. Flea beetles in the genus *Epitrix* have also been shown to transmit eggplant mosaic virus to tomato.

Seedlings can be protected from immigrating flea beetle adults with row covers of fine-mesh fabric. A single timed application of insecticide may be all that is necessary in lieu of using row covers.

Leafminers

Four species of leafmining flies are most damaging to tomato: the vegetable leafminer, *Liriomyza sativae* Blanchard; *L. trifolii* (Burgess); *L. bryoniae*

(Kaltenbach); and the pea leafminer, *L. huidobrensis* (Blanchard). Adults are 2–3 mm long and are black on the top and yellow on the heads, sides and underside. Females use their sheath-like ovipositors to make small holes in the leaf into which eggs are deposited just below the leaf surface. Females also use the ovipositor to tear larger holes, from which they feed on the exuding cell contents. The latter holes become necrotic and are often termed 'stippling'. *L. sativae, L. trifolii* and *L. bryoniae* deposit eggs primarily in the upper surfaces of leaves while *L. huidobrensis* deposits eggs primarily in the lower leaf surfaces. Larvae are yellowish maggots 2–5 mm long with black sickle-shaped mouth hooks that are used for scraping leaf tissue. Larvae of the former three species form serpentine mines in no particular pattern, while larvae of the latter species form mines that are generally associated with the midrib or lateral leaf veins. Larvae cut a hole at the end of the mine and drop to the soil, where they pupate under debris (Fig. 7.2).

Leafminers are generally held in control by large numbers of species of parasitic wasps that attack the larval stages; however, leafminer populations can increase to damaging levels if parasitoids are reduced or eliminated with applications of broad-spectrum insecticides targeted against other pests. Therefore, timed applications of selective insecticides with minimal impact on parasitic species are recommended.

Loopers

Different looper species attack tomatoes in different parts of the world but some of the most common include: the cabbage looper, *Trichoplusia ni* (Hübner); the soybean looper, *Pseudoplusia includens* (Walker); the alfalfa looper, *Autographa californica* (Speyer); and the tomato looper or golden twin spot, *Chrysodeixis chalcites* (Esper). Eggs are deposited singly on the undersides of leaves. Cabbage looper eggs resemble fruitworm eggs, except that they are flatter and have a larger number of fine ridges. Young larvae leave the upper epidermis intact, forming 'windows', while older larvae chew irregular holes. Surface feeding of green fruit occurs only rarely. Larvae are up to 3.5 cm long and pale green, sometimes with white stripes on the sides or top, and they arch their bodies into a loop as they crawl.

Generally, looper populations are not large enough to result in economic loss. Applications of *B. thuringiensis* are usually effective in controlling looper larvae. Although these products contain bacterial endospores and can cause infection, much of the mortality is attributed to endotoxins that are produced by the bacteria during fermentation and which are usually enclosed in the endospores. Rotation of products based on different subspecies of *B. thurinigiensis* (i.e. *kurstaki* and *aizawei*) or products derived from trans-conjugation or recombinant DNA techniques will ensure that different complements of endotoxins are being applied and therefore resistance to specific endotoxins is avoided. Mass releases of the egg parasitoid, *Trichogramma pretiosum* Riley, have reduced looper populations by up to 50%.

Fig. 7.2. (a) Adult *Liriomyza trifolii* and (b) leafmining damage in tomato caused by larval feeding.

Chewing damage: fruit and foliage

Armyworms

Of the numerous species of armyworm in the genus *Spodoptera* damaging tomatoes worldwide, the most common include: the beet armyworm, *S. exigua* (Hübner); the cotton leaf worm, *S. littoralis* Boisd; the southern armyworm, *S. eridania* (Cramer) (Colour Plate 2); the yellowstriped armyworm, *S. ornithogalli*

(Guenée); the western yellowstriped armyworm, *S. praefica* (Grote); the cluster caterpillar, *S. litura* (Fabricius); the fall armyworm, *S. frugiperda* (J.E. Smith); and the tomato moth or bright-line brown-eye, *Lacanobia oleracea* (L.). All species deposit eggs on the undersides of leaves in masses of 50 or more and females of most species cover the eggs with hair-like scales. Hatching larvae feed gregariously on the undersides of leaflets, leaving the upper leaf epidermis intact and forming 'windows'. Later instars disperse throughout the same and adjacent plants, eating holes through leaves and giving plants a 'shot hole' appearance. Although fruit are not preferred feeding sites, larvae will feed on the surface of fruit when they encounter them. Because of the large number of larvae present, nearly all of the fruit on a single plant may be damaged. Secondary microorganisms often enter the fruit at the feeding sites, causing the fruit to rot. Larvae of most species appear smooth to the eye and have lateral yellow, tan or white stripes. Some have black triangular markings on their backs. Mature larvae are 2.5–5 cm long, depending on species.

Although most armyworm larvae are susceptible to the bacterium *B. thuringiensis*, spray applications should be applied to coincide with egg hatch for best results. *B. thuringiensis* products containing the Cry1C endotoxin are more effective (see additional comments for using *B. thuringiensis* products under the discussion for loopers, above). Products containing polyhedral occlusion bodies of the nuclear polyhedrosis virus are available for managing *S. exigua* larvae. The spinosyns, a group of compounds derived from the bacterium *Saccharopolyspora spinosa*, are effective against many species of armyworm larvae and also are less toxic to many natural enemy species.

Fruitworms

At least three species of fruitworm attack tomatoes: the tomato fruitworm, *Helicoverpa zea* (Boddie) (Fig. 7.3); *Heliothis armigera* Hübner; and the tobacco budworm, *Heliothis virescens* (Fabricius). Eggs are usually deposited singly on terminal leaflets of leaves near flowers or small fruit. Hatching larvae prefer to feed on fruit but also will attack buds, flowers or stems. Ripe fruit are seldom attacked and *H. virescens* particularly seems to prefer stems. Larvae bore deeply into fruit, primarily at the stem end, and can complete their entire larval development inside a fruit that is large enough. Damaged fruit often rot due to invasion of secondary microorganisms. Larvae are similar to armyworms in size and appearance but have bristle-like spines and patches of shorter spines that are usually visible with a hand lens.

Because larvae quickly bore into fruit, where they are inaccessible to insecticides, control must be targeted at the eggs or at newly hatching larvae. Eggs can be monitored on leaves immediately above or below fruit hands with open flowers or small, green fruit. Mass releases of *T. pretiosum* egg parasitoids have resulted in 50–85% parasitism. *B. thuringiensis* products are generally not as effective in controlling fruitworm larvae, because of the rapidity at which the larvae bore into fruit. Products containing the Cry1A(c) toxin are

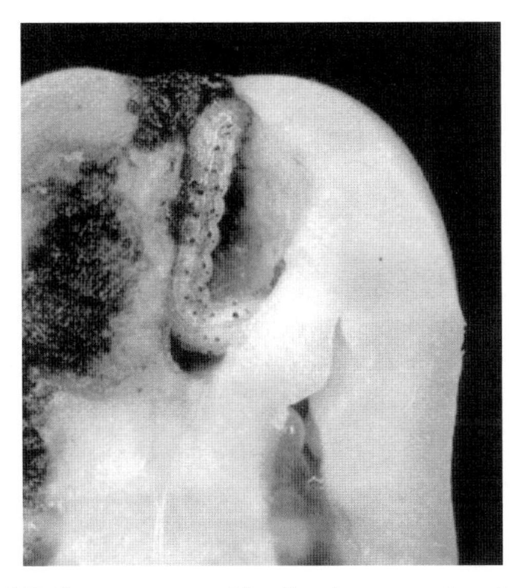

Fig. 7.3. Larva of *Helicoverpa zea* and feeding damage on tomato.

the most effective (see additional comments for using *B. thuringiensis* products under the discussion for looper control, above).

Hornworms

The tomato hornworm, *Manduca quinquemaculata* (Haworth), and tobacco hornworm, *M. sexta* (Linnaeus), usually deposit smooth, pale green eggs on the lower surfaces of leaves. Hatching larvae are pale green and have a distinctive horn or spine on the rear end. Older larvae retain the horn but also have diagonal white stripes on their sides. Larvae grow to be up to 10 cm long and are voracious feeders, consuming entire leaves and small stems. Large pieces of fruit may be consumed, but foliage is preferred.

Hornworms do not usually require control because eggs and larvae are often attacked heavily by natural enemies; however, whole plants can be defoliated if a population is left unchecked. Mass releases of the egg parasitoid *T. pretiosum* have resulted in parasitism up to 70%. Most products containing *B. thuringiensis* also are effective (see additional comments for using *B. thuringiensis* products under the discussion for looper control, above).

Potato tuberworm

The potato tuberworm, *Phthorimaea operculella* (Zeller), is more widely distributed than the tomato pinworm (see below) but attacks tomato similarly. Eggs are deposited singly on the undersides of leaves. Most larvae form blotch mines upon hatching but eventually move to the stems. Fruit may be damaged in a way similar to that by the tomato pinworm but larvae may also bore into terminal stems, causing tip dieback. Larvae are larger than the tomato pinworm, being about 13 mm long, and are greyish white to pink.

Populations of the potato tuberworm can be avoided by not planting tomatoes near infested potatoes. Insecticides should target the eggs or larvae in blotch mines.

Tomato pinworm

The tomato pinworm, *Keiferia lycopersicella* (Walsingham), is a serious pest of tomatoes in the south-eastern and south-western USA, Mexico and some Caribbean islands. Small, reticulated white eggs about 0.5 mm in diameter are deposited singly on the undersides of leaves. Newly hatched larvae are light coloured and bore into leaves to form blotch mines. Older larvae grow to about 8 mm and are reddish purple. They roll leaves or bore into green or red fruit, usually under the calyx. Fruit damage can be difficult to detect, due to feeding under the calyx and because larvae often cover their entrance holes with silk. Infested fruit seldom rot, thus causing a potential contamination problem.

New tomato fields should be set with transplants free of tomato pinworm eggs or larvae. Old crops should be destroyed immediately upon completion of harvesting. Mass releases of the egg parasitoid *T. pretiosum* may be helpful in reducing tomato pinworm populations. Mating disruption, using applications of components of the sex-attractant pheromone, has been extremely effective in managing within-field populations. Applications of pheromone can be made when pheromone-baited sticky traps capture five moths per trap per night. Because immigrating females may already have mated, spot spraying of field perimeters with insecticide may be necessary. *B. thuringiensis* products have generally been ineffective in managing the tomato pinworm.

The tomato leafminer, *Scrobipalpula absoluta* Meyrick, is similar to the tomato pinworm but most economic damage is attributed to serpentine leafmining rather than fruit feeding. It is a pest in South American countries, especially Peru, Chile, Argentina, Colombia and Brazil.

Sucking damage: fruit

Stink bugs

Tomatoes are attacked by a number of species of stink bug, including: the southern green stink bug, *Nezara viridula* (Linnaeus); the brown stink bug, *Euschistus servus* (Say); the consperse stink bug, *E. conspersus* Uhler; the Say stink bug, *Chlorochroa sayi* (Stål); and others. The green or brown adults are about 10–15 mm in length and shield-shaped. Small barrel-shaped eggs are deposited on leaves in clusters of ten or more. Nymphs are similar in shape to adults, only more rounded and without wings, and may be brightly coloured in black, green, orange and white. When handled, most kinds produce a strong odour. Both nymphs and adults feed on fruit, producing a lightened blotch beneath the fruit surface of green fruit that turns yellow as the fruit ripen.

Since stink bugs tend to increase on leguminous plants, tomatoes should not be planted near or adjacent to legume crops such as soybeans or near or adjacent to fallow fields with legume cover crops such as *Sesbania* or *Indigofera*. Some species of stink bug are predacious and so proper identification is important.

Coreid bugs

The leaf-footed bug *Leptoglossus phyllopus* (Linnaeus), *Phthia picta* (Drury) and others are plant bugs in the family *Coreidae* that are similar to stink bugs but are longer (20 mm) and darker with parallel sides. Eggs are metallic and ovate but somewhat flattened laterally and are laid in rows, sometimes along stems, or clusters, usually on leaves. As with the stink bugs, both adults and nymphs feed on fruit, but feeding punctures usually are deeper, which results in distortion of fruit as the fruit grow and expand.

Because certain leaf-footed bugs, including *L. phyllopus*, are attracted to leguminous cover crops such as beggarweed and *Crotalaria* and to solanaceous weeds such as nightshades, tomatoes should not be planted near or adjacent to fallow or weedy fields containing such plants.

Sucking damage: foliage and flowers

Aphids

The most common species of aphid attacking tomato plants are: the green peach aphid, *Myzus persicae* (Sulzer); the cotton or melon aphid, *Aphis gossypii* Glover; and the potato aphid, *Macrosiphum euphorbiae* (Thomas). All are pear-shaped with a posterior pair of tubercles projecting upwards and backwards from the top surface of the abdomen. The green peach aphid is about 1.5 mm long and light to dark green, while the potato aphid is larger (3 mm long) and either green or pink. *A. gossypii* is similar in size and shape to the green peach aphid but generally darker in colour. Winged and non-winged forms are all females and give birth to living nymphs. They feed on the undersides of leaves, though the potato aphid feeds on stems and petioles, causing spotting, general chlorosis and distortion of leaves. Large populations can cause stunting and wilting of plants and abscission of flowers. Aphids feed in the phloem, extracting more plant sap than they need. The sugary excess, called honeydew, is excreted and accumulates on the upper surfaces of leaves and fruit, where it supports the growth of sooty moulds. Aphids are most damaging as vectors of plant viruses, such as potato virus Y and tobacco etch, which cause serious debilitating diseases.

Aphids are not usually directly damaging, because they are attacked by a large complement of natural enemies. Insecticides should be selected for aphid control, as well as for control of other pests, that will conserve the natural enemies. Most insecticidal sprays do not afford protection against the primary transmission of virus by immigrating winged aphids, but they can be useful in reducing the amount of secondary spread within the field. Covering the soil underneath plants with aluminium plastic film, which reflects UV light, will reduce the number of alighting aphids and reduce or delay the incidence of plants with virus symptoms.

Mites

Two groups of mites are pests of tomatoes: the spider mites in the genus *Tetranychus*, particularly *T. urticae* Koch; and the tomato russet mite, *Aculops lycopersici* (Massee). Although both groups feed on tomatoes with piercing–sucking mouthparts, they are different in appearance and in the damage they inflict.

Spider mites are 0.3–0.5 mm long and are oval. They oviposit and feed on the undersides of lower leaves. The lower surface of infested leaflets may be covered with silken threads, while the upper surface will have small chlorotic spots or stippling. Increasing populations of spider mites move to the top leaves and produce silk webbing. Plants may appear chlorotic from a distance, mimicking nutritional deficiencies. Population increase is favoured by hot, dry conditions.

Russet mites are smaller and more elongate than spider mites and detection requires at least a 14× hand lens or dissecting microscope. Russet mites also attack the undersides of lower leaves, but the upper surfaces of damaged leaves appear silvery, chlorotic and, eventually, necrotic. As increasing populations move up the plant, stems and leaf petioles appear bronzed and lower foliage desiccates. Hot dry weather favours population increase, which may result in all foliage turning brown and desiccating. Damage may be confused with nutritional deficiencies or imbalances or with water stress.

Mites generally are kept under natural control in the field by predacious mites and numerous insect predators. Applications of pesticides for other pests can disrupt the natural control and cause outbreaks, especially in hot, dry weather. Sulphur is effective in controlling the russet mite but is less effective against spider mites. For the latter pests, other selective miticides should be used. Insecticidal soap and detergent can provide some short-term control, but thorough coverage is essential and reduced plant growth and yield can occur when detergent is applied weekly at rates above 1%.

Thrips

Numerous species of thrips are pests of tomato. Most, especially in the genus *Frankliniella*, occur primarily in flowers and include: the western flower thrips, *F. occidentalis* (Pergande); the tobacco thrips, *F. fusca* (Hinds); the flower thrips, *F. tritici* (Fitch); and *F. schultzei* (Trybom). Other species, such as *Echinothrips americanus* Morgan, the onion thrips *Thrips tabaci* Lindeman and the greenhouse thrips *Heliothrips haemorrhoidalis* (Bouché), occur in flowers but attack foliage more commonly than the *Frankliniella* species. Thrips are minute, elongate insects 0.2–0.5 mm in length. Adults are usually yellow, brown or black with slender, feather-like wings fringed with long hairs. Damage to flower parts, especially pistils, may cause or contribute to bloom abscission and reportedly results in necrotic lines or 'zippering' and deformation or 'catfacing' of fruit. Feeding or oviposition on small fruit may cause pitting, especially on the blossom end, as the fruit develop. Oviposition by *F. occidentalis* may result in white spots surrounding the ovipositional scar. Tomato spotted wilt virus is transmitted by nine species of thrips, the most important of which include *F. occidentalis*, *F. schultzei*, *F. fusca* and *T. tabaci*. The virus can only be acquired by larvae but is transmitted by feeding of adults on leaves.

Although densities of thrips greater than five per flower have resulted in an increase in bloom abscission, lower densities may actually improve pollination. Thrips migrate into tomatoes as the bloom declines in flowering plants, such as clover or fruit trees, or as weedy or grassland areas dry down. Therefore, tomatoes should be planted away from these situations in time or space. Insecticides can be helpful in reducing the secondary spread of tomato spotted wilt virus within a field but are generally ineffective in reducing the primary spread of the virus by immigrating thrips. Aluminium plastic film that reflects UV light can be used to cover the soil under plants, thereby reducing the number of alighting thrips and reducing or delaying the incidence of plants with spotted wilt symptoms.

Whiteflies

The most common whitefly pests of tomatoe are: the sweet potato whitefly, *Bemisia tabaci* (Gennadius); the silverleaf whitefly, *B. argentifolii* Bellows & Perring (formerly strain B of *B. tabaci*); and the greenhouse whitefly, *Trialeurodes vaporariorum* (Westwood) (Fig. 7.4). The latter species is more common in the greenhouse but can reach damaging populations in the field. Adults look like small white moths, about 1–1.5 mm in length. Adults prefer to settle and deposit eggs on the underside of foliage in the upper third of the plant. Immature whiteflies are scale-like and range in length from about 0.3 mm to 0.7 mm. The first instar nymph is the only mobile immature life stage, but it usually settles for feeding within a few millimetres of where it hatches from the egg. Subsequent nymphal instars are sessile and do not move. Adult feeding is usually of very little direct consequence, though adults of *B. argentifolii* can cause light spotting of small fruit. Adults of *Bemisia* spp. may transmit viruses, particularly geminiviruses, that cause very serious diseases. The most noteworthy is tomato yellow leaf curl virus, which reduces new growth so severely that there is little or no subsequent yield. Feeding of nymphs of *B. argentifolii* on foliage causes an irregular ripening disorder of tomato characterized by incomplete fruit ripening, especially in longitudinal streaks or

Fig. 7.4. Tomato plant infested by the greenhouse whitefly *Trialeurodes vaporariorum*. The leaves are covered by moulds that grow on honeydew excreted by the whitefly larvae. (Photograph by Dr W.H. Lindhout.)

sections (Fig. 7.5). Also associated with nymphal feeding is an increase in internal white tissue, which seems to be especially severe in hot weather.

Because tomato plants are not only a reservoir of the whiteflies but also a major reservoir of the viruses that they transmit, it is helpful to encourage a tomato-free period by destroying old crops as soon as harvest is completed and by delaying planting of new fields. In addition, new tomato fields should be separated in time and space from old tomato fields or from fields of other infested commodities such as potatoes, cabbages, ornamentals, etc. Transplant production facilities should be located away from production fields and should have all openings covered with a fine-mesh screen to exclude migrating adults. Plastic that absorbs UV light can be used to cover plant production facilities as well as plants in the field. Yellow plastic film or aluminium plastic film that reflects UV light can be used to cover the soil under the plants. These films will result in fewer alighting whitefly adults and thus in reducing or delaying the incidence of plants with virus symptoms.

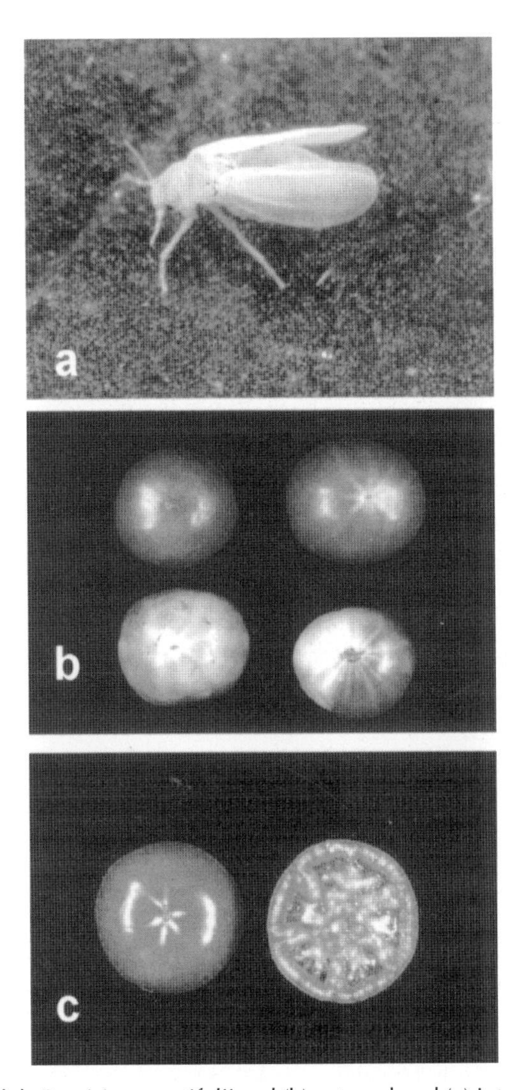

Fig. 7.5. (a) Adult *Bemisia argentifolii* and (b) external and (c) internal symptoms of irregular ripening of tomato fruit caused by feeding of nymphs.

Tomato cultivars resistant to tomato yellow leaf curl virus are available. Systemic insecticides such as imidacloprid can be applied to the soil for seedlings in transplant production facilities and to transplants in the field, providing not only whitefly control but also limited control of primary virus transmission by immigrating whitefly adults. This insecticide and insect growth regulators, insecticidal soap, oils and conventional insecticides can help to control whiteflies and reduce secondary spread of virus in the field.

TOMATO DISEASES

Plant diseases fall into two general categories: parasitic and non-parasitic. Parasitic diseases of tomato are primarily caused by bacteria, fungi, viruses and nematodes. Most fungi are microscopic organisms whose structures are elongated filaments (hyphae) and are usually branched, having cell walls and nuclei. They typically reproduce both sexually and asexually. Bacteria are microscopic organisms consisting of individual cells ranging from 0.2 to 1.0–2.0 μm in size, and contain the necessary genetic material in the cytoplasm without separation by a unit membrane. Viruses, which are tiny particles of nucleic acid (either RNA or DNA), are surrounded by a protein coat and are identified by host specificity, mode of transmission, purification properties, electron microscopy, electrophoresis, serology and genetic techniques.

Non-parasitic diseases are caused by extreme or unfavourable environmental and soil conditions, nutritional imbalances, genetic disorders (i.e. pox and fleck), pesticidal phytotoxicity and allelopathy. Symptoms may mimic those caused by pathogens and may cause difficulty in making diagnoses.

Fungal diseases

Anthracnose

Anthracnose is incited by several fungal species of the genus *Colletotrichum*, including *C. coccoides* (Wallr.) Hughes, *C. gloeosporioides* (Penz.) Sacc. and *C. dematium* (Pers. ex Fr.) Grove. *C. coccoides* is most commonly associated with the mature tomato fruit and is the only species associated with black dot root rot. Fruit may be infected when green and small, but symptoms do not appear until they begin to ripen. Symptoms first appear on ripe fruit as circular, slightly sunken lesions that become dotted with small black specks, remaining smooth and intact. Infection may occur on stems, leaves and roots. On leaves, the lesions are brown and surrounded by yellow haloes. Root symptoms, which develop on plants at the time of fruit ripening, are manifested as brown lesions and root senescence with black microsclerotia present on the root surface (hence the name black dot root rot). Larger roots are easily decorticated, while secondary roots are stunted and nearly dead.

Conditions that favour plant infection are temperatures that range from 10°C to 30°C, with an optimum of 20–24°C, combined with free moisture. Distribution of microsclerotia and conidia is favoured by splashing rain and by overhead irrigation. The fungi that cause anthracnose survive between crops in the soil and on infected plant debris. Control practices include: (i) rotating tomato crops with at least a 2-year rotation of non-solanaceous crops; (ii) minimizing overhead irrigation; (iii) using sound weed control practices; (iv) using fungicide sprays applied routinely from first fruit set to harvest; and (v) staking and mulching plants.

Late blight

Late blight, caused by the phycomycete fungus *Phytophthora infestans* (Mont.) de Bary, is a devastating disease. The fungus infects all aboveground parts of the tomato plant. Lesions begin on the foliage as indefinite water-soaked spots that enlarge rapidly into pale green to brown lesions and cover large areas of the leaf. During moist weather, lesions on the abaxial surface of the leaf may be covered with grey to white mouldy growth. On the undersides of larger lesions a ring of mouldy growth is visible. As disease develops, the foliage becomes brown, shrivels and dies. Petioles and stems are similarly affected. Dark olivaceous greasy spots develop on the fruit and a thin layer of white mycelium may be present on affected fruit during moist weather.

The pathogen survives between seasons on volunteers and in cull piles. Cool moist weather conditions favour the pathogen, with cool nights and warm days being ideal for disease development. Inoculum on infected volunteers from previous crops is transported by wind to other plants. Sporangia formation occurs at a relative humidity of 91–100% and temperature range between 3°C and 26°C, with an optimum temperature between 18°C and 22°C.

Cultural control measures include: (i) eliminating cull piles in the vicinity of tomato plantings; (ii) destroying volunteer potato and tomato plants; (iii) using transplants that have passed a certification programme; and (iv) applying fungicides when weather conditions favour disease development. Several late blight forecasting systems to predict disease outbreaks have been developed and include: (i) the Hyre system, which forecasts disease on the basis of temperature and rainfall; (ii) the Wallin system, which forecasts disease on the basis of temperature and humidity; and (iii) the Blitecast system, which integrates both systems into a computer program.

Pythium diseases

Pythium spp. cause several diseases of tomato (seed rots, pre-emergence and postemergence damping-off, stem rot and fruit rot). The species include *P. aphanidermatum* (Edson) Fitzp., *P. myriotylum* Drechs., *P. arrehenomanes* Drechs., *P. ultimum* Trow and *P. debaryanum* Hesse. Losses occur in the greenhouse and field when wet weather occurs or plants are grown in poorly drained areas of fields. Some *Pythium* species cause a fruit rot on green or ripe fruit. The disease on fruit has been referred to as 'watery rot' or 'cottony leak' of green or ripe fruit.

The fungal species may attack germinating seedlings and young plants. Untreated seed may rot in infested soil. Seedlings attacked prior to emergence often have dark brown or black water-soaked lesions that encompass the entire seedling. Plants affected in the field have dark-coloured water-soaked lesions on the root that extend up the stem. When the lesion encompasses much of the stem, the seedling collapses and dies. *Pythium* spp. can grow in the soil indefinitely. The disease is favoured by free moisture. When soil moisture is near saturation, mycelial growth and asexual reproduction are stimulated.

The disease can be controlled by: (i) using high quality seed; (ii) growing seedlings under optimal conditions, avoiding excessive watering and poorly drained soil; (iii) using sterilized planting medium in greenhouses; (iv) treating soil in the field with a broad-spectrum fumigant; (v) treating seed with a broad-spectrum fungicide; and (vi) staking tomatoes and using plastic mulch to prevent contact of the field-grown fruit with the soil.

Early blight

Early blight, incited by the fungus *Alternaria solani* (Ell. & Mart.) L.R. Jones & Grout, is present wherever tomatoes are grown, but is most destructive where frequent dews occur. Disease development in transplant houses or seedbeds results in collar rot, which may cause significant plant losses.

The early blight fungus attacks foliage, stems and fruit. Initial symptoms consist of small brownish-black lesions on the older foliage, often with a surrounding yellow halo. If the leaf is severely affected, the entire leaf may become yellow. As the spots enlarge, concentric rings are often evident on the lesion. On stems of seedlings, lesions begin as small, dark and slightly sunken, but enlarge to form circular or elongated lesions with concentric rings. The lesions may enlarge to the extent of girdling the plant and resulting in plant death. The fungus infects green or ripe fruit through the calyx or stem attachment. The fruit lesions also attain considerable size and usually have concentric ringing.

The fungus overseasons in the soil or on seed, infected debris, volunteer tomato plants or other solanaceous hosts. Environmental conditions favourable for disease development include periods of mild temperatures (24–29°C) and rainy weather.

Control is achieved by: (i) using resistant or tolerant cultivars; (ii) planting healthy transplants; (iii) applying fungicides regularly; (iv) using long crop rotations; (v) eradicating weeds and volunteer tomato plants; (vi) fertilizing properly to keep the plants growing vigorously; (vii) using pathogen-free seed; and (viii) pasteurizing soil in seed beds with steam or fumigants or using pathogen-free artificial soil mixes.

Fusarium wilt

Fusarium wilt disease, caused by the fungus *Fusarium oxysporum* Schlecht. f. sp. *lycopersici* (Sacc.) Snyd. & Hans., is of worldwide importance in warmer climates. In northern areas it is limited, by temperature, to greenhouse crops.

Infected seedlings are stunted, with the older leaves drooping and some turning yellow. The vascular tissue is dark brown. On older plants the earliest symptom is yellowing of the older leaves. The yellowing often occurs on one side of the plant and on one side of a leaf. Gradually most of the foliage yellows and then the plant wilts during the hottest part of the day. Eventually the plant collapses and dies. The vascular tissue of affected plants is usually dark brown and extends up the stem, being visible in petiole scars.

The pathogen remains in soils for several years and infects roots growing through infested soil. Factors favouring wilt development are soil and air temperatures of 28°C, soil moisture optimum for plant growth, and low soil pH. The fungus is disseminated via seed, tomato stakes, soil and infected transplants.

Control is achieved by: (i) using resistant cultivars; (ii) pasteurizing infested soil with steam or fumigants; (iii) maintaining soil pH between 6.5 and 7.0; and (iv) using nitrate-nitrogen rather than ammoniacal-nitrogen. Flooding the land or irrigating with ditch or pond water should be avoided, because this will spread the pathogen. Seedling production houses or seedbeds should be located away from production fields. A crop rotation of 5–7 years will greatly reduce losses.

Grey leafspot

Grey leafspot is incited by three fungal species of *Stemphylium*: *S. botryosum* f. sp. *lycopersici*, *S. solani* Weber and *S. floridanum* Hannon and Weber. It is one of the most destructive diseases of tomato worldwide when susceptible cultivars are grown under warm humid conditions.

Grey leafspot is almost entirely limited to infecting the leaf blades. Occasionally, when optimal conditions exist, lesions develop on the petioles and new stem tissue. On the foliage the disease first appears as circular to oblong and randomly scattered, minute brownish-black specks. The centres of the spots may dry out and crack in various patterns. When extensive defoliation occurs, yellowing of the entire leaf becomes prominent. In the seedbeds significant defoliation may occur without conspicuous yellowing.

The disease is initiated on plants in transplant production sites or on field-produced seedlings when the plants reach the first true-leaf stage. The fungus overseasons on infected debris. Abandoned fields and seedbeds, volunteer tomato plants and other weed and cultivated plant species also act as inoculum sources.

Control is best achieved by using resistant cultivars. If susceptible cultivars are grown, fungicides must be applied on a regular schedule.

Grey mould

Grey mould, incited by the fungus *Botrytis cinerea* Fr., occurs in field-planted tomato crops as well as in greenhouse culture. It consistently causes minor losses, but occasionally the disease can be a major factor in production. Its greatest effect is as a fruit rot, which can occur in the field as well as in postharvest shipments.

Grey mould affects all aboveground parts of the plant. The most characteristic sign of disease is the numerous sporophores on necrotic tissue that give the diseased tissue a fuzzy grey-brown appearance. Leaf lesions can encompass the entire leaflet, petiole and stem. The disease often occurs at the fruit's stem-end or blossom-end. Fruit lesions are characterized by a typical soft rot, with decayed areas having a whitish colour. The skin ruptures in the centre, but is unbroken over the rest of the area. Eventually grey mould develops on the entire fruit, causing it to mummify. Another phase of the disease is termed ghost-spot, in which haloes develop around a necrotic fleck on the fruit. Although no rot ever occurs with ghost-spot, many haloes on a fruit make it unsightly and unmarketable.

The pathogen has a very extensive host range. Fungal spores are wind-borne and disseminated over long distances. The fungus can survive in the form of sclerotia or as a saprophyte. The disease begins in relatively cool weather, without the requirement of prolonged periods of high humidity, and will continue during the warmer temperatures, but more slowly.

The disease can be controlled easily. Fungicides should be applied before dense canopies of plants occur and acid soils should be limed to increase the calcium content of plants.

Leaf mould

Leaf mould, caused by the fungus *Fulvia fulva* (Cooke) Ciferri., is primarily a disease of greenhouse-grown tomatoes and is most severe when tomatoes are grown under conditions of high humidity. Occasionally, it can cause serious problems on field-grown tomatoes.

The disease is generally restricted to the leaf tissue, but petioles, peduncles, stems, blossoms and fruit are attacked. Older leaves are first affected with symptoms, while younger leaves develop symptoms later. Initially, symptoms are visible on the upper leaf surface and are pale green or yellowish, eventually turning a distinct yellow. An olive-green mould is present on the abaxial surface of affected leaves and is more dense and deeper in colour toward the centre of the affected area. The leaves curl and wither and may drop from the plant. Affected green and mature fruit can have a black leathery rot at the stem-end.

Disease does not occur below 85% relative humidity. The optimal temperature for disease development is between 22°C and 24°C, but it can occur at temperatures between 4°C and 32°C. The fungus survives as a saprophyte on crop debris and in soil. Fungal spores are readily disseminated by rain or wind and various other mechanical means. Seed can also be contaminated and serve as an inoculum source.

Control is achieved by removing and destroying crop residue following the crop. Hot-water treatment of seed may help. Fungicides, when used properly, will effectively control this disease. In greenhouses following harvest, the production areas should be steamed for 6 h at 57°C. Disease can be reduced in the greenhouse by minimizing long periods of leaf wetness, by maintaining night temperatures higher than outside temperatures, and by staking and pruning of tomatoes to increase ventilation. In production areas where late crops are severely affected, planting early may reduce disease.

Southern blight

Southern blight, incited by *Sclerotium rolfsii* Sacc., is most important in tropical and subtropical areas. It has a wide host range, attacking hundreds of other species of plants, including many economically important vegetable, ornamental and field crops. The disease has several common names, including southern wilt, southern stem rot and sclerotium stem rot.

The disease is usually noticeable on plant parts in or near the soil. The typical symptom is a brown to black rot of the stem that develops near the base of the plant. Lesions develop rapidly, girdling the stem and causing a permanent wilt of the entire plant. Under conditions of high moisture, white mycelium proliferates on the affected stem tissue. Tan to reddish-brown spherical sclerotia 1–2 mm in diameter are produced on the mycelial growth. The fungus can survive many years as sclerotia, which can be disseminated in soil or plant material. The fungus uses organic matter as a substrate when infecting plant tissue. Temperatures between 30°C and 35°C and moist conditions favour pathogen development.

Control of southern blight is best achieved by rotating crops with non-susceptible grass crops, deep turning of soil to bury host debris, using effective fungicides in the transplanting solution, fumigating soils with broad-spectrum chemicals, and maintaining certain fertilization regimes (i.e. high calcium levels and ammonium-type fertilizers).

Verticillium wilt

Verticillium wilt, caused by the fungi *Verticillium albo-atrum* Reinke and Barthold and *V. dahliae* Kleb., occurs in all tomato-growing regions, causing disease on approximately 200 plant species.

Generally the first indication of this disease is the occurrence of a diurnal wilting pattern. Plants show mild wilting during the warmest part of the day, but recover at night. This is followed by some marginal and inter-veinal chlorosis on the lower leaflets. Leaflets may also have characteristic V-shaped lesions with yellowing present in a fan-shaped pattern. Vascular discoloration is evident on infected plants when the lower parts of the plant are cut open longitudinally.

Both species of *Verticillium* survive in the soil in debris from infected plants. Conditions that favour verticillium wilt are relatively cool temperatures (20–24°C) and neutral to alkaline soils.

The disease can be controlled by steam sterilization and fumigation of the soil with broad-spectrum materials. Resistant cultivars should be used where possible; however, races of *V. dahliae* overcome simple monogenic resistance. Crop rotation is limited by the wide host range of these fungi. In some cases, it may be possible to remove and destroy plant debris, thus reducing the amount of inoculum liberated into the soil.

Bacterial-incited diseases

Bacterial canker

Bacterial canker, caused by *Clavibacter michiganensis* subsp. *michiganensis* (Smith) Davis *et al.*, occurs worldwide. Early symptoms include a marginal necrosis present on leaflets, unilateral wilting of leaflets and the upward curling of leaflet margins. The lower leaves wilt first, while the upper leaves remain turgid; however, if infection occurs through wounds created when terminal buds are clipped, the disease will develop in the upper portion of the plant and move downward. Adventitious roots may develop on the stem. Light yellow to brown streaks, which later turn reddish brown, can be seen in the vascular tissues of diseased stems first. This discoloration is easiest to view at the nodes. As the disease progresses the pith becomes discoloured and 'mealy'. Symptoms of bacterial wilt and bacterial canker can be difficult to distinguish from each other. Fruit symptoms, which have been referred to as 'bird's-eye spot', are characterized by lesions with raised brown centres that are surrounded by an opaque white halo and are usually 3–6 mm in diameter.

The bacterium overseasons in soil, plant debris, weed hosts, volunteer plants, contaminated wooden stakes and seed and is spread by splashing water, contaminated equipment, pruning of staked tomatoes and clipping of transplants.

The disease is controlled by using clean seed followed by clean transplants. Soil, potting mix and pots or flats should be sterilized before use in transplant operations. In order to prevent secondary spread, tools should be disinfected frequently.

Bacterial speck

Bacterial speck, caused by *Pseudomonas syringae* pv. *tomato* (Okabe) Dye, is generally of minor concern, but the disease can be quite destructive if cool temperatures (18–24°C) and high moisture conditions exist. Severe spotting can occur on the fruit and reduce marketable yield.

Lesions on leaflets are round and dark brown to black during the early stages and develop a halo as the lesion develops. The lesion is visible on both leaf surfaces, but is most prominent on the abaxial surface. Minute lesions or specks, which are dark and rarely >1 mm in diameter, develop on fruit. Tissue around each speck appears to be a more intense green than unaffected areas. The lesions are slightly raised or flat when first observed, but in some instances are sunken.

The bacterium is seed-borne and also is disseminated by splashing rain and by mechanical means. It has been shown to survive in crop residue for an extended period (up to 30 weeks).

The disease is controlled by: (i) avoiding planting in the same field in 2 successive years; (ii) producing disease-free transplants in locations where tomato production is non-existent; and (iii) disinfecting seed to reduce seed contamination. Production fields should be kept free of weeds and volunteer tomato plants. Cull piles should not be located in or close to production fields. Appropriate bactericides should be applied where recommended.

Bacterial spot

Bacterial spot, incited by *Xanthomonas campestris* pv. *vesicatoria* (Doidge) Dye, is present wherever tomatoes are grown, but is most destructive in tropical and subtropical regions. All aboveground plant parts are infected by the bacterium. On the leaves, stems and fruit spurs, spots are generally dark brown to black and circular. On leaflets, the spots may easily be confused with bacterial speck, early blight, grey leafspot or target spot. Bacterial spot lesions do not have concentric zones as do target spot or early blight lesions, and they are generally darker in colour and less uniformly distributed than grey leafspot lesions. A prominent halo surrounding pinpoint lesions is often present with target spot and early blight but less frequently with bacterial spot lesions. General yellowing can occur on leaflets with many bacterial spot lesions. Blighting of the foliage can occur and the plants can become huddled in appearance because of severe epinasty. Often the dead foliage remains on the plant, giving it a scorched appearance. Fruit lesions begin as minute, slightly raised blisters, but increase in size, becoming brown, scab-like and slightly raised. Lesions may also be raised around the margins but sunken in the middle.

The organism can survive on volunteer tomato plants, diseased plant debris and seed. Disease development is favoured by temperatures of 24–30°C and by high precipitation. Wind-driven rain droplets, clipping of transplants and aerosols disseminate the bacterium.

The bacterial disease can be controlled by: (i) rotating crops; (ii) producing disease-free transplants; (iii) treating seed; (iv) eliminating volunteer tomato plants; (v) locating cull piles away from greenhouse or field operations; and (vi) applying bactericides or fungicide–bactericide combinations.

Bacterial wilt

Bacterial wilt, caused by *Ralstonia solanacearum*, is a serious disease of tomato in many warm temperate, subtropical and tropical regions of the world. The disease has several other common names, such as southern bacterial wilt, solanaceous wilt, southern bacterial blight, and many other common names in countries where it occurs.

Initial bacterial wilt symptoms consist of a flaccid condition on one or more of the youngest leaves. With favourable environmental conditions, rapid and complete wilting follows as soon as 2–3 days after the appearance of initial symptoms. In wilted plants, the pith and cortex are brown. Water-soaked lesions may develop on the external surface of the stem. If an infected stem is cut crosswise, tiny drops of dirty-white ooze exude from the cut surface.

R. solanacearum attacks more than 200 species of cultivated plants and weeds in 33 plant families, of which the most economically important hosts are in the family *Solanaceae*. Unlike most bacterial pathogens, *R. solanacearum* is able to survive in the soil in the absence of host plants for extended periods. Disease development is favoured by high temperatures (optimum 30–35°C) and high moisture. The bacterium is disseminated by running water or soil movement, or by movement of infected or infested transplants.

Bacterial wilt can be controlled by: (i) avoiding the use of infested fields; (ii) rotating crops with a non-susceptible crop; and (iii) using tomato cultivars with resistance to bacterial wilt, if available and acceptable.

Virus diseases

Cucumber mosaic virus

Cucumber mosaic, caused by cucumber mosaic virus (CMV), is an important disease of tomatoes in the temperate regions of the world. The virus is transmitted by aphids. The host range is extensive and includes many important vegetable crops.

Infected tomato plants in the early stages are yellow, bushy and stunted and the leaves may be mottled. Shoestring-like leaf blades are characteristic of this disease. Fruit production is greatly reduced in severely affected plants.

Numerous weed species can serve as reservoirs for CMV and it can be transmitted by more than 60 aphid species. Tomato is not generally a preferred host for aphid vectors. CMV can be transmitted mechanically, but because it is not a stable virus it is not easily transmitted by workers. The virus is seed-borne.

The virus is controlled by: (i) eradicating several key perennial or biennial species; (ii) isolating tomato fields by growing taller non-host barrier crops (e.g. maize); and (iii) applying insecticides and mineral oil sprays.

Curly top

Beet curly top or curly top of tomato, caused by the curly top virus (CTV), infects a wide range of crops. The disease occurs in arid and semiarid regions, where the leafhopper vector, *Circulifer tenellus* Baker, is confined. The disease, also called western blight, rarely affects processing tomato production, but it can cause significant problems to production of staked tomatoes.

Infection usually kills very young plants; older plants may survive but they remain yellowed, with erect terminals and a stunted plant habit. Infected leaves are thickened, crisp and rolled up, with the petioles curved down, and are dull yellow with purplish veins. One easy way to recognize this disease is to look for immature, dull and wrinkled fruit that ripen prematurely.

The virus is transmitted by leafhoppers to more than 300 dicotyledonous plant species. Monocotyledonous hosts have not been reported and seed transmission has not been shown. Virus spread into tomato fields is dependent on the seasonal cycle of the vector. Long-distance spread in weeds has been shown. Because tomato is not a desired host for the leafhopper, the insects remain in tomato fields only for short periods, thus greatly reducing secondary spread in the field.

Control can be accomplished to some extent by: (i) spraying insecticides on weeds to prevent migration of the leafhoppers (note that applying insecticides to tomato crops is ineffective); (ii) eliminating weed hosts before leafhoppers migrate into the tomato crop; and (iii) avoiding tomato production near overwintered beet fields.

Tobacco etch

Tobacco etch virus (TEV) is limited to North and South America, where it causes diseases of solanaceous crops, including tomato, pepper and tobacco. TEV symptoms include intense mottling, puckering and rugosity of the leaves. Severe stunting occurs when young plants are infected. Fruit develop a mottled appearance and do not achieve marketable size. The virus is not seed-borne, but is transmitted by ten aphid species.

Controlling the vector has not been effective in controlling the virus. TEV severity can be reduced by isolating the tomato crop from pepper production areas, because pepper is a major inoculum and aphid source. Virus infection can be reduced by applying mineral oils weekly.

Tomato spotted wilt

Tomato spotted wilt virus (TSWV) commonly occurs in the tropics, but can also be a serious problem in temperate zones. It is common in the Middle East, North Africa and India.

TSWV symptoms vary, but young leaves usually have a bronze coloration and, as the disease progresses, develop small dark spots. Streaks may develop on stems and petioles at the terminals and the growing tip may die. Fruit production is drastically reduced in plants infected at an early growth stage. Slightly raised areas with faintly visible concentric rings develop on infected green fruit. On ripe fruit, the red-and-white or red-and-yellow concentric rings are easily observed.

TSWV has a wide host range, which is predominantly dicotyledons. The virus is spread in the field by nine thrips species, but the most important vectors are *Thrips tabaci*, *Frankliniella schultzei*, *F. occidentalis* and *F. fusca*. Seed transmission has been reported.

The disease is difficult to control, because of its presence in perennials and weeds. Control of thrips does not reduce disease severity, because the virus is transmitted before the thrips are killed. UV-reflective mulches may reduce disease and some resistant cultivars are available.

Tomato yellow leaf curl

Tomato yellow leaf curl virus (TYLCV) (synonym: tomato leaf curl) is widely distributed in many Mediterranean, Middle Eastern and African countries and more recently has been established in the Caribbean and the USA.

Plants infected at early growth stages are stunted and have erect terminal and axillary shoots, small and abnormally shaped leaflets, foliage with evident vein clearing, a rosette growth habit and proliferation of axillary branch formation. Recently infected leaves are cupped downwards, while leaves that develop later after infection are chlorotic and deformed, with the leaf margins curling upwards. Plants infected at an early growth stage have significantly reduced vigour and produce no marketable fruit. Plants affected at a later growth stage contain fruit that ripen in a fairly normal manner but new fruit development does not occur, due to flower abscission.

The virus is transmitted by the sweet potato whitefly (*B. tabaci*) and the silverleaf whitefly (*B. argentifolii*) but not by the greenhouse whitefly (*T. vaporariorum*). Major outbreaks of the disease coincide with the occurrence of large sweet potato whitefly populations. Primary inoculum for infection is introduced by whiteflies from outside sources. Year-round tomato cultivation permits the virus to complete the entire disease cycle on overlapping tomato crops. Besides tomato as the primary host, TYLCV has been recovered from numerous secondary hosts, some of which (e.g. tobacco) are asymptomatic.

TYLCV may be controlled by: (i) applying insecticides with mineral oil sprays, which may reduce disease (note that the insect may develop resistance to the insecticides); (ii) changing the planting schedule to avoid the peak periods of the vector; and (iii) eliminating primary or secondary virus sources in close proximity to the susceptible crop.

Pepino mosaic virus

Pepino mosaic virus (PepMV), caused by a potexvirus, is growing in importance as a disease of greenhouse and field tomatoes in Europe and in North and South America. The virus is transmitted by plant-to-plant contact or by routine handling operations by workers. The host range includes several vegetable crops of the family *Solanaceae*.

Symptoms on infected tomato plants may vary, but usually appear first as leaf bubbling. Leaflets then become more narrow and pointed at the growing tips, causing a ‘nettle-head’ appearance. Later, the leaves have yellow spotting or a mosaic pattern and fruits show a ‘marbling’ pattern, which makes them non-marketable.

The virus is controlled by implementation of strict hygiene standards in nurseries and in greenhouses. If irrigation water is recirculated in the greenhouse, pasteurization of the water is necessary to inactivate the virus. Infected plants, and plants adjacent to them, should be removed and placed into plastic bags immediately.

INTEGRATED MANAGEMENT OF PESTS IN GREENHOUSE TOMATOES

The foregoing discussion has dealt largely with pest management in field-produced tomatoes. While pest management in the field and in greenhouses may have some aspects in common, the situation for control of damaging organisms in greenhouses is generally quite different, particularly in the large greenhouse areas of northern Europe. Weeds do not form a serious problem inside the greenhouse and are usually controlled mechanically. Nematodes, several soil diseases and soil insect pests do not play an important role either, because most tomatoes are grown in inert media and not in the soil. The pest spectrum in greenhouses is much smaller than for tomatoes grown in the open field (compare previous discussion with Table 7.2). Finally, natural enemies released in greenhouses are confined largely to the crop under production, whereas those released in the field may disperse to surrounding weeds or fields, thereby reducing or eliminating their effectiveness.

Because of the limited number of pests occurring in greenhouses, interesting IPM programmes have been developed since the 1970s for several greenhouse crops, including tomatoes (Albajes *et al.*, 1999; Van Lenteren, 2000). In addition, in temperate zones differences between greenhouse and field environments may partly explain the success of biological control in greenhouses. Greenhouses are relatively isolated units, particularly during the cold season. Before the start of a cropping period, usually during the winter, the greenhouse can be cleansed of pests and subsequently kept pest-free for several months. Later in the season, isolation prevents massive immigration of pest organisms. Furthermore, a limited number of pest species occurs in greenhouses, partly because of isolation and partly because not all pests specific to a certain crop have been imported into countries with greenhouses. Many greenhouse pests cannot survive in the field in winter or develop very slowly. Thus, biological control is easier because the natural enemies of only a few pest species have to be introduced. In addition, cultivars resistant to a number of diseases (viruses and fungi) have been developed for the most important vegetable crops. As a result, fungicides, which may lead to high mortalities of natural enemies used for pest control, do not have to be applied frequently in greenhouses. During the past 20 years, growing crops on inert media instead of in the soil has strongly decreased soil diseases and nematode problems.

Another factor easing implementation of biological and integrated control in protected crops is that cultural measures and pest management programmes can be organized for each separate greenhouse unit. Interference with pest management, or lack thereof, in neighbouring greenhouses is limited. The influence of pesticide drift on natural enemies, which is a common problem in field crops, does not play a very important role in greenhouses.

Table 7.2. Integrated pest management programme as applied in tomato in greenhouses (adapted after Van Lenteren, 2000).

	Group	Species	Methods of prevention and control
Insects and nematodes	Whiteflies	*Bemisia tabaci, B. argentifolii, Trialeurodes vaporariorum*	Parasitoids: *Encarsia, Eretmocerus* Predator: *Macrolophus* Pathogens: *Verticillium, Paecilomyces, Aschersonia*
	Spider mite	*Tetranychus urticae*	Predator: *Phytoseiulus*
	Leafminers	*Liriomyza bryoniae, L. trifolii, L. huidobrensis*	Parasitoids: *Dacnusa, Diglyphus* and *Opius* and natural control[a]
	Lepidoptera	e.g. *Chrysodeixis chalcites, Lacanobia oleracea, Spodoptera littoralis*	Parasitoids: *Trichogramma* Pathogen: *Bacillus thuringiensis*
	Aphids	e.g. *Myzus persicae, Aphis gossypii, Macrosiphum euphorbiae*	Parasitoids: *Aphidius, Aphelinus* Predators: *Aphidoletes* and natural control[a]
	Nematodes	e.g. *Meloidogyne* spp.	Resistant and tolerant cultivars, soilless culture
Diseases	Grey mould	*Botrytis cinerea*	Climate management, mechanical control and selective fungicides
	Leaf mould	*Fulvia = Cladosporium*	Resistant cultivars, climate management
	Mildew	*Oidium lycopersicum*	Selective fungicides
	Fusarium wilt	*Fusarium oxysporum lycopersici*	Resistant cultivars, soilless culture
	Fusarium root rot	*Fusarium oxysporum radicis-lycopersici*	Resistant cultivars, soilless culture, hygiene
	Verticillium wilt	*Verticillium dahliae*	Pathogen-free seed, tolerant cultivars, climate control, soilless culture
	Bacterial canker	*Clavibacter michiganesis*	Pathogen-free seed, soilless culture
	Several viral diseases		Resistant cultivars, soilless culture, hygiene, weed management, vector control
Weeds			Mechanical control

[a] Natural control: natural enemies spontaneously immigrating into the greenhouse and controlling a pest.

On the other hand, pest control is complicated by the near year-round culture of crops and by continuous heating during cold periods. These conditions provide excellent opportunities for the survival and development of a pest once it has invaded the greenhouse. Some organisms that normally show diapause when they occur in the field (e.g. spider mites) have adapted to the greenhouse climate by no longer reacting to diapause-inducing factors. As a result, rates of population growth are often much higher than in the field. These complications do not, however, create specific problems for biological control. The greenhouse climate is managed within certain ranges, which makes prediction of the population development of pests and natural enemies easier and more reliable than in field situations. The timing of introduction of natural enemies and the number and spacing of releases can be fine-tuned, resulting in season-long economic control.

Tomato production in Europe is a good example of a successful IPM programme. The programme involves ten natural enemies and several other control methods including, host-plant resistance, climate control and cultural control (Table 7.2). When tomato plants are grown in soil, soil sterilization by steaming is used shortly before planting the main crop to eliminate soil-borne diseases such as tomato mosaic virus (TMV), *Fusarium* and *Verticillium* and insects such as *L. oleracea* (tomato moth) and three *Liriomyza* spp. (leafminers). The few pest organisms that overwinter in greenhouses and survive soil sterilization are the greenhouse red spider mite (*T. urticae*) and the tomato looper (*C. chalcites*). Problems with soil diseases can also be greatly reduced by growing the crop in inert media. For 20 years, the bulk of greenhouse tomatoes in western Europe have been grown on rockwool systems, which makes soil sterilization redundant. With the cessation of soil sterilization more organisms, such as *Liriomyza* spp. and their natural enemies and *L. oleracea*, overwinter in greenhouses.

Previously, cultivars lacking TMV resistance were inoculated as young plants with a mild strain of the TMV virus to make them less susceptible; however, TMV-resistant cultivars are now available. Furthermore, many tomato cultivars in Europe are resistant to *Cladosporium* and *Fusarium* and some cultivars are also tolerant of *Verticillium* and root-knot nematodes. In tomatoes, therefore, only foliar insect pests and *Botrytis cinerea* require direct control measures. Transferring young plants free of the other pest organisms into the greenhouse is important in preventing early pest development. A recent development, which greatly stimulated the application of biological control, is the use of bumble bees for pollination.

The emergence of a new pest, the silverleaf whitefly *B. argentifolii*, has complicated IPM programmes in the USA, the Mediterranean and other parts of the world. Problems in northern Europe are fewer than initially expected after accidental importation of this pest in the 1980s. A worldwide search for new natural enemies and pathogens of *Bemisia* spp. is in progress, and as a result the pest can be controlled currently by introducing a mix of *Encarsia formosa* Gahan and *Eretmocerus eremicus* Rose & Zolnerowich (northern Europe and northern USA) or *Eretmocerus mundus* Mercet (Mediterranean). The predator *Macrolophus caliginosus* Wagner is generally added to these parasitoids, as this mix of two parasitoids and one predator results in better control over a long period.

Several expert systems, or decision-support systems, have been developed. An important factor favouring the use of such systems in the greenhouse industry is the fact that this sector is technologically highly advanced, with widespread use of computerized control of environmental conditions. Recently developed expert systems have included pest diagnosis, integrated management in specific crops, information on natural enemy release programmes, and data on side effects of pesticides on natural enemies. This type of decision-support system helps growers to manage increasingly complex production systems, though the continuous updating of decision-support software packages, which is essential for the proper functioning of the systems, is still problematic. Online services concerning biological control of greenhouse pests and the effects of pesticides on natural enemies as provided by producers of natural enemies (e.g. http://www.koppert.nl, information in English) can be updated more easily and quickly.

REFERENCES AND FURTHER READING

Agrios, G.N. (1988) *Plant Pathology*, 3rd edn. Academic Press, New York, 803 pp.

Albajes, R., Gullimo, M.L., Van Lenteren, J.C. and Elad, Y. (eds) (1999) *Integrated Pest and Disease Management in Greenhouse Crops*. Kluwer, Dordrecht, The Netherlands, 545 pp.

Beckman, C.H. (1987) *The Nature of Wilt Diseases of Plants*. APS Press, St Paul, Minnesota.

Berlinger, M.J. (1986) Pests. In: Atherton, J.G. and Rudich, J. (eds) *The Tomato Crop*. Chapman & Hall, London, pp. 391–441.

Brust, G. and Henne, R. (1995) Tomatoes. In: Foster, R. and Flood, B. (eds) *Vegetable Insect Management with Emphasis on the Midwest*. Meister Publishing, Willoughby, Ohio, pp. 77–88.

Davidson, R.H. and Lyon, W.F. (1979) *Insect Pests of Farm, Garden, and Orchard*, 7th edn. John Wiley & Sons, New York.

Dixon, G.R. (1981) *Vegetable Crop Diseases*. Avi Publishing, Westport, Connecticut, 404 pp.

Domsch, K.H., Gams, W. and Anderson, T.H. (1980) *Compendium of Soil Fungi*, Vol. 1. Academic Press, London, 859 pp.

Flint, M.L. (1990) *Pests of the Garden and Small Farm: a Grower's Guide to Using Less Pesticide*. Publication 332. Statewide Integrated Pest Management Project, Division of Agriculture and Natural Resources, University of California, Davis, California.

Francki, R.I.B. and Hatta, T. (1981) Tomato spotted wilt virus. In: Kurstak, E. *et al.* (eds) *Handbook of Plant Virus Infections and Comparative Diagnosis*. Elsevier/North-Holland Biomedical Press, New York, pp. 490–512.

Geraldson, C.M., Overman, A.J. and Jones, J.P. (1965) Combination of high analysis fertilizers, plastic mulch and fumigation for tomato production in old agricultural lands. *Proceedings Florida Soil and Crop Science Society* 25, 18–24.

Gilreath, J.P., Noling, J.W., Gilreath, P.R. and Jones, J.P. (1997) Field validation of 1,3 dichloropropene + chloropicrin and pebulate as an alternative to methylbromide in tomato. *Proceedings Florida State Horticultural Society* 110, 273–276.

Hollings, M. and Brunt, A.A. (1981) Potyviruses. In: Kurstak, E. *et al.* (eds) *Handbook of Plant Virus Infections and Comparative Diagnosis.* Elsevier/North-Holland Biomedical Press, New York, pp. 731–807.

IPM Education and Publications (1998) *Integrated Pest Management for Tomatoes,* 4th edn. Publication 3274. Statewide Integrated Pest Management Project, Division of Agriculture and Natural Resources, University of California, Davis, California.

Jones, J.B., Jones, J.P., Stall, R.E. and Zitter, T.A. (eds) (1991) *Compendium of Tomato Diseases.* APS Press, St Paul, Minnesota.

Keularts, J., Waddill, V. and Pohromezny, K. (1985) *Effect of Manual Defoliation on Tomato Yield and Quality.* University of Florida, IFAS, Agricultural Experiment Station Bulletin 847.

Lange, A.H., Fisher, B.B. and Ashton, F.M. (1986) Weed control. In: Atherton, J.G. and Rudich, J. (eds) *The Tomato Crop.* Chapman & Hall, London, pp. 484–510.

Makkouk, K.M. and Laterrot, H. (1983) Epidemiology and control of tomato yellow leaf curl virus. In: Plumb, R.T. and Thresh, J.M. (eds) *Plant Virus Epidemiology – the Spread and Control of Insect-borne Viruses.* Blackwell Scientific Publications, Oxford, pp. 314–321.

Metcalf, R.L. and Metcalf, R.A. (1993) *Destructive and Useful Insects.* McGraw-Hill, New York.

Noling, J.W. (1998) Nematodes. In: Hochmuth, G.J. and Maynard, D.N. (eds) *Vegetable Production Guide for Florida.* University of Florida, Gainesville, Florida, pp. 65–72.

Pernezny, K., Schuster, D., Stansly, P., Simone, G., Waddill, V., Funderburk, J., Johnson, F., Lentini, R. and Castner, J. (1995) *Florida Tomato Scouting Guide.* IFAS, Cooperative Extension Service, Publication SP-22. University of Florida, Gainesville, Florida.

Putnam, A.R. (1990) Vegetable weed control with minimal herbicide inputs. *HortScience* 25, 155–159.

Schuster, D.J., Funderburk, J.E. and Stansly, P.A. (1996) IPM in tomatoes. In: Rosen, D., Bennett, F.D. and Capinera, J.L. (eds) *Pest Management in the Subtropics, Integrated Pest Management – a Florida Perspective.* Intercept, Andover, UK, pp. 387–411.

Sherf, A.F. and MacNab, A.A. (1986) *Vegetable Diseases and Their Control.* John Wiley & Sons, New York, 728 pp.

Van Lenteren, J.C. (2000) A greenhouse without pesticides: facts or fantasy? *Crop Protection* 19, 375–384.

Watterson, J.C. (1986) Diseases. In: Atherton, J.G. and Rudich, J. (eds) *The Tomato Crop.* Chapman & Hall, London, pp. 444–484.

Weaver, S.E. and Tan, C.S. (1987) Critical period of weed interference in field-seeded tomatoes and its relation to water stress and shading. *Canadian Journal of Plant Science* 67, 575–583.

Production in the Open Field

A.A. Csizinszky

INTRODUCTION

The tomato is the second most widely grown vegetable crop in the world after the potato. During the 1990s the volume of the world's tomato production increased in every continent except in Europe and in North and Central America (Table 8.1). In several countries there has been an increase in average tomato yields due to the introduction of improved cultivars and production methods. Tomatoes are adapted for many environments and are grown in the open field for fresh consumption or processing where climatic conditions allow the economical production of the crop. Tomato production at low elevations in the humid tropics is limited by high disease incidence and in the temperate regions by low temperature and a short growing season. There are several different production systems in the world that are used successfully under local conditions to produce tomatoes (see Chapter 1). In this chapter the methods used in the USA will be presented but the principles involved can be applied to field production of tomatoes anywhere in the world.

ENVIRONMENTAL REQUIREMENTS FOR TIMING OF OPERATIONS

The tomato is a day-length neutral, warm-season plant that requires 90–120 days of frost-free weather with an average temperature above 16°C. Vegetative growth is limited at 12°C and prolonged periods of temperatures ≤ 12°C result in chilling injury. For vegetative growth, flowering and fruit colour development, temperatures of 25–30°C daytime and 16–20°C night-time are optimal. Depending on the growing habit, under optimum conditions the first flower trusses develop 28 days after seeding and 14 days after transplanting on determinate cultivars or up to 49 days after seeding and 21 days after transplanting on indeterminate cultivars (Geisenberg and Stewart, 1986). In fresh-market tomatoes the fruit-setting period may last for 60 days, while in

Table 8.1. World production of tomatoes, 2003 (UN-FAO, 2004).

	Area (ha × 10^3)		Yield (t/ha)		Production (t × 10^3)	
Region	1997	2003	1997	2003	1997	2003
World	3,1670.0	4,310.6	27.9	26.3	88,359	113,353
Africa	480.0	645.0	21.04	20.1	10,272	12,964
North and Central America	288.0	313.0	47.5	46.2	13,680	14,461
South America	155.0	151.0	35.4	44.0	5,487	6,644
Asia	1,623.0	2,509.0	24.8	23.1	40,250	57,958
Europe	608.0	683.0	29.6	30.4	17,997	20,763
Oceania	12.0	9.6	40.7	51.2	488	492
Leading Countries						
China	539.0	1,205.0	30.4	23.9	16,382	28,800
USA	165.0	166.1	65.1	62.5	10,762	10,381
Turkey	158.0	230.0	41.8	42.4	6,600	9,752
India	350.0	520.0	15.1	14.3	5,300	7,436
Italy	115.0	130.0	48.1	51.0	5,539	6,630
Egypt	175.0	181.0	34.0	35.1	5,950	6,353
Spain	56.0	64.5	53.1	59.7	2,984	3,851
Brazil	60.0	61.5	43.4	59.2	2,602	3,641
Iran	150.0	112.0	23.3	26.8	3,500	3,002
Mexico	72.0	67.1	26.6	32.0	1,913	2,147

processing tomatoes (which are harvested once over) the fruit-setting period lasts for approximately 30 days.

SOIL PREPARATION

Tomatoes are grown on a wide range of soil types and with various production systems. Consequently, the type of machinery used and the sequence of operations in soil preparation will depend on local conditions and cultural practices. The aim is to prepare a site that reduces the danger or will be free of waterlogging, provides pest-free soil with sufficient depth for root growth and allows the necessary field operations to be carried out during the growing season. The first step is to plough or to disc the soil to bury the vegetation that grows on the site. The optimum soil pH range for tomatoes is between 6.0 and 6.5. Where soil tests indicate that soil pH values need to be adjusted, dolomite and lime in cases of low pH (< 6.0) or gypsum and elemental sulphur in cases of neutral or high pH (> 7.0) should be applied 2–3 months before planting. If pH has to be adjusted at the time of site preparation, fine-particle gypsum, dolomite or lime should be used and incorporated into the soil during the ploughing and rotavating.

Where compact, heavy subsoils prevent water penetration and root development, chiselling may be necessary at the time of ploughing. The topsoil should have a fine tilth to provide good contact with the seeds and the developing roots, or with the roots of the seedlings if the crop is established from transplants.

The next step is the formation of plant beds. The width and height of the beds and the centre-to-centre spacing will depend on soil type, irrigation system, whether the plants will be staked and trellised and whether the plant beds will be covered with polyethylene film.

Bare ground culture

After the pre-beds have been marked out, the beds are gradually built up to the required height. Where soil tests indicate, the required amounts of phosphorus (P) and micronutrients are applied preplant in a band 10–15 cm wide or broadcast in the full width of the bed 7–10 cm below the plant row. Up to 50% of the nitrogen (N) and potassium (K) are also applied preplant where the land is furrow or sprinkler irrigated. The remaining amounts of N and K will be side-dressed, as required, during the season. The preplant N and K fertilizers may be broadcast in the full width of the plant bed and then rotavated into the soil, or band-applied at 10–15 cm to the side of the plant row. With micro- (trickle) irrigation systems, a small proportion (5–10%) of the N and K may be applied as preplant dry fertilizers just below and 5–10 cm to the side of the plant row. The remaining amounts of N and K will be applied as liquid fertilizer, injected via the micro-irrigation tubes with the irrigation water during the season. Chemicals to control plant pests and weeds may be applied and incorporated into the beds prior to banding the N and K fertilizers. The beds are then pressed with a land roller to compact the soil for good contact with the seeds, or with the roots of the transplants.

For processing tomatoes, the beds should be crowned (i.e. slightly higher in the centre) to facilitate mechanical harvest (Hartz and Miyao, 1997).

Plastic mulch culture

Where tomatoes are produced in the full-bed polyethylene mulch system, with furrow or with sprinkler irrigation, all of the required fertilizers are applied as dry fertilizers prior to planting. Phosphorus and micronutrients are banded or broadcast on the pre-bed. N and K are broadcast and then rotavated into the top 10–15 cm of the bed or applied in a groove 5 cm wide and 5–7 cm deep on both halves approximately 25–30 cm from the centre of the finished bed (Fig. 8.1). Next, the soil is treated with a broad-spectrum pesticide to control soil-borne pests and then the plastic mulch is laid by machine (Fig. 8.2). With the micro-irrigation system, the P and the micronutrients are placed on the pre-bed and up to 40% of the required N and K may be applied preplant as dry fertilizers. The remaining amounts of N and K will be injected through the micro-irrigation tubes as liquid fertilizers during the season, according to plant growth and development (Hartz and Hochmuth, 1996). The preplant dry N and K fertilizers may be broadcast in the full width of the bed or applied in a band and then incorporated into the soil.

On light sand, where horizontal movement of water from the emitter is limited, the preplant dry fertilizers should be placed in a band near or on the top of the micro-irrigation tube, in contact with moist soil. The soil is then treated with pesticides and covered with the polyethylene film (Fig. 8.3).

Fig. 8.1. Nitrogen and potassium fertilizer applied in narrow bands on the bed shoulders for tomatoes.

Fig. 8.2. Laying polyethylene film on the beds.

Fig. 8.3. Full-bed polyethylene mulch production system with micro- (drip) irrigation.

STAND ESTABLISHMENT

Planting techniques

Direct seeding

In the direct-seeding method, seeds are sown by precision mechanical and vacuum planters to meter the seed. Depth of sowing varies from 1.5 to 4 cm depending on soil characteristics. Soil crusting after irrigation or after heavy rainfall in low organic matter heavy soils or in calcareous soils is a serious impediment to seedling emergence and stand establishment. To combat crusting, growers use tillage tools that do not destroy the seedlings. Soil amendments, vermiculite and cationic polymers placed over the seed row have also been tried experimentally with good results (Hoyle, 1983) but the cost and the difficulty in mechanizing the operation on large farms has prevented the widespread acceptance of these treatments.

The number of seeds sown is usually higher than the required plant population. In California, for processing tomatoes, 0.28–0.56 kg (93,000–186,000) seeds/ha are sown in two rows per bed (Hartz and Miyao, 1997). Thinning within the row is carried out when the plants are at the three to four true-leaf stage. Spacing within the row depends on the growth habit of the cultivar, soil fertility and the availability of irrigation water. Closer spacing promotes earliness through an increased number of first flower clusters per plant and results in fewer sun-scalded fruits (Fery and Janick, 1970; Zahara, 1970).

For more uniform and speedier germination, especially under a wide range of temperature and soil moisture conditions, seeds may be 'primed' or 'osmo-conditioned' prior to sowing. Priming is a controlled hydration that permits the seeds to undergo pre-germinative metabolic activity without the emergence of the radicle. The seeds are soaked in osmotic solutions containing one or a combination of salts (KNO_3, K_3PO_4, NaCl), mannitol or polyethylene glycol (PEG). The osmotic strength of the solution is adjusted to allow the imbibition of water to initiate the germination process but to inhibit the emergence of the radicle. The treatment conditions must be determined by trial and error for the seeds, due to the variability among species, varieties and seed lots. Seedlings from primed seeds emerge earlier and more uniformly than those from unprimed seeds. Early and seasonal total fruit yields, however, were not always higher in crops from primed seeds than those from non-primed seeds (Alvarado *et al.*, 1986). Because of the mixed results in the field and the much higher cost, growers have not accepted the use of primed seed, except for greenhouse-grown transplants, where high germination percentage and uniformity of emergence are of great importance (Hartz, 1994).

Other techniques, such as gel-seeding and plug-mix planting of germinated seeds, have been developed but are not used commercially by tomato growers, due to the high cost of hybrid seeds and to the complexity of seeding operations (Gray, 1981).

Transplanting

The use of transplants for stand establishment continues to increase in the field production of tomatoes (Fig. 8.4). The high cost of hybrid seed, earlier yields, fewer pesticide applications for the seedlings in the greenhouse than for direct-seeded plants in the field, achievement of near-perfect plant stands and better weed control make transplanting a viable alternative to direct seeding. Transplants are raised in 'hotbeds' under cover, or in greenhouses.

Fig. 8.4. Loading containerized tomato seedlings on to transplanters.

Transplant production in the greenhouse is highly mechanized. The seeds are sown in moulded plastic or polystyrene production trays divided into cells. The cells are in various shapes (inverted pyramid, hexagonal or cylindrical) and cell size (volume) varies from 8.6 to 80 cm^3. The most widely used tray for tomato transplant production has 200 cells, each 27 cm^3 in volume. The cells are filled with a commercial medium mix containing essential plant nutrients and a surfactant to improve wetting of the medium. The seedlings are ready to transplant in 4–5 weeks when the roots fill out the cells and can be pulled from the cells with the root ball intact without damaging the roots. Cell size (root volume) has a significant effect on early season yields. Plants raised in 37.1 or 80.0 cm^3 root-volume cells flowered sooner after transplanting and produced higher early season yields than plants raised in smaller cells (Kemble *et al.*, 1994). However, transplants raised in the larger cells require a longer time to reach transplantable size; thus they are more expensive than plants raised in smaller cells.

There are no generalized nutrition programmes available for raising tomato seedlings. In general, seedlings should be compact and should grow rapidly when transplanted to the field. Fertilizers, irrigation, temperature and sunlight interact on plant height. Nutrients for the seedlings may be applied from a fertilizer solution that has a concentration (mg/l) of 75–100 N, 32 P and 75–100 K (Weston and Zandstra, 1989). The nutrient solution may be applied daily or every other day, depending on the condition of the seedlings. To hold back the seedlings when transplanting is delayed, a fertilizer solution of 50 mg N/l and 100 mg K/l may be applied once a week. N and K sources for the fertilizer solution are $Ca(NO_3)_2$ and KNO_3 and P may be obtained from soluble monopotassium phosphate. Ca, Mg and microelements are also applied in the growth medium.

Plant height may be controlled by irrigation and daily brushing. Reduced irrigation to control plant elongation is especially important under poor light conditions. Brushing consists of slowly moving a smooth stick, just touching the top of the plants.

Seedlings in the field should be transplanted to the depth of the first true leaves. Transplanting the seedlings to the first true leaves instead of to the top of the root ball increased early yields by 25–30% (Vavrina *et al.*, 1996). The yield increase was mainly in the extra large (≥ 70 mm diameter) size of the fruit. Seasonal total marketable yields and fruit size, however, were similar at the two planting depths.

A fully automatic transplanter has been developed for vegetables (Shaw, 1997). The machine accommodates the transplant growing trays, individually removes the seedlings from the cells and sets them into the soil without manual handling. The machine can transplant seedlings at a rate of 7000 plants/h through plastic mulch or into bare ground.

MULCHING AND PLANT COVERS

Mulches

Polyethylene film (mulch) to cover plant beds has been used for commercial production of fresh-market tomatoes since the 1960s (Geraldson *et al.*, 1965). In this production system the raised beds are covered with a low- or high-density polyethylene film 0.35 mm thick (Figs 8.2 and 8.3). The most common mulch colours are black or white-on-black. Black mulch is used when the crops are planted in the cooler part of the year, when the soil is cold. White-on-black is used in summer plantings.

The full-bed polyethylene mulch system, especially when combined with staking and trellising (Figs 8.5 and 8.6), has several advantages over bare-ground production. The mulch cover inhibits weed growth (except that of the nutsedge, *Cyperus* sp.), retains soil fumigants, improves moisture retention, reduces fertilizer leaching, prevents bed erosion and protects fruits from decay caused by soil-borne disease organisms. Mulching increased the rate of tomato basal branch appearance and led to earlier flowering and higher early yield compared with non-mulched plants (Wien and Minotti, 1987). Seasonal total yields have been increased by the large-scale adaptation of the full-bed polyethylene mulch system: in Florida, after state-wide acceptance of this system by growers, tomato yields increased by 60%.

Apart from conventional black or white-on-black plastic films, aluminium, blue, orange, red, silver and yellow mulches have been studied for growing fresh-market tomatoes (Decoteau *et al.*, 1989; Csizinszky and Schuster, 1995). The reflective and colour mulches affect plant growth and development by increasing soil or aboveground temperatures and the quality

Fig. 8.5. Staked tomatoes with the full-bed polyethylene mulch and furrow irrigation system.

Fig. 8.6. Trellised tomatoes with the full-bed polyethylene mulch and micro- (drip) irrigation system before harvest.

and quantity of light reflected on to the plant. The wavelength of light, especially near the UV (390 nm) region, affects the orientation of insects that visit the plants (Kring, 1972). In spring, planting on red mulch resulted in earlier yields and greater fruit size due to warmer soil temperatures than on white or black mulches. In summer/autumn, under stress from tomato mottle virus transmitted by the whitefly *Bemisia argentifolii* (Bellows and Perring), early yields were lower on the red than on the white mulch. Thus, under high insect stress, mulch colour should be selected for its effect on insects in addition to its effect on soil temperature and plant morphology.

Among the disadvantages of using mulch in the field production of tomatoes are: (i) the polyethylene film requires specialized equipment to lay it on the raised bed; (ii) the preplant cost of growing tomatoes is increased due to the cost of purchasing the polyethylene film; and (iii) the mulch has to be removed and disposed of after the last harvest (Brown and Splittstoesser, 1991). Plastic film may be disposed of by discing, burning, physical removal, or by removal and storage. Long-term degradable film builds up in the soil and interferes with field operations. Burning *in situ* with a flame thrower is expensive and in many areas burning the plastic is prohibited by environmental laws. Physical removal is labour intensive and finding a disposal site is difficult. To avoid these disposal problems, the feasibility of sprayable mulching using biodegradable latex sprays and bituminous residues has been investigated (Stapleton, 1991). Black-pigmented latex materials, similar to polyethylene film mulches, retained soil moisture, increased soil temperature and reduced numbers of soil-borne *Pythium* spp.

Row covers

Under adverse, cool, windy or rainy environmental conditions early in the season, row covers are used to protect the young developing plants and to enhance growth. Row covers are flexible, transparent or semitransparent materials used to enclose plant rows. They are made of either polyethylene film or porous, non-woven (spun-bonded) polypropylene or polyester. Polyethylene film, clear or painted white, is used as a tunnel on wire hoops. The width and height of the tunnels depend on the local conditions. The spun-bonded non-woven fabric-like covers can be used either as tunnels or as floating covers. Spun-bonded covers have low light reflectance and transmit approximately 80% of the light.

Thickness of row covers depends on the expected temperatures during the period for which young and growing plants are to be covered. For growth enhancement, polyethylene film 0.028 mm thick or spun-bonded covers of 17 g/m^2 weight are used. For frost protection, spun-bonded covers of heavier weight (37–60 g/m^2) are recommended (Geisenberg and Stewart, 1986). Under rapidly changing weather conditions, proper ventilation is very important, or the plants will be damaged by heat. In any case, the covers have to be removed from the field when anthesis begins or when cold temperatures and wind are not expected to injure the plants. As with other production methods, the cost of investment compared with the expected benefits has to be considered by growers when deciding to use row covers in the field production of tomatoes.

NUTRIENT MANAGEMENT

For acceptable economic yield with good fruit size and colour, the crop has to be provided with adequate nutrients throughout the season. Plants require 16 elements for normal growth and development. The actual quantity of each element that has to be supplied from fertilizer sources will depend on the concentration of available nutrients in the soil prior to fertilizer application. Determination of available nutrients in various soil types depends on the extraction method and on the type of extractants used. In Florida, the Melich I (double acid) index is used for fertilizer recommendations for tomatoes (Hochmuth, 1998). Prediction of fertilizer requirements based on the results of analyses also has to take into consideration the usable rooting volume of the plant (Adams, 1986). Fertilizer sources for the crop depend on the irrigation system and on the soil type. In general, on soils with pH values $\geqslant$ 7.0, fertilizers that reduce the soil pH are applied for the crop; and in soils with pH values $\leqslant$ 5.9, fertilizers that increase the soil pH are applied.

Nitrogen

To produce 1 t of tomato fruit, approximately 2.5 kg of available N should be applied in the soil (at 75% N uptake efficiency) (Feigin and Shakib, quoted by Geisenberg and Stewart, 1986). The amount and source of N applied prior to planting will depend on the production and irrigation system. In non-mulched systems, 50% of the N is applied prior to planting and the remaining amount of N is applied in two to four applications during the season, depending on the rainfall. The fertilizer is applied on the bed shoulders just ahead of the expanding roots. With the full-bed polyethylene mulch production system and furrow irrigation, all of the N fertilizers are applied prior to planting. With micro-irrigation, 25–40% of the N is applied as dry fertilizer prior to planting and the remaining amount of N is injected from liquid sources via the drip tube according to plant growth and development (Table 8.2).

During the growing season, N and K concentrations in the plant sap can be determined by ion-specific electrodes or by test strips. Nitrogen concentration in the leaves should be 4.0–4.5% early in the season, 3.0–3.5% during fruit set and 2.0–2.5% during harvest. Very high N concentration prior to or during full bloom may result in excessive shoot growth and reduced fruit set. In neutral and alkaline soils, the N source is ammonium sulphate or urea. In acid soils, calcium-ammonium nitrate or potassium nitrate should be applied. On sandy soils and under low soil temperatures or on fumigated soils, the ratio of NO_3-N to NH_4-N should be 0.7 to 0.3. In warm soils, or in non-fumigated soils, the ratio of NO_3-N to NH_4-N is not as critical.

Table 8.2. Application schedule of N and K fertilizers for micro- (drip) irrigated transplanted tomatoes.

Stage	Daily % of season's total	Day number
Stand establishment	0.6	14
Pre-bloom	0.9	21
From 1st bloom to 1st harvest	1.6	35
Early harvests	0.8	14
Late harvests	0.4	14

Nitrogen may be applied from controlled (slow)-release fertilizers. However, under high soil temperatures the N is released at a rapid rate from coated fertilizers (Csizinszky, 1993; Shaviv *et al.*, 1993). Therefore, where high soil temperatures are expected early in the season, controlled-release fertilizers have little or no advantage over soluble N fertilizer sources.

Phosphorus

Tomatoes have the greatest demand for phosphorus (P) at the early stages of development. In soils where preplant analyses indicate low P (< 31 P by the Melich I index), phosphorus fertilizer should be applied prior to planting. For soils that test very low (< 10 P), up to 90 kg P/ha may be applied. For a fresh-market tomato crop when more than three harvests are expected, P at 44–55 kg/ha should be applied if the soil tested low (< 10) or medium (≤ 30) for P. On soils that tested high (> 30) or very high (≥ 60) for P, only a small amount (10–15 kg P/ha) may be applied just under the transplant or seed row. Phosphorus applications for the crop during the season are justified only if the plants are diagnosed with P deficiency. In tomato leaves and fruits, P at 0.2–0.4 g/kg (dry weight) is adequate for good growth, yield and quality.

Phosphorus for the tomato crop can be applied from superphosphate, mono- and diammonium phosphate, monopotassium phosphate and rock phosphate. Ammonium and potassium phosphates, due to their high water solubility, are applied on calcareous soils. Rock phosphate may be applied on acid soils, but the availability of P for the plants is very low (Hausenbuiller, 1972).

Potassium

For good fruit colour, taste and firmness, potassium (K) should be present in the soil in an N:K ratio of 1:1.15–1.66. The amount of K applied will depend on preplant soil tests, cultivars and the production system. On light sandy soils that are low (20–35) or very low (< 20) in K by the Melich I index, 235–360 kg K/ha are necessary for fresh-market tomatoes that are harvested more than three times during the season (Everett, 1976). On loamy soils that are medium (36–60) for exchangeable K, from 60 to 90 kg K/ha is adequate for good yields. The amount of K applied for the crop also depends on the cultivar: those that use K efficiently produce satisfactory yields even under K stress; those that are inefficient in K use will produce satisfactory yields only under adequate K supply (Makmur *et al.*, 1978).

Potassium concentrations in the leaves at the first flowering stage should be 2.5–4.5% (dry weight) and at the early harvest 2.0–3.0% for a good quality crop. Potassium sources for tomatoes include KCl, KNO_3, K_2SO_4, K_2SO_4•$MgSO_4$ and KH_2PO_4 (potassium monophosphate). Any of these K fertilizers can be used for the crop with good results but KCl, because of its high salt index, should be used with caution (Csizinszky, 1999).

Calcium, magnesium and sulphur

In soils with pH ≤ 5.6 or when the calcium (Ca) is ≤ 300 by the Melich I index, Ca has to be applied to the soil for normal plant growth and development and for good fruit quality. To increase the pH value from 5.5 to 6.5, $CaCO_3$ is applied in a sandy soil at 1300 kg/ha and on clay loam at 5000 kg/ha prior to planting. In tomato leaves, 0.8–2.0% Ca (dry weight) is in the adequate range for growth and development.

Magnesium (Mg) should be applied to the soil when its Melich I index is < 15. In the leaves, 0.35–0.60% Mg (dry weight) is a sufficient concentration for the crop. Mg may be applied to the soil with the Ca from dolomite ($CaCO_3$•$MgCO_3$) or from $MgSO_4$, $MgCO_3$ and K_2SO_4•$MgSO_4$. The $MgSO_4$ and the K_2SO_4•$MgSO_4$ may be mixed with other fertilizers when applied to the soil at 35–50 kg Mg/ha.

Sulphur (S) deficiency is seldom observed in tomatoes. In the leaves, 0.3–0.8% S (dry weight) is adequate for the crop. Under most conditions, a sufficient quantity of S is applied for the tomatoes when S-containing fertilizers, such as superphosphate, potassium sulphate, magnesium sulphate, ammonium sulphate or potassium magnesium sulphate, are used in the fertilizer programme. The application of S at 34–45 kg/ha from these fertilizer sources should provide sufficient S for the tomato crop.

Micronutrients

Micronutrient deficiencies may occur in tomato plants in soils at high or low pH and in fine sandy soils. Where the soil is low in micronutrients, then (in kg/ha) 2.0–2.5 B, 2.0–2.5 Cu, 5.0–6.0 Fe, 2.5–3.0 Mn, 0.02–0.025 Mo and 2.0–2.5 Zn are applied during land preparation. When micronutrient deficiency is observed on plants during the season, Cu, Fe, Mn and Zn can be applied by foliar sprays.

IRRIGATION

Crop water requirement

Tomatoes require proper water management to attain high yields and good quality fruit; therefore, where natural rainfall is lacking, irrigation is necessary. Fresh-market tomatoes have to be provided with water continuously throughout the growing season. For processing tomatoes, it is common practice to cut off irrigation 2–4 weeks before harvest to increase the soluble solids concentration in the fruits (Hartz and Miyao, 1997). The actual amount of irrigation will depend on crop water requirements and the irrigation or production system (Smajstrla, 1998).

The crop water requirement depends on evaporative demand, termed as daily evapotranspiration (ET), which in turn depends on crop development stage and climatic factors (rainfall, temperature, wind). Irrigation requirements depend on crop water requirements and on the irrigation system, soil characteristics and production system. Thus the crop's ET can be calculated from the daily open-pan evaporation, the crop coefficient (k_c) and the application efficiency of the irrigation system. The k_c depends on the growth stage of the crop and on the production system (bare ground or full-bed mulch). For example, tomatoes at the fruit sizing stage on plastic mulch with drip irrigation have a k_c of 0.9, and with furrow irrigation a k_c of 1.15. The drip irrigation system's efficiency (see below) is 0.85; therefore, on a day when the open-pan evaporation is 5 mm (5 l/m^2), a drip-irrigated crop at the fruit sizing stage with the plastic mulch cultural system on a per hectare (10,000 m^2) basis will require $(0.9 \times 50{,}000\ l)/0.85 = 53{,}000$ l/ha irrigation.

Frequency of irrigation depends on the soil characteristics. On soils with a high water-holding capacity, the crop may be irrigated at intervals of 5–7 days. On coarse sand, the crop (especially during fruit setting and fruit development) has to be irrigated daily. Soil moisture can be monitored by tensiometers or by portable soil moisture meters. Placement of the tensiometers will depend on the rooting depth. For example, with a full-bed polyethylene mulch and furrow irrigation production system on light sand, the tomato roots seldom penetrate deeper than 25–28 cm. Tensiometers are placed 12–15 cm deep in the plant row and soil moisture tension is maintained at $\leq$ 10–12 kPa.

Irrigation systems

The main types of irrigation systems used in the tomato production are: (i) the overhead sprinkler; (ii) furrow (ditch) irrigation; and (iii) micro- (drip) irrigation. Irrigation systems are rated with respect to application efficiency, which is the fraction of the water that is available for the plant from the total applied. The efficiency of the sprinkler system is 0.6–0.8, that of furrow irrigation 0.4–0.6, and that of drip irrigation 0.75–0.9.

Overhead irrigation is often used in conjunction with other irrigation systems for stand establishment and for cooling the foliage during hot periods. One disadvantage of the overhead sprinkler system is its potential for spreading foliar diseases.

The furrow irrigation system has the lowest initial cost compared with other systems. The centre-to-centre spacing of irrigation furrows depends on the soil type. On light sand the horizontal movement of water is limited; consequently, the plant rows have to be close to the furrows. In many countries, runoff of water from the furrows has to be controlled in order to save water and to reduce nutrient or pesticide contamination of surrounding areas adjacent to the production fields.

Drip irrigation is expensive to install and requires skill and technical knowledge to operate. However, it has the highest application efficiency among the irrigation systems; it can be used on uneven land surfaces; and it can deliver plant nutrients and pesticides during the growing season. Under dry climatic conditions, furrow irrigation is used to establish optimum soil moisture for land preparation and planting. When the crop is established, the crop water requirement is supplied by the drip system.

The distance of the drip tubes from the plant rows and the depth of the tubes in the plant bed depend on the soil's hydraulic characteristics, the water discharge pattern from the emitters and the production system. Drip emitters should be 10–15 cm to the side of the plant row. On sand, the tubes are placed 2.5–5.0 cm deep and on heavy soils not more than 15 cm deep.

In order to prevent clogging of the emitters, the residue of injected chemicals has to be purged by running water through the drip tubes. The tubes also have to be cleaned periodically by injecting chlorine to obtain 5 ppm free Cl^- in the runoff water. Phosphoric and sulphuric acid, or special commercial cleaning products, may also be used to prevent clogging of the emitters (Clark and Smajstrla, 1996).

Under saline conditions, where tomatoes are planted in soils with an electrical conductivity (EC) of $\geq$4 dS/m, or irrigated with water that has EC $\geq$ 4 dS/m, the crop requires different management and irrigation methods than plants in non-saline soils or irrigated with good quality water. Land preparation includes deep chiselling to facilitate drainage. Between seasons, or in a crop rotation system, deep-rooted cover crops are planted to help water penetration. Prior to planting, the salts are leached from the top soil layer with water. In sodic soils that have ≥ 15% exchangeable Na^+, plants require soluble Ca, gypsum, elemental S and acid-forming fertilizers (K_2SO_4, calcium and ammonium polysulphide) to remedy ionic imbalances (Sanchez and Silvertooth, 1996). The soil should be flood-irrigated before planting. to leach salts from the top soil layer (Geisenberg and Stewart, 1986). Plants at the salt-sensitive stage (during seed germination and as young transplants) should be irrigated with good quality water, followed during later growth stages by saline water when good quality water is scarce for irrigation purposes (Hamdy *et al.*, 1993). Saline water, with or without acid treatment, may be used with the drip irrigation system especially on light sand, where water can be applied frequently to keep soil moisture high and leach salts from the root zone. Overhead irrigation of tomato plants with saline water should be avoided, due to salt accumulation on the leaves.

CROP MANAGEMENT

Crop management during the growing season involves cultivating the soil to control weeds, protecting the crop from plant pests, operating and maintaining the irrigation system, applying nutrients, and pruning, staking and trellising the plants. The actual operations will depend on the production system and whether the fruit is used for fresh consumption or processing.

Fresh-market tomatoes are pruned early in the season in order to increase fruit size and earliness. Pruning involves removing from the stem a number of axillary shoots that grow between the soil level and the first flower cluster (Olson, 1989). The axillary shoots are removed when 7–10 cm long. The number of shoots removed depends on the cultivar. Indeterminate cultivars with a large shoot mass can be pruned heavily by removing all the axillary shoots up to the one immediately below the first flower cluster (first fork). Determinate cultivars, which seldom grow taller than 90–100 cm, need light pruning: only the first two or three axillary shoots are removed from the stem, because heavier pruning will reduce marketable yields.

In the staked production method, wooden stakes 140 cm long and 2.5 × 2.5 cm thick are driven into the soil between the plants 2–3 weeks after planting (see Fig. 8.5). The growing plants are supported by tightly stretched twine running horizontally along the plant row. The function of staking and tying is to hold plants off the ground to achieve better pesticide spray cover, cleaner fruits and easier harvest than with ground tomatoes. The first tying is performed when the plants are 30–40 cm high, or before the plants fall over. At the first tying the twine is about 25 cm above the bed surface and is run horizontally along one side of the plant row and looped around each stake until the end of the row. It is then run back on the opposite side of the row, again being looped around each stake. It is important to stretch the twine tight. The plants are usually tied three times during the season and each subsequent tying is approximately 25–30 cm above the previous tying.

Under adverse climatic conditions, when endogenous hormone production and nutrient uptake by the plants are disrupted, the supply of plant hormones and essential nutrients from exogenous sources will benefit plant growth and fruit yield (Coombe, 1976). The biostimulant products available for the growers contain seaweed extracts, plant growth hormones, humates, α-keto acids and nitrophenolates. In addition, the biostimulants may contain one or more of the essential macro- or microelements. It is important to use the products according to the manufacturer's instructions. Over-application of biostimulants, or combination of biostimulants with frequent applications of nutrient sprays, will result in the accumulation of toxic levels of micronutrients in tomato shoots (Csizinszky, 1996). Under favourable growing conditions and with good fertilizer and irrigation practices, biostimulants and nutritional sprays had little or no beneficial effect on tomato yields and fruit size; furthermore, biostimulant effects on tomato yields were inconsistent from season to season (Csizinszky, 1996).

HARVESTING AND HANDLING

Tomatoes for fresh consumption are harvested at various stages of ripeness, depending on labour costs, availability of controlled atmospheric storage facilities, distance of the production area from the markets and market demands. Careful handling of the fruit to prevent bruising and injury during harvest and the subsequent sorting, grading and packing operations is very important. Mechanical injury to the fruit provides infection sites for decay organisms that render the fruit unsuitable for human consumption.

In jointed varieties, the stem and the calyx have to be removed from the fruit in the field since stems will puncture the adjacent fruit. Some markets, however, prefer fruit with calyx and stem intact, because there is a perceived association of calyx and stem with vine-ripe fruit and freshness. Fruit harvested with calyx and stem are placed and marketed in containers having only a single layer of fruit.

Fruit harvested at the light red or red stage are packed directly into cartons in the field. Carton size will depend on the particular market demand. Firm ripe fruit can be stored at temperatures from 8°C to 10°C and at 85–96% relative humidity (RH) for 1–3 weeks.

Fruit harvested at the mature-green stage, i.e. with jelly-like placenta in all locules and hardened seeds, are harvested into buckets. The buckets are emptied into large bulk container boxes holding up to 450 kg fruit. In some farms, mature-green fruit from the buckets are emptied into a large metal tub (gondola) that is filled with water to reduce injury to the fruit when they are emptied from the buckets. The tomatoes are transported to the packinghouses, where they are washed in chlorinated water (100–150 mg Cl^-/l and pH 6.8–7.2), sorted, graded and placed into cardboard boxes that have holes in the sides for ventilation. The fruit are then stored at 13–15.5°C and 90–95% RH. Under these temperatures and relative humidities, mature-green fruit can be kept in storage for 4–7 weeks, depending on the shipping schedule. When the shipping schedule requires, the fruit are treated with ethrel at 21°C, maintaining an ethylene concentration of 100–150 ml/m^3 in the storage atmosphere, until the first external colour development at the blossom end of the fruit is noted. Further storage temperatures vary from 15.5°C to 18°C, depending on the length the fruit have to be kept in the storage.

Processing tomatoes are harvested mechanically in a once-over destructive harvest when at least 90% of the fruit is red ripe or pink (Hartz and Miyao, 1997). To maximize the percentage of coloured fruit, the tomato field is sprayed with chemicals that enhance fruit ripening. Commercially available ethephon is applied to the plants when 15–20% of the fruit is ripe, usually 2–3 weeks before the harvest date (Dostal and Wilcox, 1971). Fruit from the mechanical harvesters are loaded into tandem bulk trailers or bins and transported to a government-administered inspection station. At the inspection station fruit colour, soluble solids content and defect levels (insect damage, mould, green fruit) are evaluated, then the fruit are transported to the processing plant.

REFERENCES

Adams, P. (1986) Mineral nutrition. In: Atherton, J.G. and Rudich, J. (eds) *The Tomato Crop*. Chapman & Hall, London, pp. 281–334.

Alvarado, A.D., Bradford, K.J. and Hewitt, J.D. (1986) Osmotic priming of tomato seeds: effects on germination, field emergence, seedling growth, and fruit yield. *Journal of the American Society for Horticultural Science* 112, 427–432.

Brown, J.E. and Splittstoesser, W.E. (1991) Plasticulture: disposal problems limit its commercial uses. In: Brown, J.E. (ed.) *Proceedings 23rd National Agricultural Plastics Congress*, Mobile, Alabama, pp. 35–38.

Clark, G.A. and Smajstrla, A.G. (1996) Design considerations for vegetable crop drip irrigated systems. *HortTechnology* 6, 155–159.

Coombe, B.G. (1976) The development of fleshy fruits. *Annual Review of Plant Physiology* 27, 507–528.

Csizinszky, A.A. (1993) Evaluation of controlled-release N and K fertilizers for bell pepper and fresh-market tomato on sandy soil. In: Hagin, J., Mortvedt, J.J. and Shaviv, A. (eds) *Proceedings Dahlia Greidinger Memorial International Workshop on Controlled/Slow Release Fertilizers*, Haifa, Israel, 7–12 March 1993, pp. 273–290.

Csizinszky, A.A. (1996) Foliar- and soil-applied biostimulants for fresh-market tomatoes in Florida. *Proceedings Interamerican Society for Tropical Horticulture* 40, 246–252.

Csizinszky, A.A. (1999) Yield response of polyethylene mulched tomato to potassium source and rate on sand. *Journal of Plant Nutrition* 22, 669–678.

Csizinszky, A.A. and Schuster, D.J. (1995) Color mulches influence yield and insect pest populations in tomatoes. *Journal of the American Society for Horticultural Science* 120, 778–784.

Decoteau, D.R., Kasperbauer, M.J. and Hunt, P.G. (1989) Mulch surface color affects yield of fresh-market tomatoes. *Journal of the American Society for Horticultural Science* 114, 216–219.

Dostal, H.C. and Wilcox, G.E. (1971) Chemical regulation of fruit ripening of field-grown tomatoes with (2-chloroethyl) posphonic acid. *Journal of the American Society for Horticultural Science* 96, 656–660.

Everett, P.H. (1976) Effect of nitrogen and potassium rates on fruit yield and size of mulch grown staked tomatoes. *Proceedings Florida State Horticultural Society* 89, 159–162.

Fery, R.L. and Janick, J. (1970) Response of tomato to population pressure. *Journal of the American Society for Horticultural Science* 95, 614–624.

Geisenberg, C. and Stewart, K. (1986) Field crop management. In: Atherton, J.G. and Rudich, J. (eds) *The Tomato Crop*. Chapman & Hall, London, pp. 511–557.

Geraldson, C.M., Overman, A.J. and Jones, J.P. (1965) Combination of high analysis fertilizers, plastic mulch and fumigation for tomato production in old agricultural lands. *Proceedings Florida Soil and Crop Science Society* 25, 18–24.

Gray, D. (1981) Fluid drilling of vegetable seeds. *Horticultural Reviews* 3, 1–27.

Hamdy, A.S., Abdel-Dayem, S. and Abu-Zeid, M. (1993) Saline water management for optimum crop production. *Agricultural Water Management* 24, 189–203.

Hartz, T.K. (1994) Minimizing environmental stress in field establishment of vegetable crops in the south western United States. *Horticultural Technology* 4, 29–33.

Hartz, T.K. and Hochmuth, G.J. (1996) Fertility management of drip irrigated vegetables. *HortTechnology* 6, 168–176.

Hartz, T.K. and Miyao, G. (1997) *Processing Tomato Production in California*. Publication 7228, University of California Division of Agriculture and Natural Resources, Davis, California.

Hausenbuiller, R.L. (1972) *Soil Science*. Wm C. Brown Co., Dubuque, Iowa, 504 pp.

Hochmuth, G.J. (1998) Soil and fertilizer management for vegetable production in Florida. In: Hochmuth, G.J. and Maynard, D.N. (eds) *Vegetable Production Guide for Florida*. SP 170, University of Florida, Gainesville, Florida.

Hoyle, B.J. (1983) Crust control aids seedling emergence. *California Agriculture* 37, 25–26.

Kemble, J.M., Davis, J.M., Gardner, R.G. and Sanders, D.C. (1994) Spacing, root cell volume, and age affect production and economics of compact-growth habit tomatoes. *HortScience* 29, 1460–1464.

Kring, J.B. (1972) Flight behavior of aphids. *Annual Review of Entomology* 17, 461–492.

Makmur, A., Gerloff, G.C. and Gabelman, W.H. (1978) Physiology and inheritance of efficiency in potassium utilization in tomatoes grown under potassium stress. *Journal of the American Society for Horticultural Science* 103, 545–549.

Olson, S.M. (1989) Pruning method effects on yield, fruit size, and percentage of marketable fruit of 'Sunny' and 'Solar Set' tomatoes. *Proceedings Florida State Horticultural Society* 102, 324–326.

Sanches, C.A. and Silvertooth, J.C. (1996) Managing saline and sodic soils for producing horticultural crops. *HortTechnology* 6, 99–107.

Shaviv, A., Zlotnikov, E. and Zaidel, E. (1993) Mechanisms of nutrient release from physically protected controlled-release fertilizers. In: Hagin, J., Mortvedt, J.J. and Shaviv, A. (eds) *Proceedings Dahlia Greidinger Memorial International Workshop on Controlled/Slow Release Fertilizers*, Haifa, Israel, 7–12 March 1993, pp. 120–132.

Shaw, L.N. (1997) Automatic transplanter for vegetables. *Proceedings Florida State Horticultural Society* 110, 262–263.

Smajstrla, A.G. (1998) Irrigation management. In: Hochmuth, G.J. and Maynard, D.N. (eds) *Vegetable Production Guide for Florida.* SP 170. University of Florida, Gainesville, Florida, pp. 31–34.

Stapleton, J.J. (1991) Behavior of sprayable polymer mulches under San Joaquin valley conditions: potential for soil solarization and soil sealing. In: Brown, J.E. (ed.) *Proceedings 23rd National Agricultural Plastics Congress*, Mobile, Alabama, pp. 254–259.

Vavrina, C.S., Olson, S.M., Gilreath, P.R. and Lamberts, M.L. (1996) Transplant depth influences tomato yield and maturity. *HortScience* 31, 190–192.

Weston, L.A. and Zandstra, B.H. (1989) Transplant age and N and P nutrition effects on growth and yield of tomatoes. *HortScience* 24, 88–90.

Wien, H.C. and Minotti, P.L. (1987) Growth, yield and nutrient uptake of transplanted fresh-market tomatoes as affected by plastic mulch and initial nitrogen rate. *Journal of the American Society for Horticultural Science* 112, 759–763.

Zahara, M. (1970) Influence of plant density on yield of process tomatoes for mechanical harvest. *Journal of the American Society for Horticultural Science* 95, 510–512.

GREENHOUSE TOMATO PRODUCTION

M.M. PEET AND G. WELLES[†]

IMPORTANCE OF THE INDUSTRY

Although definitive numbers are not available on the extent of greenhouse vegetable production worldwide, Table 9.1 provides recent estimates from a number of sources. Combining plastic and glass greenhouses and large and small plastic tunnels, protected cultivation covers 1,612,380 ha worldwide. The largest area of protected cultivation occurs in Asia, with China having almost 55% of the total world's plastic greenhouse acreage (including large plastic tunnels) and over 75% of the world's small plastic tunnels (Costa *et al.*, 2004). The next largest area is in Europe, with 23% of the total plastic greenhouse and large tunnel acreage, mostly in Italy and Spain. For (non-plastic) glasshouses, the largest concentration is in The Netherlands, which has more than a quarter of the total 39,430 ha under glass worldwide.

In comparing different areas and types of protected cultivation systems, climatic conditions (light intensity and temperatures), greenhouse construction and equipment, as well as technical expertise, differ considerably. This results in yield differences between regions when expressed on a per plant or per unit area basis. While it might be expected that regions with higher light would have higher production, the level of greenhouse technology used may be a more important factor. For example, average tomato yields in a high-light area (Almeria, Spain) are lower (28 kg/m^2) than in The Netherlands or Canada (60 kg/m^2) even though light intensity on a daily basis averages five times higher in Spain in the winter and 60% more on an annual basis (Costa and Heuvelink, 2000) compared with The Netherlands. Total production under protected cultivation is still much greater in Almeria than in The Netherlands, however, because the production area is much larger.

[†] Deceased before publication.

Table 9.1. Total protected cultivated area in the main horticultural countries (Costa *et al.*, 2004).

Location	Area (ha)		
	Greenhouses and large tunnels (plastic)	Small tunnels (plastic)	Glasshouses
Europe			
Italy	61,900	19,000	5,800
Spain	46,852	17,000	4,600
France	9,200	20,000	2,300
The Netherlands	400	–	10,500
UK	2,500	1,400	1,860
Greece	3,000	4,500	2,000
Portugal	1,177	450	–
Ex-Yugoslavia	5,040	–	–
Poland	2,031	–	1,662
Hungary	6,500	2,500	200
Total	160,000	90,000	–
Africa and Middle East			
Egypt	20,120	17,600	–
Turkey	17,510	26,780	4,682
Morocco	10,000	1,500	500
Israel	5,200	15,000	1,500
Total	55,000	112,000	–
Asia			
China	380,000	600,000	–
South Korea	43,900	–	–
Japan	51,042	53,600	2,476
Total	450,000	653,600	–
Americas			
USA	9,250	15,000	1,000
Canada	600	–	350
Colombia	4,500	–	–
Mexico	2,023	4,200	–
Equator	2,700	–	–
Total	22,350	30,000	–
WORLD TOTAL	687,350	885,600	–

COSTS OF PRODUCTION

Greenhouse production is more expensive than producing the same crop in the open field. The most important factors determining costs are depreciation of the structure and equipment, labour, energy and variable costs such as plant material, substrate and fertilizer. In British Columbia (BC), Canada, direct capital investment for a high-tech greenhouse operation in 2003, including utility hook-up, computerized environmental control system, heating and irrigation systems and basic site preparation, but not land costs, was US$1.8 million (CAN$2.5 million) (BC MAFF, 2003). For just the greenhouse structure and equipment, 1998 estimates for California (Hickman, 1998) were US$52/m^2. In 1993 (Jensen and Malter, 1995), the cost of a modern greenhouse, exclusive of land, was estimated at US$90–100/m^2 when the hydroponic plant growing system was included. This included frame, construction labour, heaters, fan cooling, irrigation, pump and well, electrical equipment and tools. In The Netherlands, costs of a modern greenhouse, exclusive of land but including total climate control, transport and fertilization, is about US$75/m^2 (Woerden and Bakker, 2000). This lower price per unit area in The Netherlands compared with Arizona is a consequence of the number and the high degree of specialization of Dutch greenhouse manufacturers.

Greenhouse vegetable production is very labour intensive, requiring 7–12 workers/ha in North America (Jensen and Malter, 1995) but only 5–8 workers/ha in The Netherlands (Woerden and Bakker, 2000) when transplants are purchased rather than grown. In BC, Canada (BC MAFF, 2003), the main operating costs are labour (25%), heating (28%) and marketing (25%), with larger units having 9–10% lower operating costs because of lower heating and labour costs and other economies of scale. Economic feasibility of mechanization generally increases with the size of the greenhouse. The estimated minimum economical commercial greenhouse area was estimated at 1.5 ha in The Netherlands (Woerden and Bakker, 2000). Table 9.2 details production costs for cluster and beefsteak tomato production in The Netherlands. Although production is highly efficient, with 45–71 kg of fruit produced per man-hour, labour still accounts for 37% of production costs. Marketing costs are relatively low in The Netherlands at US$0.21/kg, compared with US$0.32/kg in Canada and Spain (Boonekamp, 2003).

GREENHOUSE STRUCTURES

Tomatoes can be grown in every type of greenhouse, provided it is sufficiently high to manage and to train the plants vertically. High light transmission is very important and this varies between 70% and 81% in modern greenhouses. In many countries above 50°N latitude, Venlo-type glasshouses, consisting of a 1.5 ha block of 3.2 m spans, are used (Atherton and Rudich, 1986). Gutter height is 4–6 m to accommodate high wire planting systems, thermal screens and supplementary lighting. In other countries other greenhouse dimensions, structures and coverings may be used, as described below. For example, in China most of the greenhouse structures are unheated (Jensen, 2002).

Table 9.2. Cost of production, yield, price received, labour and natural gas consumption for two types of greenhouse tomatoes grown in The Netherlands (personal communication, N.S.P. de Groot, LEI, The Hague).

	Cluster	Beefsteak
Costs (US$/m^2)		
Labour	13.22	9.60
Depreciation and interest	4.62	4.18
Energy	6.31	5.94
Plant material	2.01	1.61
Other materials (excluding plants)	2.27	2.48
Delivery costs	1.11	2.81
Other costs	3.43	2.81
Total costs	32.98	29.43
Yield		
Fruit (kg/m^2)	46.7	53.4
Price received (US$/m^2)	31.94	30.00
Costs (US$/kg)		
Labour	0.26	0.18
Plants	0.04	0.03
Delivery	0.02	0.05
Total	0.70	0.54
Labour (h/m^2)	1.09	0.77
Fruit produced (kg/h labour)	45	71
Natural gas used (m^3/m^2)	60.8	56.9

Frame types and greenhouse orientation

Greenhouse frames are generally made of aluminium or galvanized steel, though the ends of double-poly houses may be wood-framed. The shape varies (Fig. 9.1) depending on: (i) expected snow load; (ii) use of natural ventilation; (iii) whether a number of houses are to be joined at the gutters; (iv) whether the covering is to be glass or plastic; (v) the growing system; and (vi) whether screens or artificial lighting are used. The straight sidewall greenhouse with arch roof is probably the most common shape, because it can be covered with double layers of plastic and connected to other houses at the gutter to create a large open growing area. Sidewall heights have been increasing and range from 3.5 m to 6 m in most new greenhouses. High sidewalls (Fig. 9.2) allow for a taller crop and for more climate control equipment (such as horizontal airflow fans, screens for shading or energy conservation, lights and heaters) to be installed above the crop. High sidewalls also increase the effectiveness of natural ventilation in open-roof systems. Space near the sides can be used more efficiently in straight sidewall than in Quonset-style structures. Gothic-arch frame structures (Fig. 9.1), which have a peak at the top but curving

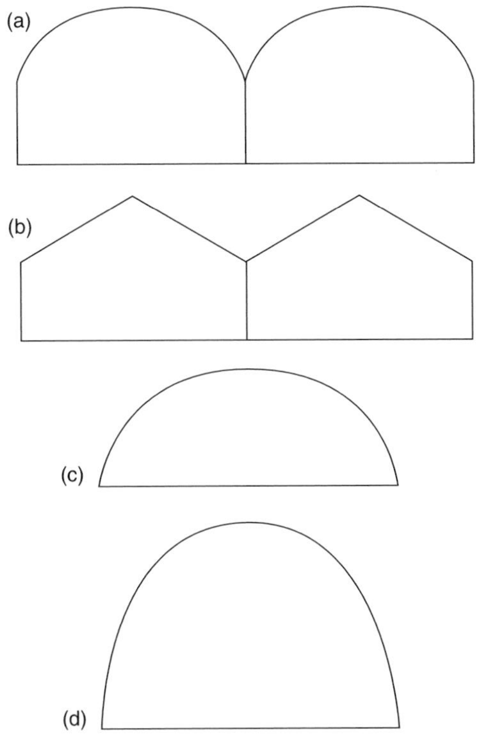

Fig. 9.1. Shapes of greenhouse frames: (a) gutter-connected straight sidewall with arch roof; (b) ridge-and-furrow style straight sidewall with gable roof; (c) hoop or quonset style; (d) Gothic-arch frame.

Fig. 9.2. Greenhouse in Leamington area of Canada showing high sidewalls, hanging gutters, leaning and lowering, vine clips and leaf pruning.

sides, also provide adequate sidewall height without loss of strength and can be free-standing or part of a range of multi-span, gutter-connected units. The advantage of Gothic-arch structures in double-polyethylene greenhouses is better runoff of condensation from the inner layer of plastic. The runoff can be channelled outside the greenhouse, reducing greenhouse humidity.

Greenhouses have traditionally been oriented north/south to optimize light, with cooling pads and sometimes insulation placed on the north side where they intercept less light. In large multi-bay greenhouses, which are almost square, orientation for light may be less critical than optimizing wind direction if the greenhouse is to be naturally ventilated (roof opening), so that the greenhouse is perpendicular to the direction of the prevailing winds during the hottest times of the production cycle. An additional consideration is that the distribution of shaded and non-shaded areas should be uniform over the course of the day. That is, all areas of the greenhouse should receive uniform illumination over a 24 h period in order for plant growth to be uniform throughout the greenhouse.

Coverings

There are three main types of greenhouse covering: glass, rigid plastics and polyethylene plastic film. Plastic film coverings can be either double or single. In cold climates, double layers are separated by a insulating layer of air, usually about 10 cm thick, to conserve energy. Traditionally, greenhouses have been made from glass; hence the use of the terms glasshouse and glazing (covering). Glass maximizes light transmission and requires only regular cleaning and sealing of the edges. Temperature extremes, dust or sand, ultraviolet (UV) radiation and air pollutants reduce life expectancy of all plastics, and polyethylene coverings are generally replaced every 2–4 years to maintain acceptable light transmission. Variations in plastic formulations and advanced extrusion technologies for polyethylene coverings make it possible to extend life and combine different types of plastic layers to modify thermal properties or to reduce condensate dripping. Newer plastics can reduce heat loss by 20% (Jensen, 2002). Some manufacturers also offer wavelength-selective plastics said to reduce disease or insect pressure or to control plant height growth. UV-blocking films developed in Israel are said to reduce populations of flying insects such as whiteflies, aphids and thrips (Jensen, 2002). However, issues such as pollination by bumble bees in greenhouses covered with wavelength-selective plastics have not been extensively evaluated. At this time, because of their higher cost wavelength-selective plastics are not widely used by growers.

Glass can be used in large panels (up to 1.8 m × 3.6 m), reducing structural shading (Giacomelli and Roberts, 1993). Glass is expensive compared with polyethylene plastic film, but is generally less expensive than rigid plastics with comparable properties. Tomato and cucumber yields in Ontario, Canada, were reported to be similar under all three types of coverings (Papadopoulos and Hao, 1997a,b). Presumably this was because light was less limiting than in earlier studies in The Netherlands that compared single and double glass greenhouses. In these studies, 1% light loss during the production stage resulted in about 1% loss of production (Van Winden *et al.*, 1984). Whatever the type of glazing, increasing light transmission in greenhouses is always a priority for manufacturers of greenhouses and greenhouse coverings.

Rigid plastics used in greenhouse construction include fibreglass-reinforced polyester, polycarbonate, acrylic (polymethylmethacrylate) and polyvinyl chloride (Giacomelli and Roberts, 1993). Some are energy efficient, have good light transmission in the first year of usage and last at least 10 years, but rigid plastics are more expensive than polyethylene films (Giacomelli and Roberts, 1993). Additional disadvantages are that acrylic and fibreglass panels deteriorate from dust more rapidly than glass and are fire hazards. Like glass panes, rigid plastics are strong and can be installed as large panels to reduce shading from support structures. Compared with glass, plastic panels shade the house less, because they are generally stronger and so require less support. Insulated rigid double-walled plastic panels are sometimes used to conserve energy, but they reduce the rate of snowmelt compared to glass or plastic film, so more snow accumulates, which reduces light and can potentially collapse the greenhouse. Double-layer polyethylene greenhouses are also energy efficient, but the double layers can be collapsed when snow accumulates to increase melting rate. Thus, in Canada, Mexico and the USA, it is more common for new greenhouses to be covered with a double layer of plastic film than with glass or rigid plastic panels. In north-west Europe use of glass is common because of the economic value of light transmission.

Double-poly houses often have quonset-style rounded roofs (Fig. 9.1), which contributes to condensate dripping on to leaves and makes it more difficult to design ridge openings for natural ventilation. However, roof opening designs (Fig. 9.3) are now available for plastic film greenhouses as well as acrylic (Giacomelli and Roberts, 1993).

Greenhouse installations

Modern glasshouses include automatic irrigation, ventilation and heating systems, and accommodate movable screens for shading or energy conservation. With flower crops, fogging systems and artificial lighting may be included and movable benches are often included for pot plants. For cooling, a new greenhouse type being tested in The Netherlands (ECOFYS, 2002; Armstrong, 2003) replaces ventilation with an aquifer-based cooling system. Solar energy is ‘harvested’ in summer and stored in the aquifer to be used in winter. With this method, energy savings of 30%, production

Fig. 9.3. Plastic greenhouse in Quebec, Canada, with roof opening for natural ventilation. Note also tomato hooks used to provide more twine during the leaning and lowering process.

increases of 20% in tomatoes and 80% reductions in pesticides can be realized, compared with most vented greenhouses. The greatest potential of these greenhouses is the ability to control humidity and CO_2 concentrations independently of temperature throughout the year. High CO_2 concentrations in summer increase yields by at least 20% (Nederhoff, 1994). However, the system costs 50% more than a traditional greenhouse and has an 8-year payback period, even taking into account higher yield and greater efficiency.

CROPPING SCHEDULES

The tomato plant is a short-lived perennial and can be maintained for periods of a year or more in favourable environments. However, most production schedules allow at least a month between crops for clean-up and pest control. The time chosen to be out of production is usually based on unfavourable prices or environmental conditions. By seeding in late summer or autumn and carrying the crop until early summer of the next year, growers in southern latitudes avoid the high costs of summertime cooling, poor fruit set and quality, pest build-up and competition from field tomatoes. In northern areas and for large commercial operations, greenhouses produce almost year-round in order to lower costs per kilogram of produce and to avoid the problem of buyers switching to alternative sources in southern countries such as Spain. In some cases a second crop is interplanted (intercropped) (Fig. 9.4) within the existing crop to ensure minimal interruption in supply during summer. There are some

Fig. 9.4. Willcox Arizona greenhouse showing new intercropped cluster tomato plants with older vines on either side. Note severe leaf pruning on older vines with only four remaining leaves, in order to provide additional light to the new crop. Vines of the older crops are turned at the walkway by wire supports. A retractable screen is used to shade the walkway, and others can be used to shade the crop if temperatures are excessive. Note also strips of yellow sticky tape over the crop to trap whiteflies.

examples where artificial lighting has been used successfully in The Netherlands (Marcelis *et al.*, 2002) but the overall economics of artificial lighting have not been established. In the UK (Ho, 2004), lighting is not currently considered cost-effective, but this situation may be changed by the introduction of new combined heat and power (CHP) units.

A few growers, especially in the south-eastern USA and in some cases also in The Netherlands (approximately 100 ha), grow separate autumn and spring crops, leaving short production breaks both midwinter and midsummer. Where both autumn and spring crops are grown in the south-eastern USA, a separate transplant production house reduces carryover of pest and diseases. Additional advantages of a transplant house in those areas are the ability to maintain different temperatures and to add supplemental lighting or CO_2 enrichment. In The Netherlands, such transplant houses are not economical and all growers buy high quality plants at specialized plant nurseries.

There is considerable interest in organic greenhouse tomato production, but at this point, it is only a very minor segment of the industry. With organic systems, crop production must be certified by international organizations that regulate the types of material that can be used. Biological factors, such as soil condition and fertility and the use of beneficial insects, are the main factors used to assure a vital, healthy crop and good fruit quality. Use of high-analysis chemical fertilizers and most chemical pesticides is prohibited. The total area of organic production in the main tomato production areas in The Netherlands is < 1%. Because of the restrictions and the necessity in the European Community to grow the crop in soil, instead of a soilless substrate such as rockwool, yields are 20% lower than normal. Consequently market prices have to be at least 20% higher than for conventionally grown produce (Anonymous, 2003; Welles, 2003). In the USA, the National Organic Program does not preclude soilless or hydroponic production, but rockwool is not allowed. All guidelines relative to organic fertilizers and pesticides must still be followed and the choice of cost-effective soluble organic fertilizers is very limited.

TRANSPLANT PRODUCTION

Transplant quality is defined as a plant free from pests and diseases, quickly grown with no suppression of yield due to poor quality roots. Transplant production requires 3–6 weeks, depending on temperature and light conditions. Tomato seed germinates best at 25°C, while seedling growth is optimal at 18°C night-time minimum and 27°C daily maximum. Germination rates are at least 80% and so only one seed needs to be planted per container. The ideal transplant size is 15–16 cm tall with a weight of 100 g (not including roots). A good transplant is one that is as wide as it is tall and is not yet flowering. A larger transplant means greater height, fresh and dry weight, and earlier yield so growers try to use a transplant as large as it is possible to handle (Atherton and Rudich, 1986) except in winter, when large plants may flower too early. Supplemental light (15,000 lx) and CO_2 enrichment (800–1000 ppm) during transplant production increase plant quality by increasing plant growth rates.

Seedlings for perlite or peat bag culture are generally started in small pots filled with the same medium, which are then planted directly into the container. Fertilization practices are similar to those of the production phase. Seedlings for rockwool systems are generally started in a sterile inert medium, such as rockwool plugs, and then moved into progressively larger volumes of media. Plugs should be presoaked with electrical conductivity (EC) 0.5 dS/m nutrient solution before seeding, then re-wetted the day after seeding and 4 days later. After emergence, EC can be increased to 1.0–1.5 dS/m. At emergence of the first true leaves, seedlings can be transplanted into 75–100 mm rockwool blocks (Fig. 9.5). If seedlings are somewhat 'leggy', with long stems, they can be transferred into blocks with their stems bent 180 degrees, so that the original cube is upside-down inside the larger block and the main stem forms a 'U' shape, emerging vertically upwards from the block. Adventitious roots grow readily from the bent stems. If stems are to be inverted, water should be withheld for 24 h prior to transplanting to avoid stem cracking.

Fig. 9.5. Rockwool plug with small seedling on right. This will be inserted into the rockwool cube on the left.

During the transplanting stage, plant density should be 20–22 plants/m^2. A rule of thumb is that transplant leaves should not touch. EC in the blocks should be raised to 3.0 dS/m before the final transplanting to rockwool slabs (Fig. 9.6). These slabs should have been leached and moistened according to the manufacturer's instructions and warmed to greenhouse air temperature before planting. Plants should be irrigated with nutrient solution immediately after transplanting. Table 9.3 summarizes temperature, EC, pH and irrigation recommendations from germination through harvest.

Table 9.3. Growing recommendations for tomato cropping (adapted from OMAFRA, 2001).

	Germination	Plant raising	Transplanting	Harvesting	Full harvest
Temperature (°C)					
Day	25	19–21	24	19	20–22
Night	25	19–21	24	19	17–19
EC (dS/m)	0.0–0.1	2.5–3.0	2.5–3.0	2.7–3.5	2.7–4.0
pH	5.8	5.8	5.8	5.8	5.8
Volume of feed (l/day)	–	0.2–0.3	0.2–0.3	0.5–1.5	0.5–2.5

Fig. 9.6. Rockwool cubes containing two seedlings each are placed on rockwool slabs in a greenhouse in Leamington, Ontario, Canada. Emitters for irrigation are placed both in the slab and in the cube in this system. The crop is a cluster tomato (tomatoes on the vine, TOV) with some kinking of the peduncle. Hot-water heating pipes on the floor in the background can be used as a rail for equipment.

PLANT SPACING AND EXTRA STEMS

Generally, tomatoes are set out in double rows, normally around 0.5 m apart with 1.1 m access pathways between the double rows. In a typical 3.2 m Venlo house span, there are four rows of plants and two pathways. Plant populations are adjusted at the start of the crop by altering the in-row spacing and later in the season by allowing extra heads (side shoots) to develop. At the end of the season, plant populations can be increased by intercropping (see Fig. 9.4). Atherton and Rudich (1986) gave detailed information on the relationship between plant density and yield per plant and the consequences of spacing for mean fruit weight and harvest costs. In general, under north European conditions a plant density of 2.5 plants/m^2 has been found to give the best financial margin. In more southern latitudes, a higher plant density (3.3–3.6 plants/m^2) may be used, because of higher light intensity. Similarly, the number of plant stems or side shoots allowed to develop should be based on light intensity. This ensures not only a high yield but also optimal quality, including taste. Uniformity of fruit size is also improved when the number of side shoots is matched to incident light (Ho, 2004). For example, Canada has about two times higher radiation in winter and 40% higher radiation in spring compared with The Netherlands, and so optimal plant spacing in mid December is 50 cm rather than 55 cm and extra stems are left starting in week 5 rather than week 9. For spring, recommended spacings in Canada are 40 cm in the row, compared with 44 cm in The Netherlands.

CULTIVARS

The large red multi-locular 'beefsteak' type of tomato with an indeterminate growth habit is the industry standard in North America. 'Trust' has been the most popular red multi-locular type in North America, reportedly planted on 80% of the beefsteak tomato acreage, but 'Quest' and 'Rapsodie' are becoming more widely grown. 'Rapsodie' is also the most widely grown beefsteak tomato in Europe.

Smaller two- to three-locule round fruits (47–57 mm) or cherry tomatoes (< 15 mm) are the most common types grown in Europe, and a few areas also produce some 'pink' or yellow tomatoes. 'Eclipse', 'Prospero' and 'Aromato' are widely grown cultivars in Europe. Recently, there has been increased production of the smaller cluster tomato types, also called truss tomatoes or on-the-vine (TOV) tomatoes (Figs 9.4 and 9.6), in North America. In BC, Canada, for example, beefsteak tomatoes represent 24% of all greenhouse vegetables sold, and cluster tomatoes 34% (BC MAFF, 2003). Cluster tomatoes can be sold loose, in plastic clamshells, in single-layer boxes or in net bags, but usually still have the vine attached. Yields of cluster and cherry types may be lower than those of beefsteak types, but this is not always the case (e.g Hochmuth *et al.*, 1997, 2002). Quality attributes for cluster types include uniformity of fruit size within the cluster, maintaining a fresh green calyx and vine after harvest, simultaneous ripening of all the fruit on the cluster and fruit staying on the vine after harvest. In Arizona and Canada, 'Campari' is widely grown in large greenhouse operations, but seed availability is limited. In BC, recommended cluster types included 'Jamaica', 'Aranca', 'Tradiro' and 'Vitador'. In The Netherlands, standard cluster tomatoes varieties are 'Clotilde', 'Aranca' and 'Cedrico'.

Greenhouse tomato seeds are relatively expensive (US$0.20–0.25 or more) compared with seeds of open-field cultivars. However, cultivars designed for outdoor production do not do well in greenhouses. Their determinate plant growth habit makes them hard to maintain over extended periods and they require higher light and lower humidity than greenhouse cultivars. Most greenhouse cultivars have a number of disease resistances. For production in the soil, or where cultivars lack resistances or sufficient vigour, greenhouse cultivars can be grafted on rootstocks. This practice is discussed further in the section on root grafting, below.

CROP MANAGEMENT

Training systems

In the 1970s and early 1980s, plants were trained up to the wire and then allowed to drape down the other side (the up-and-down system). The negative effects on yield of plant heads hanging down in the shade is now understood and the most widespread training system in The Netherlands and the northern USA at the present time is the high-wire system (Figs 9.2, 9.4 and 9.6), which allows one crop to be carried over several seasons. In this system, the growing tip remains at the top of the canopy, but the stem is lowered and trails along the base of the plants. This system combines the yield-improving advantages of maximum light interception by young leaves with increased labour efficiency resulting from easier removal of leaves and fruit. However, it requires a high enough greenhouse structure to accommodate the height of the horizontal wires used in training the plants and any screening materials used for shading or energy conservation. Foggers and CO_2 injection equipment are also sometimes installed above the crop.

The high-wire system requires early training of the main stem. As soon after transplanting as possible, plant stems should be secured to plastic twine hung from horizontal wires that run at a height of 3.2 m above the ground. The end of the twine is attached to the base of the stem with a non-slip loop. The twine is then wound around the stem in two or three spirals for each truss (Figs 9.4 and 9.6). The length of the supporting twine should allow an extra 10–15 m to unwind. Usually this extra twine is held in a winding hook placed near the wire (Fig. 9.3). As an alternative to winding twine around the stem, the stem can be clipped to the twine every 18 cm (Fig. 9.2). Clips can be sterilized and reused, but twine should be discarded after each crop. If the vines are to be 'leaned and lowered' (see below), it is useful to wind twine around the lower stem at least, as the twine provides a better support than clips, and it is also useful to start out with the stems angled in the direction in which they are to lean in the row.

The objective of 'leaning and lowering' is to keep the head of the plant upright for pollination and light interception and still have the clusters at a convenient height for workers even when crops are in the greenhouse for long periods (Fig. 9.2). When plants are near the overhead wire, the twines are unwound from the twine hook hangers and the twine and plant are both moved sideways down the horizontal wire. This process is called 'lowering' and is a delicate operation in order to avoid breaking the stems. It should be performed every 7–10 days. Flowering of the fourth cluster is a good developmental stage to start leaning and lowering, as the stem is relatively vigorous and should resist breakage. In some greenhouses, especially those using upright bags, the vines rest on special holders designed to give additional support. At the end of the double row, the vines are wound around a corner and back down the next row. Upright rods or wire supports are placed at the corners to turn the vines (Fig. 9.4) and this is another spot where vines frequently break. Black plastic drain tubes or other types of 'bumpers' are sometimes placed on the rods to protect the vine as it is turned. Vine breaks can sometimes be successfully mended with duct tape.

Intercropping or interplanting (Fig. 9.4) is a variation of the high-wire system that minimizes down time between crops and allows production of two crops per year. One row of old plants is removed in preparation for the new crop. Leaves of plants in the remaining row are removed except for the top four leaves. The height of the leafy part of the old canopy is adjusted to increase light penetration to the new plants. Young plants are then placed next to the existing plants in the row for the last month or so of a 6-month cropping cycle. The combination of intercropping and hanging gutters offers high production and year-round cropping. Disadvantages of intercropping are that greenhouse clean-up is more difficult and the foliage of the new transplants is high in nitrogen, stimulating whitefly feeding. Diseases may also be carried over, and if ethylene is used to promote ripening of the old crop of cluster tomatoes, transplants may be adversely affected.

Side-shooting and trimming

All greenhouse cultivars have an indeterminate growth habit, but if vines are not pruned, side shoots will develop between each compound leaf and the stem. These side shoots should be removed weekly, leaving only one main stem as a growing point (Figs 9.2 and 9.6). Workers must be careful not to remove the main stem accidentally, rather than the side shoot. If this happens, a side shoot can be left to form a new main stem, but yield is reduced and harvest delayed. For this reason, side shoots are usually not pruned until they are a few inches long, at which time they are easier to distinguish from the main stem. Knife pruning weekly will reduce the size of pruning scars and thus the risk of botrytis.

As is indicated in the section on plant spacing and extra stems (p. 268), extra side shoots may be maintained when light intensity is high compared with available leaf area. Sometimes an extra head is left (twin-heading, side-shoot taking) when a gap is left in the row by removal of a neighbouring plant. In The Netherlands, management of side shoots is an important tool for optimizing the fruit load of the crop and hence yield (De Koning, 1994). Leutscher *et al.* (1996) presented an economic evaluation of the optimal number of additional side shoots, based on a modelling approach.

In the UK (Ho, 2004), uniform fruit size is maintained by increasing the number of fruit left on the truss and letting a side shoot develop as light increases. During the winter and spring season in the UK, more than 50% of the fruit falls into the 40–47 mm class and only 35% into the preferred 47–57 mm class. However, during the summer, too many fruits fall into the less desirable 57–67 mm size. To address this problem in the UK (Cockshull and Ho, 1995), crop densities are adjusted upwards from a relatively low density of 8000 plants/acre (20,000 plants/ha) in the winter, which gives > 70% fruit in the 47–57 mm class. The crop density is then increased during the summer to 12,000 plants/acre (30,000 plants/ha) by letting a side shoot develop on every other plant, resulting in 80% of fruit falling within the 47–57 mm class for the summer crop as well.

Trimming of the stems is done more or less weekly, depending on growth rate of the crop. With the high-wire system, side-shooting and other operations may be done standing on an electrical lift (Fig. 9.7).

Pollination

Before the early 1990s, each flower cluster had to be vibrated with an electric pollinator at least three times weekly to release pollen. Poor pollination results in flower abortion and/or small, puffy or misshapen fruit. It is particularly important to get good fruit set on the first three clusters to establish an early pattern of generative growth. Pollination should take place at midday, when humidity conditions are most favourable (50–70%). If humidity is too high in winter, temperatures can be raised by 2°C at midday to reduce humidity, but in summer too high a mean daily temperature reduces pollen development and release (Sato *et al.*, 2000). Temperatures that are too low (night temperatures below 16°C) (Portree, 1996) have the same effect. Compensation with high day temperatures is possible.

Commercially, bumble bees are now used for pollination. Generally, one worker bee can service 40–75 m^2 (Portree, 1996) and so 5–7.5 hives/ha are required. As well as saving labour, increases in yield and quality have been reported (Portree, 1996) compared with manual vibration. Hives are placed on stands 1.5 m above the ground along the centre aisle (Fig. 9.8) and protected from ants with sticky bands or water troughs. Hives should be shaded by foliage or covers, and marked distinctively above the hive and around the entrance so that bees can return to the correct hive (Portree, 1996). Bees are docile, unless the hive is disturbed or an individual is squeezed, but it is still a good idea to maintain first aid supplies on site.

Some additional management of the bees throughout the season is also required. Bees harvest pollen from the tomato flowers to maintain their young, but early in the season a small amount of pollen by the exit hole helps to establish the hive. Tomato flowers do not produce a source of carbohydrates and so the grower must supply a sugar-water solution. Usually there is an indicator so that the solution can be replaced as it is used. Solution levels should be monitored daily and the solution should be replaced if it becomes cloudy from contamination. No broad-spectrum insecticides or those with residual action should

Fig. 9.7. Greenhouse worker in British Columbia, Canada, standing on an electrical lift to sucker and train plants. The lift moves down the row on the rails from the hot-water pipes. Note also the hanging gutters. Medium consists of sawdust in most BC greenhouses.

be used once a hive is in place. All pesticides should be checked for effects on bees and, if compatible, application should be done at night with the hive closed and covered. Within 2 months or less, most hives will need to be replaced. Some pesticides may be used if the hives remain closed for 3 days after treatment. The health of the hive can be monitored by observing activity (Fig. 9.9) and looking for brown bruise marks on the anther cone as evidence of flower visitation. At least 75% of withered flowers should have evidence of bee visits.

Fig. 9.8. Beehives are placed on stands in an Arizona greenhouse. Note slide on the top front of the box, which can be used to open or close off access to the hive.

Fig. 9.9. Bumble bee pollinating a tomato flower.

De-leafing

When vines are lowered, leaves are removed to prevent disease development (see Fig. 9.2). To avoid introducing *Botrytis*, leaves should be cut with a knife or pruned flush to the stem (Fig. 9.6). The amount of de-leafing that occurs higher up the plant varies. Typically 18 compound leaves are left on 'Trust' but only 14 on 'Quest', which is a more vigorously growing cultivar. A vigorously growing plant will produce 0.8–1 truss and three leaves per week. When total leaf numbers reach the maximum desired, from that point on the bottom two to three leaves are removed each week, depending on the average 24 h temperature. If the plant is still too vigorous, the middle one of the three leaves between each cluster can also be removed. Also the leaf under the truss can be removed at the same time as the sucker. Pruning may be less severe during the final months of a crop, leaving 18–21 leaves. The purpose of de-leafing higher up the plant stem is to increase light penetration and air circulation. Typically, all leaves are removed below the bottom fruit cluster, but de-leafing may be more severe when a new crop is intercropped next to the old (Fig. 9.4). Effects of de-leafing on light interception and yield are discussed in Chapter 4.

Another factor to be considered in de-leafing is the effect on diseases, pests and beneficials. Removing lower leaves from the greenhouse and then destroying them will remove whitefly immatures developing on lower leaves. However, if beneficials have been introduced, they will have parasitized the immatures, and removing and destroying leaves will also prevent the beneficials from emerging. If parasitized pupae are known to be present, leaves are sometimes removed but left piled in the greenhouse for a few days to allow emergence. In this case, de-leafing and leaf removal represent a trade-off between emergence of whiteflies, emergence of beneficials and spread of disease from the discarded leaves. If botrytis is present, however, leaves should be removed from the greenhouse immediately after pruning.

Fruit pruning and development

The purpose of fruit pruning is to increase fruit size and fruit quality and to balance fruit load. Pruning can also be used to maintain uniform fruit size. Misshapen fruits and small, undersized fruits at the end of a cluster are always removed, as these will generally not grow to marketable size and are thought to reduce the size of other fruits on the cluster. In some cases, all clusters are pruned to leave only the four fruits nearest the plant (proximal fruit).

Whether or not clusters are pruned depends on the expected fruit size for that cultivar, how many fruits normally form on the cluster, growing conditions and the size demanded by the market. A typical benchmark for beefsteak tomatoes is to have no more than 18 fruits present on the plant at any one time. Yield prediction may be achieved with the help of an expert system, such as that used commercially in The Netherlands (http://www.LetsGrow.com). During periods of good fruit set, flowers can be removed before setting (within 3–6 days of opening). If fruit set is poor, only misshapen fruits are removed. The recommended fruit pruning practice for large-fruited cultivars such as 'Trust' is to prune the first three trusses to a total of eight to nine fruits and subsequently leave four fruits per truss. If fruit loads are low, this can be increased to five fruits per truss, or, if expected fruit size is low, decreased to three per truss. With cluster tomatoes it is most important that fruits develop regularly in each cluster. With some cultivars, up to eight to ten fruits can be allowed to develop on each cluster. Effect of fruit pruning on fruit size is also discussed in Chapter 4.

When some greenhouse tomato cultivars are grown under relatively low light conditions, the peduncles of the inflorescences (trusses) are too weak to support the weight of fruit they bear and are liable to bend (Horridge and Cockshull, 1998) or 'kink'. Another reason sometimes given for kinking is too high a temperature during the vegetative phase, which causes the truss to become almost vertical ('stick trusses'). As fruit develop on these trusses, they may become kinked (Fig. 9.6). Truss hooks suspended from the tomato stem prevent heavy trusses from pulling off the vine and keep the cluster from bending sharply under the weight of the fruit. Truss support devices, which also include peduncle clamps, are thought to prevent a reduction in fruit size on kinked trusses. There is some evidence for this in the scientific literature, though results are not conclusive (Horridge and Cockshull, 1998). The standard practice to prevent kinking is to apply truss braces for the first eight to ten trusses. Truss braces are applied to the cluster before fruit development, when the stem is still flexible. Some growers rub the underside of the truss with a roughened piece of PVC piping to create a scar, but this method requires experienced labour and heavy application to reduce truss kinking.

Topping plants at the end of the crop

The growing point is removed 5–8 weeks before the anticipated crop termination date. A week later, all remaining flowers are removed. An individual fruit requires 6–9 weeks from anthesis to harvest, and so flowers or small fruit present after topping will not have enough time to develop to maturity. It may be helpful in summer to leave some shoots or leaves at the top of the plant to shade the fruit and prevent sunscald. Leaving shoots at the top (or not topping at all) is also thought by some growers to provide shade to top fruit and increase transpiration, thereby reducing risks of fruit cracking and russeting. For further discussion of factors contributing to cracking and russeting, see Chapter 6. With the high-wire system used in northern Europe and in the USA, topping during the growing season is practised infrequently and plant stems continue to grow from December of one year until November of the following year.

SUBSTRATES AND SUBSTRATE SYSTEMS

Except for organic growers, there is relatively little commercial tomato production directly in the soil. In Europe and Canada, and in large greenhouse complexes in the USA, 95% of greenhouse tomatoes are grown on inert artificial substrates, a system usually referred to as soilless culture. The term 'hydroponic' can refer to soilless culture or to systems such as nutrient film technique (NFT), in which no solid substrate is used and water flows almost constantly down troughs holding plant roots.

Rockwool is the most common substrate for soilless culture (Sonneveld, 1988). Manufactured by heating basaltic rock, rockwool is usually provided as plastic-wrapped slabs of spun wool. A distinctive characteristic of rockwool is its high air-holding capacity even when fully saturated. Slabs are available as high or normal density. High-density slabs have good structural stability, high water-holding capacity and good capillarity. Normal-density slabs are less compact and have a slightly lower water-holding capacity. They are still structurally stable and have good capillarity (ability of water to rise to the surface through channels in the rockwool, thus becoming available to the plant). If the slabs are to be reused after pasteurization, higher-density slabs should be chosen.

For tomato crops, rockwool of density 10–12 l/m^2 is recommended (Sonneveld and Welles, 1984). Sizes up to 16 l/m^2 and down to 5 l/m^2 can be used, but higher volume increases cost, while lower volume leaves little buffer for errors in irrigation. The size selected will depend on the amount of water to be applied to the crop, the plant density and the row centre spacing. Typically, either $90 \times 15 \times 7.5$ cm slabs hold two plants, or $120 \times 15 \times 7.5$ cm slabs hold three plants. The greenhouse floor should be covered with white polythene to suppress weeds and increase light to the crop. If the greenhouse floor is not heated, rockwool slabs may be placed on polystyrene for insulation. In closed systems, return gutters are placed under the slabs to recapture excess water (overdrain). The slabs should either be placed flat or with a 2% slope towards the drainage ditch. One advantage of the hanging gutter system (see Figs 9.2, 9.6 and 9.7) is that it is possible to control the slope more accurately than when bags are placed on the floor, and thus irrigation can be more uniform. In addition, gutters can be moved so that plants are at a convenient height for workers. Finally, hanging gutters allow installation of hanging tubes for cooling and control of humidity in closed greenhouses, as discussed in the section on greenhouse installations (above).

Although rockwool is the most widely used substrate in soilless culture, perlite, peat and to some extent also pumice (rock and limestone) are also used. Perlite is a volcanic glass formed when lava cools very rapidly, trapping small quantities of water. The glass is crushed and heated, vaporizing the trapped gas, which expands the material into foam-like pellets. Initial pH of perlite is near neutral. Typically 0.03 m^3 of medium to coarse perlite is sealed into opaque white-on-black polyethylene bags treated with enough UV inhibitors to last 2 years. Bags are typically 1.1 m long by 0.2 m wide and contain three plants. Some growers buy perlite in bulk and fill bags themselves to cut costs. Drainage slits are cut about 25 mm from the bottom of the bag to provide a shallow reservoir for nutrient solution. As with rockwool slabs, water rises from the reservoir through capillary action to replace that lost by plant transpiration, maintaining a constant moisture profile as long as the reservoir is maintained. Perlite can also be placed in buckets (Fig. 9.10).

Rockwool and perlite have many similar advantages: (i) excellent aeration and water-holding capacity; (ii) sterile and lightweight when dry; (iii) easily installed and cleaned up; and (iv) both types of medium can be unwrapped, steam sterilized, rebagged and used again once or twice. Successful reuse of the medium without sterilization has also been reported. Yields in Florida were comparable for new rockwool (two brands), 1- and 2-year-old rockwool, upright peat bags and lay-flat peat bags (Hochmuth *et al.*, 1991), as has also been seen in numerous experiments in The Netherlands (Sonneveld and Welles, 1984). There may be increased commercial interest in perlite, pumice or other substrates in the near future, since disposal of rockwool is difficult, reuse is limited and some consider peat to be a non-renewable resource. In a trial of growing media at the University of Arizona (Jensen, 2002), there were no

Fig. 9.10. Bato® Buckets containing perlite in a North Carolina greenhouse. Note white PVC pipe on the floor to collect drainage from the buckets, to be recirculated with the addition of some fresh water and nutrients as needed. Black plastic clips have been used to attach the vine to the twine. Spools near the greenhouse roof allow the vines to be lowered. A horizontal airflow fan for air circulation and yellow sticky cards to monitor whiteflies can also be seen above the plants.

significant differences in yield of greenhouse tomatoes between five different media (coconut coir, perlite, peat–vermiculite mixes, coir/perlite and rockwool).

NUTRITION AND IRRIGATION

Nutrition

Symptoms of nutrient deficiency and excess, pH, EC and ion ratios are discussed individually in Chapter 6. Tables 9.4 and 9.5 provide fertilizer, pH and EC recommendations for tomato production in rockwool and other soilless systems in the south-eastern USA and in Canada. Compared to the Canadian recommendations (Table 9.5), Florida recommendations (Table 9.4) are lower and increase more gradually. This is based on findings by Florida researchers (Hochmuth and Hochmuth, 1995) that higher fertility levels result in excessive vegetative growth (bullish plants) under high light and temperatures. Voogt (1993) discussed nutrient uptake of tomato crops in The Netherlands. In systems with drip irrigation, nutrients are usually injected into the irrigation water (fertigation) from concentrated solutions in stock tanks. The fertilizers must be separated into at least two tanks (Fig. 9.11) to avoid precipitation of calcium phosphate and calcium sulphate. Some greenhouses have duplicate

Table 9.4. Final delivered nutrient solution concentration (ppm) and EC recommendations for tomatoes grown in Florida in rockwool, perlite or nutrient film technique (Hochmuth and Hochmuth, 1995). Numbers in bold denote changes from previous stage.

	Stage of growth				
Nutrient	Transplant to first cluster	First cluster to second cluster	Second cluster to third cluster	Third cluster to fifth cluster	Fifth cluster to termination
N	70	**80**	**100**	**120**	**150**
P	50	50	50	50	50
K	120	120	**150**	150	**200**
Ca[a]	150	150	150	150	150
Mg	40	40	40	**50**	50
S[a]	50	50	50	**60**	60
Fe	2.8	2.8	2.8	2.8	2.8
Cu	0.2	0.2	0.2	0.2	0.2
Mn	0.8	0.8	0.8	0.8	0.8
Zn	0.3	0.3	0.3	0.3	0.3
B	0.7	0.7	0.7	0.7	0.7
Mo	0.05	0.05	0.05	0.05	0.05
EC (dS/m)	0.7	**0.9**	**1.3**	**1.5**	**1.8**

[a]Ca and S concentrations may vary depending on Ca and Mg concentrations in well water and amount of sulphuric acid used for acidification.

Table 9.5. Final delivered nutrient solution concentrations (ppm) recommended for greenhouse tomato production in rockwool in Ontario, Canada (OMFRA, 2001).

	Stage of growth			
Nutrient	Saturation of slabs	For 4–6 weeks after planting	Normal feed	Heavy fruit load
N	200	180	190	210
NH_4	10	10	22	22
P	50	50	50	50
K	353	400	400	420
Ca[a]	247	190	190	190
Mg	75	75	65	75
S[a]	120	120	120	120
Fe	0.8	0.8	0.8	0.8
Cu	0.05	0.05	0.05	0.05
Mn	0.55	0.55	0.55	0.55
Zn	0.33	0.33	0.33	0.33
B	0.5	0.5	0.5	0.5
Mo	0.05	0.05	0.05	0.05
Cl	18	18	18	18
HCO_3	25	25	25	25

[a] Ca and S concentrations may vary depending on Ca and Mg concentrations in well water and amount of sulphuric acid used for acidification.

Fig. 9.11. Injectors used to control two nutrient solutions (A and B tanks) and pH in a North Carolina greenhouse.

sets of stock tanks, so that irrigation will not be interrupted while solutions are remade. A third tank may be added for pH correction, and some large commercial greenhouses inject out of six or more separate tanks to give better control of the nutrient solution.

Controlling growth

As well as avoiding nutrient deficiency or excess, it is also important to control the balance between vegetative and reproductive (or generative) growth in the tomato crop. A well-balanced plant (OMAFRA, 2001) has a thick stem, dark green leaves and large, closely spaced flower clusters that set well. Specifically, the stem should be 1 cm thick 15 cm below the growing tip. Thicker stems indicate excessive vegetative growth and are usually associated with poor fruit set and low productivity. Thinner stems usually indicate carbohydrate starvation, slow growth and, ultimately, low overall productivity. There are a number of ways to control this balance, including the environmental controls summarized in Table 9.6. EC, water supply and the ratio of nitrogen to potassium in the feed also influence plant balance. EC influences plant growth through its effect on plant water relations (Heuvelink *et al.*, 2003). High salinity in the root environment, infrequent irrigation and low volume of irrigation water reduce water availability to plant roots, thereby decreasing water uptake and overall growth rate, and steering the plant towards generative growth. High temperature and low relative humidity also have a generative effect, because they make water less available, resulting in 'hard' plants and slow growth. Lowering nitrogen or maintaining a high potassium/nitrogen ratio in the fertilizer feed is another technique to reduce the rate of growth and steer plants towards generative development (OMAFRA, 2001).

Recirculating systems

There has been increasing interest over the past decade in nutrient-recycling systems with provision for disinfection of the water, and/or replenishment of nutrients before reuse. Recirculation can decrease fertilizer costs by 30–40% and water usage by 50–60% (Portree, 1996). In an open (non-recirculating) system, in order to compensate for variations in drippers, 20–50% excess irrigation is applied and plants draw from a small reservoir in the individual bag or slab. The main problem with open systems is that in areas of intensive production there may be significant discharge of nutrients into the environment. Provisions for reuse or at least recapture of greenhouse runoff should be designed into new greenhouses, as they are already required in many countries and recapture systems are not easy to retrofit.

Table 9.6. Regulating plant growth by adjusting environment and nutrition. Adapted from *Reading the Plant* (Portree, 1996).

Plant part	Observation	Recommendation
Plant head	Thick head	Too vegetative. Increase day temperature 1–2°C, especially during peak light period; increase spread between day/night temperature settings by 1–8°C (the bigger the difference, the stronger the 'generative' signal to the plant)
	Thin head	Too generative. Bring day/night temperatures closer together. Reduce the 24 h average in low radiation situation, e.g. early spring, late autumn. Target 10–12 mm diameter head measured approximately 15 cm from growing tip or at 1st fully expanded leaf before the flowering truss
	Head is 'tight' – leaves do not unfold until late in the day	Vegetative imbalance. Increase 24 h average by increasing temperature between midnight and sunrise. Curl should be out between 11 a.m. and 4 p.m. Target slightly higher temperature in afternoon (+1 to 21°C)
	Heads are purple	Vegetative imbalance. Slight purpling acceptable. Increase night temperature
	Grey head	High tissue temperatures in combination with high CO_2 levels or high temperature and low light. Can be observed in early spring when venting is limited. Reduce CO_2 levels and shut off CO_2 earlier in day
	Chlorosis in head	Chlorosis in head can occur if media water/air ratio not in balance. If slab is dry, increase EC. If slab is wet, increase only the micronutrients 10%. Maintain temperature differential between head temperatures and root temperatures of >5°C
Flower/ truss	Flower colour pale yellow, especially in a.m.	Colour should be bright 'egg yolk' yellow. If climate is humid, low vapour pressure deficit (< 2) will often occur in a.m. Recommend increase VPD, especially early in morning, from 3 to 7 VPD. Create active climate with minimum pipe heat 40°C and limited venting (1–2%). Flowering rate should be 0.8–1.0 truss/week
	Long, straight flower trusses (kink trusses)	Aggravated by low light and high temperature. Decrease 24 h temperature by decreasing day temperature. Promote active climate 3+ VPD. Avoid increasing plant density too early in season when light levels low (< 600 J/cm^2/day)
Flowers	'Sticky flowers' in which sepal does not roll back	Caused by too humid climate, especially a.m. Activate plant in a.m. with minimum pipe and crack the vent. If left unchecked, these flowers result in poor quality fruit. Higher day temperature = higher VPD = less sticky
	Flowers too close to head, < 10 cm below growing tip	Too generative. Go down with day, up with night, i.e. bring day/night temperatures closer together. Late April/early May: close flowers may be desirable in order to get enough fruit on plant for summer fruit loads
Leaves	Short leaves in head, e.g. < 35 cm in length	Occurs in late spring. Plant is in vegetative imbalance. Fruit load is low (< 85 fruit/m^2). Increase differential between day and night

All NFT systems recirculate nutrient solution. In the original NFT systems, the starter nutrient solution was replenished by a 'topping-up' solution (Jones, 1999) as it was used up by plant transpiration. The solution was then discarded after some period of time, usually 2 weeks, and a fresh solution made up to correct nutrient imbalances. In modern systems, the solution is monitored for salts and water, and specific nutrients may be replenished. There are many different types of closed recirculating systems available and experimentation continues in this area. In The Netherlands (Voogt and Sonneveld, 1997) the most common system to capture runoff is that of plastic gutters with rockwool slabs (Fig. 9.12). In the southern USA, Bato Buckets® can be used to collect drain-water (Fig. 9.10), using drainage outlets that suction excess water from the reservoir in the bottom of the Bato Bucket® into a PVC pipe.

The main problem with these systems is preventing contamination by pathogens. Additional considerations are oscillations in nutrients caused by plant uptake and autotoxicity. Ikeda *et al.* (2001) listed the following types of control methods: (i) physical and cultural: heat treatment, UV radiation, membrane filtration, nutrient solution temperature, pH and EC control, and

Fig. 9.12. Recirculation with rockwool slabs in a gutter in a greenhouse in The Netherlands.

sanitation; (ii) chemical methods: ozone, chlorination, iodination, hydrogen peroxide, metal ions, inorganic elements, non-ionic surfactants and chitosan (a bioactive product derived from a polysaccharide found in the exoskeleton of shellfish such as shrimp or crabs); and (iii) biological: slow filtration, rhizosphere bacteria and antagonistic fungi, mycoparasitic fungi, suppressive substrates, and biosurfactants.

Irrigation

Large amounts of high quality water are needed for plant transpiration, which serves both to cool the leaves and to trigger transport of nutrients from roots to leaves and fruits. Water consumption of 0.9 m^3/m^2/year is estimated for greenhouses in The Netherlands (Anonymous, 1995) and 0.8 m^3/m^2/year for BC, Canada (Portree, 1996). Before building a greenhouse, it is important to ensure adequate water availability and quality. EC should be < 0.5 dS/m, pH from 5.4 to 6.3 and alkalinity < 2 meq/l. Water treatment to lower alkalinity and adjust pH is usually possible, if expensive. Lowering EC by reverse osmosis is usually not economically feasible but in The Netherlands it has been shown that an alternative water treatment consisting of mixing a poor quality water supply with rainwater increases irrigation water quality.

Frequency of irrigation varies with substrate rooting volume and water-holding capacity. In rockwool slabs, rooting volume is very restricted, and slabs may be watered five to six times per hour, or up to 30 times a day under summer conditions. Gieling (2001) developed a basic controller for water supply. Watering frequency in perlite systems is usually less than in rockwool systems, but more frequent than in peat bag systems, which may be watered only three to four times a day. The amount of water needed by plants varies from 1 to 14 l/m^2/day (0.4–5.6 l/plant/day), depending on stage of growth and season. Daily timing of irrigation cycles also varies with water demand. For example, when the heating system is on during the winter, up to 50% of total daily transpiration can take place at night, compared with only 5–8% of the daily total during summer nights (Portree, 1996). In rockwool systems, fertigation should begin 1–2 h after sunrise and end 1–2 h before sunset to decrease diseases as well as summertime russeting and fruit cracking. Night watering may be needed during the winter, when night-time heating decreases relative humidity, and in summer if conditions are hot and dry (OMAFRA, 2001). In The Netherlands a well-balanced irrigation model has been developed for recirculating systems, based on leaf area, air temperature and season (De Graaf, 1988). Very accurate weighing units are being introduced, so that moisture content of the slab, as well as plant transpiration, can be monitored every hour to avoid stress to plants.

ENVIRONMENTAL CONTROL

Because of their increasing sophistication, ease of use and affordability, even in relatively small greenhouse ranges, computers are used to control temperature, relative humidity, CO_2 concentrations and light intensity. They are also very useful in providing a history of the crop environment over time and alerting operators to malfunctions. Computers can control many mechanical devices within a greenhouse (vents, heaters, fans, evaporative pads, CO_2 burners, irrigation valves, fertilizer injectors, shade cloths, energy-saving curtains) based on preset criteria, such as outside or inside temperature, irradiance, humidity, wind and CO_2 levels. More importantly, they can integrate the results of different sensors and process all the data to achieve a desired result, such as maintaining a particular temperature or humidity regime. It is much easier to balance plant growth using environmental controls in a computerized greenhouse. Computer control of irrigation and fertilization regimes based on environmental conditions is discussed in Chapter 6.

Relative humidity

Humidity in a greenhouse is a result of the balance between transpiration of the crop and soil evapotranspiration, condensation on the greenhouse cover and vapour loss during ventilation. In winter, humidity is generally low because of low transpiration and high levels of condensation, but humidity levels may be high in late spring and autumn. Energy conservation features, such as the use of double layers of polyethylene films, have increased relative humidity (Hand, 1988).

Although computer control programs can be very sophisticated, there are limitations on the effectiveness of humidity control. For example, as vents are opened and closed to control temperature, the humidity and CO_2 levels also change. If humidity levels become too high, while temperatures remain in an acceptable range, some combination of heating and ventilation may be necessary to maintain acceptable humidity and temperature. In glasshouses with vents, the heat should be turned on and the vents opened. In houses with fans, the fans should be turned on for a few minutes and then the heater turned on to maintain air temperature. Venting for humidity control is most effective when outside air is significantly cooler and drier than that inside the greenhouse. As cool, dry air heats up in the greenhouse, it absorbs moisture and lowers the humidity. Humidity reduction by bringing in outside air can be somewhat effective even if the outside air is very humid, as long as it is significantly cooler than the inside air. In practical terms, however, outside air should be significantly cooler and drier to justify the cost of ventilation. With a 'closed' glasshouse, humidity control may be achieved without influencing other climatic factors because cooling is provided by stored groundwater rather than ventilation (see section on greenhouse installations, above).

Relative humidity is sometimes discussed in terms of the corresponding vapour pressure deficit (VPD) of the air, i.e. the amount of moisture in the air compared with the amount of moisture the air could hold at that temperature. This topic is discussed further in Chapter 6. In a study in The Netherlands (Bakker, 1990), high humidity (low VPD) reduced leaf area because of calcium deficiency, and also increased stomatal conductance, reduced final yield and reduced mean fruit weight. This study was conducted over a fairly limited range of VPDs, however: 0.35–1.0 kPa in daytime and 0.2–0.7 kPa at night. It is unclear to what extent low humidity (high VPD) is deleterious to the plant if adequate water is available, but in general, VPDs > 1.0 are considered potentially stressful. In northern Europe, VPDs > 1 kPa are rarely seen, but they will sometimes exceed this range in parts of North America and certainly in southern Europe. A greenhouse temperature of 26°C and relative humidity of 60% would result in a VPD of 1.35, for example. In arid climates, greenhouse VPD can be as high as 3–5 kPa. If plants transpire more water than can be supplied through the roots, fruit may develop blossom-end rot (BER) and stomates may close, resulting in poor growth. Chapter 6 gives a fuller discussion of humidity and temperature interactions and recommended VPD levels.

The most important reason for reducing humidity and keeping leaf surfaces dry is disease prevention. Diseases spread rapidly when VPD is 0.2 kPa or less, and germination of fungal pathogen spores increases on wet leaf surfaces. This is most likely when warm sunny days increase leaf transpiration and evaporation but moisture is held as water vapour until air cools to the dewpoint at night. Water vapour then condenses on to cool surfaces, such as the leaves and the inside skin of the greenhouse, and drips from the greenhouse skin on to the leaves. Wetting agents, either sprayed on the inside film or incorporated into the plastic, prevent condensation from dripping onto the plants because moisture remains as a film, which slides off in a sheeting action rather than dripping off on to the leaves. The problem of condensate dripping on leaves is most severe in quonset-style double-poly greenhouses, because the rounded arch makes it hard to collect and remove drainage. Glass and acrylic panel greenhouses are less humid to start with and the roof is more steeply pitched, as is also the case with Gothic-arch greenhouses, so moisture runs off rather than accumulating. It can then be collected and drained to the outside. Increasing air movement to 1 m/s (leaves move slightly) in the greenhouse reduces condensation on the leaves by reducing temperature differences between the leaf surface and the air, thereby preventing leaf surfaces from cooling below the dewpoint. Air movement can be increased either by running the fans on hot-air furnaces or by horizontal airflow (HAF) fans. These small fans (Fig. 9.10) are placed along the sides of the house to push air in one direction on one side of the greenhouse and in the opposite direction on the other side, and they operate continuously, except when the exhaust fans are turned on. This creates a slow horizontal air movement, which also makes temperatures more uniform.

In The Netherlands a condensation model has been developed (Rijsdijk, 1999) that enables growers to modify the heating regime during sunrise (the typical period when condensation is formed) as a function of measured fruit temperature rather than by use of ventilation fans. Condensation can form on the fruit because at sunrise air heats up faster than the fruit and so the surface is colder than the air. Fruit heats more slowly than leaves and so fruit temperature is measured rather than leaf temperature. If no condensation forms on the fruit, it should also not occur on the leaves.

Temperature: heating and cooling

Maintaining optimal temperatures

Optimal day and night temperatures for different crop developmental stages are shown in Table 9.3. As temperatures increase within the range 10–20°C, there is a direct linear relationship between increased growth and development. If daytime temperatures are warm, night-time temperatures can be allowed to fall to conserve energy as long as the means remain in the optimal range (see also Chapter 4).

In The Netherlands, temperature integration strategies in tomato production in glasshouses target mean daily temperature rather than maintaining specific day and night temperatures (De Koning, 1990). Energy savings of 10–15% have been realized, compared with maintenance of regimes of high day and low night temperature regardless of the 24 h mean. Work summarized in Papadopoulos *et al.* (1997) suggested that, over periods ranging from 24 h to several days, plants tolerate some variation about the optimal temperature. For example, tomatoes can tolerate a deviation of 3°C below standard for 6 days, provided the following 6 days are 3°C above standard and as long as the average temperature over the 12-day period stays the same. Even a deviation as high as 6°C can be tolerated if the 6-day temperature average is unaffected (De Koning, 1990; Portree, 1996).

Energy conservation measures are widely used in greenhouses in northern latitudes to reduce heating costs. Pulling thermal curtains of porous polyester or an aluminium foil fabric over the plants at night reduces heat loss by as much as 20–30% on a yearly average. Dual-purpose lightweight retractable curtains are sometimes used for energy conservation at night and for shade in daytime. In most southern growing regions, such as Spain and the southern part of the USA, however, thermal curtains will not provide enough energy saving to justify their high cost; and even rolled to the side, shading during the day gives rise to production losses. They are also not practical in most Quonset-style greenhouses, because there is not enough space overhead to pull them back and forth.

Maximal temperatures for greenhouse tomato production are less well established but are considered to limit summertime production in southern latitudes, especially where evaporative cooling is less effective because of high humidity. There are three common methods of greenhouse cooling: (i) natural ventilation; (ii) mechanical fan-and-pad cooling; and (iii) fog cooling. In The Netherlands, a new system is being tested at tomato nurseries that utilizes heat storage, heat pumps, heat exchangers and cooling plates to control greenhouse temperatures. With these closed greenhouses, temperatures can be kept below 26°C (De Gelder *et al.*, 2005). (See section on greenhouse installations, p. 263.)

Natural ventilation from ridge vents is popular in areas with relatively few days with high ambient temperatures. For natural ventilation, some part of the greenhouse (usually at the peak or ridge) is opened and air movement created by wind pressures or by gradients in air temperatures draws cooler air into the house. Side curtains that can be rolled up either manually or automatically can easily be installed in double-polyethylene film houses. New designs also allow natural ventilation in double-poly greenhouses (Giacomelli and Roberts, 1993). With any type of natural ventilation system, however, insect netting in the ventilation opening to prevent pest entry and escape of pollinators or beneficials reduces ventilation capacity by approximately 20%.

Cooling is difficult in humid climates, because plant transpiration, fan-and-pad evaporative cooling and fogging are all less effective than in arid climates. While natural ventilation is used in some warm-climate greenhouses, hot conditions outside and lack of wind reduce its effectiveness. In the south-eastern USA, pest pressures, high humidity and high temperatures force most growers to invest in active mechanical cooling, usually with a combination of fans and pads. With mechanical cooling, low-pressure propeller-blade fans are placed opposite the air intake, which is covered by cellulose evaporative cooling pads. Louvres or other types of covers are placed outside the cooling pads and are closed when the greenhouse is not venting. Ventilation fans are normally sized to allow one air exchange per minute, although researchers in North Carolina and Israel have documented increased cooling at higher rates, especially when combined with evaporative cooling (Willits, 2000).

Presumably, the temperature averaging method described above as a way of reducing heating costs can also be applied to warm conditions where cooling costs are a concern. This question has not been addressed directly, but Peet *et al.* (1997) reported that, over the range 25–29°C, the actual day and night temperatures and the day/night differential were less important than the daily average in accounting for declines in fruit set, yield, fruit number and seediness. This suggests that, in areas where summer night temperatures are low, day temperatures can be allowed to exceed the normal maximal levels. Similarly, in areas where night temperatures are excessive, lowering daytime temperatures may be useful, but at a cost of higher energy consumption for cooling. Although the applicability of temperature averaging to above-optimal conditions for plants setting fruit has not been tested directly, data collected in the south-eastern USA in an experiment with night-time air conditioning (Willits and Peet, 1998) suggested the potential feasibility of such an approach during the fruit-set period. After fruit set, plants were much less sensitive to high temperature.

Using temperature to control plant growth

Temperatures can also be used to steer the plant towards a particular growth pattern (Table 9.6). In the long cropping cycles typical of greenhouse production, tomatoes tend to cycle between being overly vegetative at the beginning of cropping (too much growth and too little fruiting) and later being overly generative (too little growth and excessive fruit loads). Where uniform production is desired, it is helpful to be able to moderate these swings in crop productivity. Temperature is considered to be the most important tool to control flowering and fruit growth, and thus to determine the yield over a particular period. Quantitative data on the effects of temperature on flowering, fruit set, fruit growth and yield (De Koning, 1994) have been used to develop yield prediction models for The Netherlands. In 2002 these models were further developed into an Internet-based expert system (http://www.LetsGrow.com).

Carbon dioxide

CO_2 can be added to the greenhouse in several ways. Natural gas or propane burners hooked up to sensors can be used to generate CO_2. Different fuel sources provide different amounts of CO_2. Burning 1 m^3 natural gas, 1 l kerosene or 1 l propane provides 1.8 kg, 2.4 kg and 5.2 kg of CO_2, respectively (Portree, 1996). Flue gases from a hot-water boiler burning natural gas can be captured and recirculated. All these sources of CO_2 will add heat and water vapour to the greenhouse, as well as potential pollutants. Low-NOx (nitrous oxide, nitrogen dioxide) burners are available to minimize risks of pollutants reducing yield. The most expensive but safest option is compressed or liquid CO_2, which is unlikely to contain combustion gases as contaminants and does not add heat or water vapour to the greenhouse. CO_2 sensors should be calibrated periodically and located near the top of the plant. CO_2 distribution within the greenhouse should also be as uniform as possible, to avoid yield differences and for efficient utilization. CO_2 is heavier than air and so it is important that it should reach the plant canopy, rather than remain near the floor.

CO_2 enrichment to 750–800 μmol/mol increases yields by 30% compared with standard outside conditions (about 340 μmol/mol). A standard approach to enrichment (Nederhoff, 1994) is to inject CO_2 as a by-product of combustion of natural gas, at a level of 800 μmol/mol during heating. At low ventilation rates (< 10% opening), this level is reduced to 500 μmol/mol. With further vent opening, the goal is to maintain a base level of 350 μmol/mol but this is not always possible. Currently in The Netherlands, maximum levels of CO_2 enrichment are used except for the period from May until September. Even during this period, CO_2 is maintained at ambient levels.

Computer control models (Aikman *et al.*, 1996) have been developed to convert the rate of carbon assimilation in CO_2-stimulated photosynthesis into an anticipated financial return from fruit sales by linking biological processes with a fruit price model. Since fruit prices are very difficult to predict, use of such models is limited. Bailey (2002) considered strategies for CO_2 enrichment both with liquid CO_2 and with CO_2 from greenhouse heaters or CHP units. On the basis of the financial margin between crop value and the combined costs of CO_2 and natural gas, it was shown that the most economic CO_2 control point with liquid CO_2 depended on its price. With exhaust gas CO_2 and CHP units, financial margin depended on whether there was a heat store for excess daytime heat.

In southern latitudes, greenhouses are vented so frequently that CO_2 enrichment is not practical. In Raleigh, North Carolina, tomatoes could only be CO_2 enriched for 2–3 h daily for most of the growing season (Willits and Peet, 1989). In any case, when temperatures are above 25°C, CO_2 enrichment may not be cost-effective in North American conditions (Portree, 1996) and may cause stomatal closure, which reduces transpiration. In the 'closed' glasshouse concept, maximum levels of CO_2 will be used year round.

Light intensity

Light intensity is reduced by 20–30% compared with outside, depending on the covering and the greenhouse structure and is further reduced within the plant canopy. Therefore, in almost all regions, CO_2 and irradiance (light intensity) are the most limiting factors for maximizing yield. Economic use of supplemental light is not feasible except in areas with very short days in winter, although in The Netherlands increases in yearly production of 55% have been reported (Marcelis *et al.*, 2002). One problem is that, when placed overhead, the bulb assembly (reflector, transformer and starter) reduces the interception of natural light by the crop during periods when artificial light is not needed. Assemblies designed to intercept less light have now been developed and in northern Europe and Canada there is interest in systems where supplemental high-intensity discharge (HID) lighting is combined with hanging gutters and intercropping to maximize productivity.

Shade cloths and screens are used in southern production areas to protect fruit at the top of the canopy from sunscald, russeting, and cracking caused by high temperatures and to reduce greenhouse temperatures (see Chapter 6 for additional discussion of the causes of these disorders). Shade cloths also reduce leaf cooling through transpiration, because stomata close, and so reductions in leaf temperatures are less than reductions in air temperature. In North America, a 3% light reduction from screening resulted in a 1% yield reduction, but fruit quality was increased (Portree, 1996). In Dutch growing conditions, Van Winden *et al.* (1984) showed that 1% light loss even in summer conditions gave rise to almost 1% yield loss.

Air pollutants

The most common and serious forms of greenhouse pollution are combustion gases generated by faulty heat exchangers, dirty fuel openings and incomplete fuel combustion. Well-sealed, energy-efficient greenhouses have added to the problem by reducing outside air exchanges. At low concentrations, carbon monoxide (CO) can cause headaches and dizziness for workers; and injury

Fig. 9.13. Plant exposed to 10 ppm ethylene overnight. Note twisting vines, down-turned leaves, yellowing and flower abortion.

and death can occur above 50 ppm (0.005%). Leaks of fuels such as propane and methane must be fairly large to be hazardous for human health, but even small leaks can adversely affect plants. Similarly, ethylene is dangerous to humans at 5 ppm, but ethylene levels of < 0.05 ppm can make tomato leaves bend downward (epinasty) (Fig. 9.13). With chronic exposure to levels as low as 0.02 ppm, stems may thicken, branching may increase, and flower buds may abort or develop into malformed fruit. Symptoms of chronic low exposure may be hard to recognize, especially if plants grown in clean air are not available for comparison. Diagnosis is also difficult because of the time lag between the period of ethylene exposure and the time when damage is noted. Equipment to detect most pollutants directly is not practical for use in the greenhouse, but some North American growers use inexpensive CO monitors in the flue gas. CO levels > 30 ppm may indicate incomplete combustion and potential pollutants. Table 9.7 indicates potentially harmful levels for people and plants of air pollutants likely to be found in the greenhouse.

Most problems are noted when greenhouses are first started up in winter. It may be useful to bring a few potted tomato plants into the greenhouse before transplanting. If symptoms of epinasty are noted on these, the system should be thoroughly checked before the crop is brought into the greenhouse. Air pollution can result from other sources as well, such as paint on heating pipes, cleaning agents and new PVC (Portree, 1996). The safest practice is to maintain proper ventilation, even at the expense of energy conservation, and observe plants closely for signs of damage when heaters first come on in the autumn and during periods of unusually cold weather in the winter.

Proper maintenance also prevents problems: cleaning the unit heater and fuel orifice at least twice a year and regularly inspecting the flame for changes in appearance. Propane flames should have a small yellow tip, while natural gas flames should be soft blue with a well-defined inner cone. Heater adjustment and checking for gas leaks is best done by professionals before the start of the heating season.

Table 9.7. Maximum acceptable concentration (ppm) for humans and plants of common greenhouse air pollutants (various original sources; table adapted from Portree, 1996).

Gas	Humans	Plants	Plants (long-term exposure)
Carbon dioxide (CO_2)	5000	4500	1600
Carbon monoxide (CO)	47	100	
Sulphur dioxide (SO_2)	3.5	0.1	0.015
Hydrogen sulphide (H_2S)	10.5	0.01	
Ethylene (C_2H_4)	5.0	0.01	0.02
Nitrous oxide (NO)	5.2/5.0	0.5/0.01–0.1	0.250
Nitrogen dioxide (NO_2)	5.0	0.2–2.0	0.100

Pest and disease management

Insect and disease problems

Chapter 7 provides an overview of pests and diseases in tomato. Typical insect problems in greenhouses include whiteflies, mites, thrips, aphids, pinworms, caterpillars, psyllids and leafminers. Typical fungal pathogens include damping-off (root rot, *Pythium* sp.), botrytis grey mould (*Botrytis cinerea*), powdery mildew (*Erysiphe* sp.), leaf mould (*Fulvia fulva*, previously known as *Cladosporium*), and Fusarium crown rot (*Fusarium oxysporum*). A number of viral diseases can also be found in the greenhouse, including tomato spotted wilt virus (TSWV), tomato (ToMV) and tobacco (TMV) mosaic viruses, beet pseudo-yellows virus and various gemini viruses. Pepino mosaic virus has recently been observed in North America, but is not yet widespread. The best prevention for pseudo-yellows and gemini viruses is to exclude the vector (silverleaf whiteflies) as is also the case for TSWV, vectored by western flower thrips. Leafhoppers, planthoppers, psyllids and possibly whiteflies are vectors of phytoplasmas, previously known as mycoplasmas.

Corynebacterium may spread in some production areas with high temperatures. Disease and insect problems typical of field production, such as early blight and beet armyworms, can also occur in greenhouses, particularly those with open sidewalls as in the southern USA and Mediterranean regions.

Biological control

Tomatoes can be grown virtually insecticide-free in North America and northern Europe through the use of biological controls. A number of beneficial insects are available for use against greenhouse pests. Success with biological control requires experience, patience and a good supplier. Problems can arise before the beneficials even enter the greenhouse. Shipments may be the wrong amount or at the wrong stage of development. If packaging was inadequate, or shipping conditions too hot or too cold, the beneficials may have perished during shipment.

Keeping records of lot numbers and date and location released can be helpful. With experience and a good hand lens, samples can be inspected on arrival and for a few days after placement in the house to determine viability, but, with each type of beneficial, different characteristics are important. Treatment after arrival is also crucial. Beneficials should be released immediately, and not left in hot conditions or exposed to temperatures below 10°C (as in a refrigerator). It is also not easy to determine how well the beneficials are establishing, since results are not immediate as with conventional pesticides.

The main groups of biocontrol agents are: (i) parasitoid wasps to control whiteflies; (ii) aphid parasite (*Aphidius matricariae*) and predator (*Aphidoletes aphidimyza*, a small midge); (iii) predatory mites (*Phytoseiulus persimilis*, *Amblyseius cucumeris* and *Hypoaspis*) to control spider mites; (iv) nematodes to control fungus gnats; and (v) some general predators (e.g. lacewings, minute pirate bug). Within these general groups, not all types of beneficials will control all types of pests. For example, predatory mites will control spider mites but not russet mites. Whitefly control is also complicated, unless the only species present is the greenhouse whitefly (*Trialeuroides vaporariorum*), which can be controlled by *Encarsia formosa* (Fig. 9.14). Control of the silverleaf whitefly (*Bemisia tabaci*) is more difficult and requires *Eretmocerus eremicus* (formerly called *E. californicus*) or *Eretmocerus mundus*. Control of silverleaf populations is critical, because they not only reduce plant vigour and excrete honeydew, but also are vectors of viruses and cause uneven fruit ripening. Honeydew from either type of whitefly, or from aphids, supports development of sooty moulds and other fungi on the leaf and fruit surface (Fig. 9.15). With mixed whitefly populations, it may be necessary to release both types of predators; and with all types of pests, it is often more effective to use more than one type of biocontrol agent, as environmental conditions, or shifts in pest populations, may favour one over the other at any particular time.

Beneficial insects tend to be more sensitive to pesticides than pests. Once beneficials have been introduced into the greenhouse, the types of pesticide that can be used are very restricted. Lists of allowable pesticides are usually available from suppliers.

Biopesticides

Biopesticides represent a new category of products (sometimes also called biorationals or reduced-risk pesticides), which are safer for humans and have fewer off-target effects. This category includes microbial pesticides, such as *Bacillus thuringiensis* (Bt), insect protein toxins, entomopathogenic nematodes, baculoviruses, plant-derived pesticides and insect pheromones used for mating disruption. The registration process for these products is streamlined compared with conventional pesticides, making them more likely to be registered for a speciality crop such as greenhouse tomatoes. Material derived

Fig. 9.14. Underside of tomato leaf, showing a group of adult greenhouse whiteflies. Dark spots are whitefly immatures parasitized by *Encarsia formosa*.

Fig. 9.15. Leaf covered with sooty mould growing on honeydew deposited as a result of whitefly or aphid feeding.

from *Beauveria bassiana*, an entomopathogenic fungus which serves as a bioinsecticide, is available in several formulations to control whiteflies. These materials can be sprayed on the crop just like conventional pesticides but, like conventional pesticides, applications must also be repeated. Insect growth regulators and products derived from natural insecticides, such as neem oil, are other types of biopesticides, but in some cases these can be quite toxic to bumble bees and beneficials.

Conventional pesticides

Few conventional pesticides are registered specifically for greenhouse tomato production. Some materials registered for field production of tomatoes specifically prohibit use in the greenhouse on the label, but others may be used. Growers need to consult the local or national pesticide registry for clarification on which materials can legally be used on greenhouse tomatoes and restrictions as to re-entry interval and days before harvest. There are also detrimental effects of most pesticides on bumble bees and introduced beneficial insects. Although pesticides may be useful for clean-up after the crop or to reduce populations before introducing beneficials, growers should adopt integrated pest management practices rather than rely exclusively on insecticides. In Spain, the response to pests and diseases is mainly preventive spraying; biological control is used on < 5% of the acreage. In Almeria (as shown in Table 9.8), three to four times the amount of active ingredient per square metre is used compared with The Netherlands, where pesticide use is integrated with biological pest control. Per kilogram of product, the use of active ingredient in Almeria for tomato crops is about 20 times higher than in The Netherlands, even without taking into account soil-applied chemicals in Spain.

Table 9.8. Comparison of yields and pesticide utilization in greenhouses in Almeria, Spain, and in The Netherlands (ai, pesticide active ingredient) (source: Van der Velden *et al.*, 2004).

		Pesticide usage	
Region	Yield (kg/m^2)	Per unit area (kg ai/ha)	Per unit crop produced (mg ai/kg)
Almeria, Spain	9	26.0	289
The Netherlands	50	7.7	15

Cultural practices to avoid insect and disease problems

It is of the highest importance to start with plant material that is clean from insects or has a well-balanced system of predators and pests. Great care should also be taken to avoid introducing pests through transplants or ornamentals, which can be sources of thrips, mites and whiteflies. Preventing pest entry through air inlets and entry-ways is also essential. Placing screens directly over the air inlets would reduce circulation too much. With forced air systems, such as fan-and-pad evaporative cooling systems, plenums (screened air entry areas) can be constructed to reduce the pressure drop caused by the screening material. This excludes insects, but still allows adequate ventilation rates. Charts available from screening manufacturers allow calculation of the volume of these screening boxes based on air intake. Double-entry doors with positive pressure are the best way to prevent pests coming in with workers. Within the greenhouse, pest populations should be monitored with yellow sticky cards or tape (see Fig. 9.10) and control measures should be instigated as soon as adults are detected. Blue sticky cards are especially attractive to western flower thrips and may be a better choice if these have been a problem in the past. However, yellow cards attract whitefly, thrips, leafminers, fungus gnats and winged aphids.

Another cultural practice that can affect pest control is de-leafing, which may remove the parasitized immature stages of beneficials, especially parasitic wasps. Sometimes piles of leaves are left between the rows in order to let the adult beneficials emerge, but this is not advisable when disease inoculum is present.

The best way to prevent diseases is to maintain a good greenhouse environment, as discussed under environmental control: good air circulation, optimal plant temperatures, low humidity and no dripping of condensate on plant leaves. In addition, plant wastes should be removed and destroyed promptly.

Diseases in the root environment

Problems of pathogen spread must also be overcome in recirculating systems, as discussed earlier. Heat treatment (30 s at 95°C) and UV radiation are currently the most widely used disinfection methods, but both are expensive and not always effective. In most cases, the return is only partially sterilized.

Recently, there has been an interest is the use of slow sand filtration as an alternative to complete disinfection of the nutrient solution in recirculating systems. With slow sand filtration, the resident microflora may have the ability to suppress pathogens such as *Pythium* and *Phytophthora*. Thus, disease outbreaks are prevented, without the expense and difficulty of complete sterilization. Soil-borne diseases, such as Fusarium crown rot and Pythium, can also be a problem in non-recirculating systems, though they are mostly controllable through sanitation and resistant cultivars (in the case of Fusarium crown rot).

Root grafting

In soilless production, the main objective of grafting tomato plants on rootstocks is to obtain a higher yield. Since the late 1920s, grafting has been carried out in production in soil to reduce infection by soil-borne diseases caused by pathogens such as *Fusarium oxysporum* (Chapter 7). For soilless cultivation in greenhouses, therefore, plants were not grafted. However, recently it has been found that grafting results in a higher resistance against virus and fungal diseases such as *Verticillium*, as grafting results in more vigorous plants (Heijens, 2004). Grafting of tomato cultivars on rootstocks has also been found to increase high-temperature tolerance (Rivero *et al.*, 2003), drought tolerance (Bhatt *et al.*, 2002) and salinity tolerance (Fernández-García *et al.*, 2002).

Plant performance depends on the combination of root and scion. For example, the rootstock effect on the tomato salinity response depends on the shoot genotype (Santa Cruz *et al.*, 2002). These interactions make it difficult to predict the rootstock effect in a particular rootstock–scion combination. Zijlstra and Den Nijs (1987) tested the contribution of the roots to growth and earliness under low-temperature conditions for nine genotypes by making reciprocal grafts using the same genotypes as scion and as rootstock (81 combinations). Although they reported little interaction between rootstock and scion, they observed that tomato genotypes selected for growth and early production under low temperature, performed very poorly when used as a rootstock at low temperature.

The use of rootstocks is common in greenhouse production in The Netherlands. The most widely used rootstock is 'Maxifort' (De Ruiter Seeds), but 'Eldorado' (Enza Zaden), 'Beaufort' (De Ruiter Seeds) and 'Big Power' (Rijk Zwaan) are also used (Heijens, 2004). Rootstocks are sown 7–10 days earlier than the scion cultivar. The scion is sown about 5 days earlier than the non-grafted plants, to obtain equal-sized plants on the desired planting date, as grafting results in a small delay. Also the number of leaves below the first truss is usually one more than for non-grafted plants (Heijens, 2004). Grafted plants are 50–100% more expensive and therefore often two stems per plant are kept. The second stem is obtained as a side shoot below the first, second or third truss, or grafted plants are decapitated above the cotyledons. The latter procedure will only result in two equal stems when light levels are high enough (planting from mid-March onwards). Decapitation above the cotyledons results in a delay of 14 days compared with retaining a side shoot on a standard plant. A third method is decapitation above the second leaf. In that case there is the risk of unequal shoots, as the highest one will 'dominate' the lower one (Heijens, 2004).

MARKETING

Because greenhouse tomatoes are harvested riper than field-grown tomatoes, they are highly perishable, and shippers and buyers must be located well in advance. Since production costs and product quality are higher compared with field production, special attention must be given to receiving a price that will offer a sustainable rate of return on investment. The key is to sell a product clearly superior to field-grown tomatoes in appearance and flavour. Many growers attach stickers or provide promotional material to attract customers and build name recognition (Figs 9.16 and 9.17). The rapid rise in US sales is testimony to consumer willingness to pay more for a high quality product. In northern Europe, growers' groups sell their tomatoes under a specific brand, e.g. 'Tasty Tom' or 'Nature's choice'. They guarantee their brand by adhering to strict growing strategies, focused on optimizing shelf-life and taste. For that reason a specific 'taste' model is used, manipulating cultural practices such as temperature control, feeding regime and cultivar to

Fig. 9.16. Single-layer boxes of harvested tomatoes have individual stickers applied automatically as they go through the packing line. These tomatoes have the calyx intact.

Fig. 9.17. Automated packing line separates tomatoes at varying degrees of ripeness and packs them into a single-layer box.

optimize taste attributes (Verkerke *et al.*, 1998). For smaller growers, direct sales at their greenhouse, grower cooperatives, farmers' markets, or speciality outlets, such as organic and health food stores, are all viable options.

HARVEST

The superior taste and texture of greenhouse tomatoes is often attributed to the fact that they remain longer on the vine and reach the consumer sooner than field tomatoes. For direct marketers, fruit are harvested virtually red-ripe. For large operations shipping cross-country or even overseas, beefsteak types may be harvested earlier, but never before the breaker stage (first show of colour at the blossom scar). For both types of sales, attractive packaging and presentation are critical. The thin skins and fruit walls of greenhouse cultivars contribute to their appeal to consumers but expose them to injury during harvest and packing. Generally they are packed in a single layer, rather than being stacked (as is the practice with field tomatoes).

Many large greenhouses have systems designed to protect the fruit and reduce labour costs during harvest. These include pipe-rail systems for moving picking carts along the rows and hydraulic lifts to allow workers to work the plants and harvest. In beefsteak tomatoes, the calyx and stem are usually removed at harvest to prevent puncture wounds, although those shown in Fig. 9.16 have the calyx intact. In cluster tomatoes, the entire cluster is cut off at the main stem and kept together either by placing in a single layer in boxes, clamshells, plastic sleeves or mesh bags.

POSTHARVEST PACKING AND STORAGE

Most large greenhouses have automated packing lines, similar to those used for packing the field crop except that greenhouse tomatoes are usually packed in stackable single-layer boxes (Figs 9.16 and 9.17). Greenhouse tomatoes are generally not stored, or in the case of North America for not more than the few days necessary to move the crop to market, but even in this short period optimal temperature and humidity should be maintained (10–13°C and 90–95% relative humidity). Because tomatoes are picked at the breaker stage or later, artificial ripening with ethylene is unnecessary. Ethrel is sometimes used by cluster-tomato growers to promote simultaneous ripening of all fruit on the cluster. Normally this is only done when the grower wants to finish up the harvest in late autumn before the new crop comes into production. In this case, care should be taken not to expose the transplants to such high levels of ethylene as to promote flower abortion and growth abnormalities (see Fig. 9.13). Furthermore, shelf-life might be affected negatively. Naturally ripening tomatoes are also a source of ethylene and should not be stored or transported with ethylene-sensitive crops such as broccoli, lettuce and cucumber.

POTENTIAL PRODUCTION

Yields are increasing worldwide because of better cultivars and more intensive use of technology. However, in comparing production figures, it is important to note the length of the production season, plant density and the number of crops per year. In general, greenhouse tomato yields are much higher than for outdoor production: 375 t/ha/year compared with 100 t/ha/year (Jensen and Malter, 1995). Yields in soilless greenhouse systems average higher than yields in soil-based greenhouse systems, though meaningful comparisons of productivity are difficult. Soil-based greenhouse systems are usually managed less intensively, with lower overhead, capital, marketing, production and operating costs and a shorter growing season. Growers in soil-based greenhouse systems also often retail locally, or sell organic produce, rather than selling their crop wholesale or through a broker. Thus, smaller growers can also operate profitably, especially if they do not compete in the same markets with large operations.

For an individual plant in a high light environment, such as the south-western USA, 18 kg/plant over a 7–8-month cropping period represents an excellent yield (Jensen and Malter, 1995). In the south-eastern USA, yields of 9–10 kg/plant are more common, representing shorter seasons, lower light and less intensive production techniques. In BC, Canada, target yields are 65 kg/m^2, or 275–350 fruit/m^2 over the entire production cycle (Portree, 1996). Tomato production in Dutch greenhouses increased from 250 t/ha in 1985 to 400 t/ha in 1994 (Anonymous, 1995). Recent estimates in The Netherlands range from 47 kg/m^2 for cluster tomatoes to 53 kg/m^2 for beefsteak tomatoes (see Table 9.2). Systems with hanging gutters and HID lights may push the production barrier even higher. In the UK, the best growers already achieve annual yields of 800 t/ha compared with average national yields in 2000 of 440 t/ha (Ho, 2004). Further large increases are most likely to result from a longer harvest season, which will require supplemental lighting and interplanting. This will only be practical if a relatively cheap year-round supply of CO_2 and electricity is available from CHP or other energy sources. In that case, an annual yield of more than 1000 t/ha could be achieved by a year–round, non-stop picking production system (Ho, 2004). In the meantime, breeders, engineers, horticulturalists and physiologists are working on ways to increase yield and quality of greenhouse tomatoes, while decreasing costs, including labour, and adverse environmental impacts.

CONCLUSION

Production of greenhouse tomatoes is demanding in terms of capital, energy, labour and management. Although production levels might be raised further to as much as 100 kg/m^2 in the coming decades, profitable operation requires excellent management and tight integration of the various production processes.

REFERENCES

Aikman, D.P., Fenlon, J.S. and Cockshull, K.E. (1996) Anticipated cash value of photosynthate in the glasshouse tomato. *Acta Horticulturae* 417, 47–54.

Anonymous (1995) *Kwantitatieve informatie voor de glastuinbouw 1995–1996.* Informatie en Kennis Centrum Landbouw, Afdeling Glasgroente en Bloemisterij, Aalsmeer/Naaldwijk.

Anonymous (2003) *Feitelijke prestaties biologische landbouw.* PPO report no. 47. PPO, Wageningen, The Netherlands, 95 pp.

Armstrong, H. (2003) Shut the roof and save energy. *Fruit&Veg Technology* 3, 69. Available at: http://www.HortiWorld.nl

Atherton, J.G. and Rudich, J. (1986) *The Tomato Crop. A Scientific Basis for Improvement.* Chapman & Hall, London, 661pp.

Bailey, B.J. (2002) Optimal control of carbon dioxide enrichment in tomato greenhouses. *Acta Horticulturae* 578, 63–70.

Bakker, J.C. (1990) Effects of day and night humidity on yield and fruit quality of glasshouse tomatoes (*Lycopersicon esculentum* Mill.). *Journal of Horticultural Science* 65, 323–331.

BC MAFF (2003) *An Overview of the BC Greenhouse Vegetable Industry*. British Columbia Ministry of Agriculture, Fisheries and Food. Available at: http://www.agf.gov.bc.ca/ghvegetable/

Bhatt, R.M., Srinivasa Rao, N.K. and Sadashiva, A.T. (2002) Rootstock as a source of drought tolerance in tomato (*Lycopersicum esculentum* Mill.). *Indian Journal of Plant Physiology* 7, 338–342.

Boonekamp, G. (2003) Spaanse kopgroep vergroot voorsprong. *Groenten en Fruit* 28, 14–15.

Cockshull, K.E. and Ho, L.C. (1995) Regulation of tomato fruit size by plant density and truss thinning. *Journal of Horticultural Science* 70, 395–407.

Costa, J.M. and Heuvelink, E. (2000) *Greenhouse Horticulture in Almeria (Spain): Report on a Study Tour, 24–29 January 2000*. Horticultural Production Chains Group, Wageningen University, Wageningen, The Netherlands, 119 pp.

Costa, J.M. and Heuvelink, E. (eds) and Botden, N. (co-ed.) (2004) *Greenhouse Horticulture in China: Situation and Prospects*. Ponsen & Looijen BV, Wageningen, The Netherlands, 140 pp.

De Gelder, A., Heuvelink, E. and Opdam, J.J.G. (2005) Tomato yield in a closed greenhouse and comparison with simulated yields in closed and conventional greenhouses. *Acta Horticulturae* (in press).

De Graaf, R. (1988) Automation of water supply of glasshouse crops by means of calculating the transpiration and measuring the amount of drainage water. *Acta Horticulturae* 229, 219–231.

De Koning, A.N.M. (1990) Long term temperature integration of tomato: growth and development under alternating temperature regimes. *Scientia Horticulturae*, 45, 117–127

De Koning, A.N.M. (1994) Development and dry matter distribution in glasshouse tomato: a quantitative approach. PhD Dissertation, Agricultural University, Wageningen, The Netherlands, 240 pp.

ECOFYS (2002) Closed greenhouse concept. Available at: http://www.Innogrow.nl

Fernández-García, N., Martínez, V., Cerdá, A. and Carvajal, M. (2002) Water and nutrient uptake of grafted tomato plants grown under saline conditions. *Journal of Plant Physiology* 159, 899–905.

Giacomelli, G.A. and Roberts, W.J. (1993) Greenhouse covering systems. *HortTechnology* 3, 50–58.

Gieling, T.H. (2001) Control of water supply and specific nutrient application in closed growing systems. Dissertation, Agricultural University, Wageningen, The Netherlands, 172 pp.

Hand, D.W. (1988) Effects of atmospheric humidity on greenhouse crops. *Acta Horticulturae* 229, 143–155.

Heijens, G. (2004) Enten niet meer weg te denken bij tomaat. *Groenten & Fruit* 2, 22–23.

Heuvelink, E., Bakker, M.J. and Stanghellini, C. (2003) Salinity effects on fruit yield in vegetable crops: a simulation study. *Acta Horticulturae* 609, 133–140.

Hickman, G.W. (1998) *Commercial Greenhouse Vegetable Handbook*. Publication 21575. University of California, Oakland, California.

Ho, L.C. (2004) The contribution of plant physiology in glasshouse tomato soilless culture. *Acta Horticulturae* 648, 19–25.

Hochmuth, G. and Hochmuth, B. (1995) Challenges for growing tomatoes in warm climates. Greenhouse Tomato Seminar, Montreal, Quebec, Canada. American Society for Horticultural Science, ASHS Press, Alexandria, Virginia, pp. 34–36.

Hochmuth, G.J., Hochmuth, R.C. and Carte, W.T. (1991) Preliminary evaluation of production media systems for greenhouse tomatoes in Florida. Suwannee Valley Research and Education Center, Suwannee Valley, Live Oak, Florida. Research Report 91–17.

Hochmuth, R.C., Leon, L.L. and Hochmuth, G.J. (1997) Evaluation of several greenhouse cluster and beefsteak tomato cultivars in Florida. Extension Report, Suwannee Valley Research and Education Center, Suwannee Valley, Live Oak, Florida. NFREC-SV Research Report 97–3. Available at: http://nfrec-sv.ifas.ufl.edu/Reports%20HTML/97–03_report.htm

Hochmuth, R.C., Leon, L.L. and Tillman, N. (2002) Evaluation of greenhouse beefsteak and cluster tomato varieties for North Florida 1999–2000. Extension Report, Suwannee Valley Research and Education Center, Suwannee Valley, Live Oak, Florida. NFREC-SV Research Report 2000–02. Available at: http://nfrecsv.ifas.ufl.edu/Reports%20HTML/2000–02_report.htm

Horridge, J.S. and Cockshull, K.E. (1998) Effect on fruit yield of bending ('kinking') the peduncles of tomato inflorescences. *Scientia Horticulturae* 72, 111–122.

Ikeda, H., Koohakan, P. and Jaenaksorn, T. (2001) Problems and countermeasures in the re-use of the nutrient solution in soilless production. *Acta Horticulturae* 578, 213–219.

Jensen, M. (2002) Controlled environment agriculture in deserts, tropics and temperate regions – a world review. *Acta Horticulturae* 578, 19–25.

Jensen, M.J. and Malter, A.J. (1995) *Protected Agriculture. A Global Review*. The World Bank, Washington, DC.

Jones, J.B. Jr (1999) *Tomato Plant Culture in the Field Greenhouse and Home Garden*. CRC Press, Boca Raton, Florida, 199 pp.

Leutscher, K.J., Heuvelink, E., Van de Merwe, R.A. and Van den Bosch, P.C. (1996) Evaluation of tomato cultivation strategies: uncertainty analysis using simulation. In: Lokhorst, C., Udink ten Cate, A.J. and Dijkhuizen, A.A. (eds) *Information and Communication Technology Applications in Agriculture: State of the Art and Future Perspectives*. Proceedings of the 6th International Congress for Computer Technology in Agriculture (ICCTA '96). VIAS, Wageningen, The Netherlands, pp. 492–497.

Marcelis, L.F.M., Maas, F.M. and Heuvelink, E. (2002) The latest developments in the lighting technologies in Dutch horticulture. *Acta Horticulturae* 580, 35–42.

Nederhoff, E.M. (1994) Effect of CO_2 concentration on photosynthesis, transpiration and production of greenhouse fruit vegetable crops. Dissertation, Wageningen Agricultural University, Wageningen, The Netherlands, 213 pp.

OMAFRA (2001) *Growing Greenhouse Vegetables*. Publication 371. Ontario Ministry of Agriculture, Food and Rural Affairs, Toronto, Canada, 116 pp.

Papadopoulos, A.P. and Hao, X.M. (1997a) Effects of greenhouse covers on seedless cucumber growth, productivity, and energy use. *Scientia Horticulturae* 68, 113–123.

Papadopoulos, A.P. and Hao, X.M. (1997b) Effects of three greenhouse cover materials on tomato growth, productivity, and energy use. *Scientia Horticulturae* 70, 165–178.

Papadopoulos, A.P., Pararajasingham, S., Shipp, J.L., Jarvis, W.R. and Jewett, T.J. (1997) Integrated management of greenhouse vegetable crops. *Horticultural Reviews* 21, 1–39.

Peet, M.M., Willits, D.H. and Gardner, R. (1997) Response of ovule development and post-pollen production processes in male-sterile tomatoes to chronic, sub-acute high temperature stress. *Journal of Experimental Botany* 48, 101–112.

Portree, J. (1996) *Greenhouse Vegetable Production Guide.* British Columbia Ministry of Agriculture, Fisheries and Food, Abbotsford, British Columbia, 117 pp.

Rijsdijk, A. (1999) Risico van condensatie bepalen met de computer. *Groenten en Fruit* 20, 12–13.

Rivero, R.M., Ruiz, J.M. and Romero, L. (2003) Can grafting in tomato plants strengthen resistance to thermal stress? *Journal of the Science of Food and Agriculture* 83, 1315–1319.

Santa-Cruz, A., Martinez-Rodriguez, M.M., Perez-Alfocea, F., Romerio-Aranda, R. and Bolarin, M.C. (2002) The rootstock effect on the tomato salinity response depends on the shoot genotype. *Plant Science* 162, 825–831.

Sato, S., Peet, M.M. and Thomas, J.F. (2000) Physiological factors limit fruit set of tomato (*Lycopersicon esculentum* Mill.) under chronic high temperature stress. *Plant, Cell and Environment* 23, 719–726.

Sonneveld, C. (1988) Rockwool as a substrate in protected cultivation. In: Takadura, T. (ed.) *Horticulture in High Technology Era, Proceedings of the International Symposium on High Technology Protected Cultivation* (Tokyo), pp. 173–191.

Sonneveld, C. and Welles, G.W.H. (1984) Growing vegetables in substrates in The Netherlands. *Proceedings 6th ISOSC International Congress on Soilless Culture*, pp. 613–632.

Van der Velden, N.J.A., Janse, J., Kaarsemaker, R.C. and Maaswinkel, R.H.M. (2004) Duurzaamheid van vruchtgroenten in Spanje: proeve van monitoring. LEI Report 2.04.04, The Hague, 52 pp.

Van Winden, C.M.M., Van Uffelen, J.A.M. and Welles, G.W.H. (1984) Comparison of the effect of single and double glass greenhouses on environmental factors and production of vegetables. *Acta Horticulturae* 148, 567–573.

Verkerke, W., Janse, J. and Kersten, M. (1998) Instrumental measurement and modelling of tomato fruit taste. *Acta Horticulturae* 456, 199–206.

Voogt, W. (1993) Nutrient uptake of year round tomato crops. *Acta Horticulturae* 339, 99–112.

Voogt, W. and Sonneveld, C. (1997) Nutrient management in closed growing systems for greenhouse production. In: Goto, E. *et al.* (eds) *Plant Production in Closed Ecosystems.* Kluwer Academic, Dordrecht, The Netherlands, pp. 83–102.

Welles, G.W.H. (2003) Vliegende start biokas. *Ekoland* 1/2, 26–27.

Willits, D.H. (2000) *The Effect of Ventilation Rate, Evaporative Cooling, Shading and Mixing Fans on Air and Leaf Temperatures in a Greenhouse Tomato Crop.* ASAE Paper No. 00–4058. ASAE Meeting Presentation, Milwaukee, Wisconsin.

Willits, D.H. and Peet, M.M. (1989) Predicting yield responses to different greenhouse CO_2 enrichment schemes: cucumbers and tomatoes. *Agricultural and Forest Meteorology* 44, 275–293.

Willits, D.H. and Peet, M.M. (1998) The effect of night temperature on greenhouse grown tomato yields in warm climates. *Agricultural and Forest Meteorology* 92, 191–202.

Woerden, S. and Bakker, J. (2000) *Quantitative Information.* Applied Plant Research, Naaldwijk, The Netherlands, 320 pp.

Zijlstra, S. and Den Nijs, A.P.M. (1987) Effects of root systems of tomato genotypes on growth and earliness, studied in grafting experiments at low temperature. *Euphytica* 36, 693–700.

10

Postharvest Biology and Handling

M.E. Saltveit

INTRODUCTION

Originally, the tomato plant (*Lycopersicon esculentum*) had an indeterminate growth habit, continuously producing flowers and fruit during an extended period of time. Cultivars with this indeterminate growth habit are used where multiple picks are economically justified, e.g. in the greenhouse, or in the field when plants are supported by poles or on a rope or wire trellis. These fresh-market tomato cultivars have fruit at various stages of maturity on the plant and hand harvesting is often necessary to maximize yield through multiple harvests. While this labour-intensive practice is economically justified for some fresh-market field- and glasshouse-grown fruit, processing tomatoes must be culturally managed to produce maximum yields through once-over mechanical harvests.

Breeding and cultural practices have modified processing cultivars and most field-grown fresh-market tomato cultivars into a determinate growth habit in which the bush-like form produces side shoots that terminate in inflorescences. Determinate varieties lend themselves to once-over mechanical harvesting and are therefore well suited for processing. Foliar applications of chemicals that release or break down to ethylene are used before harvest to hasten the maturity of processing tomatoes and to reduce variability among the ripening fruit.

Processing tomatoes are harvested when fully ripe and are quickly transported from the field to the processing plants. The minimal postharvest care they receive during the hours for which they are held before processing is often limited to shading them from the sun or removing field heat by precooling. In contrast, fresh-market tomatoes are often harvested when mature but less than fully ripe, and their ripening is carefully managed before, during and after transport to distant markets.

Greenhouse tomatoes can be grown in soil, or in peat moss, sawdust, rockwool or other soilless and/or inert media, or using a nutrient film technique or other hydroponic systems. The protected greenhouse environment allows manipulation of many environmental parameters to maximize yield and quality (Dorais *et al.*, 2001). Fruit are picked two or three times a week as they reach the proper stage of development. More harvests are needed during warmer weather. Fruit picked before they are fully developed are more prone to mechanical injuries because of poorly developed epidermis and waxy cuticle. Maturity should be uniform, to minimize handling during grading and packing and to enable the uniform application of postharvest treatments. Greenhouse tomatoes are usually harvested when they are riper than fresh-market field-grown fruit.

HARVESTING

Tomatoes are harvested at different maturities for different purposes (Hardenburg *et al.*, 1986; Kader, 2002; Sargent and Moretti, 2002). Processing tomatoes are harvested red-ripe and immediately transported to a processing plant. Fruit destined for the fresh market can be harvested at mature-green through to red-ripe stages of maturity (Cantwell, 2000). Mature-green fruit are firm enough and have sufficient postharvest-life to survive the stress of being shipped considerable distances and still arrive at the market with enough shelf-life to be marketed at the wholesale and retail level. Partially to fully ripe fruit can be harvested for regional markets where transportation is shorter and less damaging to the softer, riper fruit. Greenhouse tomatoes are usually harvested riper than field-grown fruit and are therefore more prone to mechanical injuries because they are softer and have a shorter shelf-life than fruit harvested at the mature-green stage. The thin skins and locular walls of greenhouse cultivars are valued by consumers, but make them more prone to mechanical injury. Some greenhouse-grown fruit are harvested at the breaker stage 'on the vine' and marketed in clusters of five to six fruit. Tomatoes that quickly reach the pink stage soon after harvest are called 'vine-ripened'. Fewer mature-green and more vine-ripe fruit will probably be marketed in the future because of the development of non-toxic gaseous inhibitors of ethylene action (e.g. 1-methylcyclopropene, 1-MCP) (Sisler *et al.*, 1996) and fruit genetically engineered to retain firmness during the early stages of ripening.

Processing tomatoes are machine harvested in most industrialized countries (Fig. 10.1). The once-over, destructive harvest starts when at least 90% of the fruit are ripe. This is determined by periodic visual inspections of the field. Chemicals to stimulate ripening (e.g. an ethylene-releasing compound) can be applied a few weeks before harvest to maximize the percentage of coloured fruit. The entire plant is lifted from the soil to a mobile platform, where the fruit are mechanically removed. The vines are discarded and the fruit are quickly inspected on the platform before they are dumped into large gondolas or field bins. The mechanical injuries inflicted on the fruit as a result of machine harvesting are not severe enough, given the rapidity of processing, to reduce the quality of the processed product significantly. In comparison, the often long period from harvest to consumption of fresh-market tomatoes allows sufficient time for water loss and pathogenic infections to occur as the consequence of injuries to the surface of the fruit and render most mechanically harvested fruit unacceptable for the fresh market. Also, small bruises and skin injuries that do not affect internal quality are visually apparent and render the fruit unacceptable to many consumers. Processed tomato products include peeled whole, quartered and diced fruit, and concentrates for ketchup, juice, pulp, paste and soup. The concentrate market is by far the largest.

While appearance, texture and flavour are the important quality attributes of fresh-market tomatoes, the major quality components of processing tomatoes are soluble solids, pH, titratable acidity, viscosity and colour. Other quality characteristics include uniform colour (i.e. high lycopene and low chlorophyll content), freedom from decay, and firmness. Soluble solids include sugars, organic acids and other dry matter constituents, such as pectic fragments, that remain in solution. The production of tomato concentrates and paste requires removal of water, which is an energy-intensive process; therefore, it is less expensive to produce concentrates from fruit with a high content of soluble solids and dry matter. Viscosity, or consistency, is a complex physical property that is influenced by the amount and suspension of tomato solids, the size and linkage of pectins, and the solution of salts, proteins, sugars and organic acids in the processed product. Colour in tomato products is predominantly the result of two lipid-soluble pigments: lycopene and β-carotene.

Over the years, a great deal of work has been done by traditional plant breeders (and recently by genetic engineering) to increase the soluble solids,

Fig. 10.1. Fully ripe processing tomatoes being mechanically harvested.

viscosity and colour of processing tomatoes. There is difficulty in significantly improving soluble solids content because the trait is polygenetic. Environment exerts a large effect on expression of the trait, and there is a negative relationship between soluble solids and other desirable characteristics such as high yield and concentrated ripening. For these reasons, the product of yield multiplied by soluble solids content (expressed as °Brix) is more useful in estimating the productivity of processing tomatoes. In general, yields have been significantly increased, but the soluble solids content of the fruit has at best increased only slightly (Stevens, 1994; Zamir *et al.*, 1999). The main reason for this relatively slow progress in increasing solids is the lack of genetic variability for high soluble solids in the germplasm of processing tomato. Attempts to increase soluble solids by introducing high solids potential from wild species of *L. chmielewskii* and *L. cheesemanii* have been unsuccessful. This is because most of the high solids potential has disappeared by the time the appropriate genes have been introgressed into an acceptable horticultural type (Stevens, 1994). Modern genetic engineering may be able to introduce specific genes or cluster of genes that will increase soluble solids, but it will require a better understanding of the relation between partitioning of photosynthate within the plant, mechanisms of soluble solids accumulation within the fruit, and the various components of yield.

For the fresh market, multiple hand harvests are often needed to produce an economic yield when indeterminate cultivars are grown. Development of determinate, bush-type plants almost eliminated the use of field-grown indeterminate tomatoes that were supported by a pole or trellis. Plant decline is rapid after harvest of the major set from determinate plants, and few additional harvests are economical. However, the demand for high quality fruit in regional upscale markets has made cultural practices that favour multiple hand harvest of riper fruit economically justifiable in some instances.

Harvested greenhouse-grown tomatoes are commonly carried by hand to packing stations within the greenhouse and then moved on small carts to collection centres (Dorais *et al.*, 2001). In some large greenhouse operations, transportation systems are incorporated into the structures to protect the fruit from damage and to reduce labour costs during harvest. Canals have been built under main greenhouse floors to float the fruit from areas of production to the packing area. The pickers empty their baskets into the troughs; the flowing water carries the tomatoes to a roller elevator, where the fruit passes over a dryer prior to being graded and packed. This can be an efficient system, but skin cracking can be a problem and shelf-life can be decreased if the fruit remain in the water for more than 6 h.

A number of rail systems have been developed that are economical and efficient. The rail system can be temporarily installed during harvest, or permanently installed so that it can be used during the rest of the production cycle. A cart or trolley can run on heating pipes or specially made rails positioned among the rows. In the over-the-crop system, the heating pipes or rails are suspended between the rows and the suspended trolley runs along them. In both rail systems, the baskets of harvested fruit are transferred to a motorized car that is hauled to the packing area. In the suspended system, the rails often continue to the packing area. Overhead wire cables can also be installed in a continuous loop to bring picking carts or baskets from the picking areas to the packing area. Some systems use self-propelled 'robotic' carts that follow sensing strips embedded in the greenhouse floor and can be programmed to go from one specific location within the greenhouse to another without manual intervention. When properly designed, such systems reduce labour and damage and shorten the time from harvest to packing.

Most field-grown fresh-market tomatoes are hand harvested at the mature-green stage of maturity. Fruit grown on indeterminate plants supported by a pole or trellis, whether in the field or greenhouse, are usually harvested at more advanced stages of maturity. Mechanical harvesters are being developed to assist in harvesting fresh-market fruit, but none is yet commonly used in a commercial setting. Pickers put the fruit into buckets or bins, or on to conveyer belts that move the fruit into large field bins or gondola trailers (Fig. 10.2). The fruit should be shielded from the sun during transport to the packing facility, to prevent excessive heating. Without the plant canopy to shade them, the fruit can be heated by direct sunlight to a high enough temperature to damage the exposed tissue and adversely affect subsequent ripening.

In most cultivars, ripe fruit abscise at the stem scar, but mature-green fruit may retain part of the pedicle in jointed fruit. Joints or knuckles form when a zone of abscission develops in the pedicle. With fruit harvested in clusters or with a portion of the stem or pedicle attached, care should be taken so that the protruding tissue does not puncture adjacent fruit during harvesting, handling, packing and transport. However, retention of the

Fig. 10.2. Fresh-market mature-green tomato fruit being hand harvested into buckets and field bins.

pedicle is seen as a positive quality aspect in some markets (Fig. 10.3). For example, in cluster tomatoes, the entire cluster is cut off at the main stem and kept together either by being placed in a single layer in boxes or by being put into mesh bags.

Fruit quality is not only influenced by cultivar practices, but also by postharvest handling procedures (Kader *et al.*, 1978; Sargent and Moretti, 2002). Trauma induced during harvesting and packaging, and ripening and temperature exposure during marketing can markedly affect postharvest-life and quality.

PACKAGING

Upon reaching the packing shed, the fruit can be flumed out of the trailer with chlorinated water. Smaller field bins or boxes can be unloaded by hand or gently dumped into water or on to a padded table. Although it would seem advantageous to start to remove field heat as soon as possible by using cold water, in reality the water should be a few degrees warmer than the fruit. Cold water will cool the gases inside the fruit and facilitate water intrusion through the stem scar, as will hydrostatic pressure resulting from immersing the fruit

Fig. 10.3. Breaker fruit (with pedicle attached) in small plastic field trays.

too deeply in water. Water entering the fruit carries chemical and biological contaminants with it and should be avoided.

Pre-sizing is sometimes used to eliminate undersized tomatoes. The fruit are then washed in chlorinated water and dried. A visual inspection is often used to eliminate cull, over-mature, misshapen and otherwise defective fruit. If allowed, the fruit may then be sprayed with an edible food-grade wax. However, consumers in many markets discourage the application of waxes and fungicides.

After being sorted for defects and colour differences, the fruit are segregated into several size categories by weight or diameter. Moving belts with holes of increasing diameter (Fig. 10.4) or diverging rollers can sort by size, while spring-loaded pans can sort by weight. Other belts running perpendicular to the sorting belts convey the sized fruit to an area for one additional quality inspection, where unmarketable fruit are eliminated (Fig. 10.5).

These sorting and grading steps can be injurious to softer fruit and are therefore done primarily by hand when more mature fruit are being packed. In small packing sheds, breaker and turning fruit can be unloaded from small field bins on to sorting lines, where they are segregated into ripeness classes and are hand packed (Fig. 10.6). The minimal handling in this type of packing house permits riper fruit to be picked, packed and marketed.

Instruments that measure external colour have been developed, but lycopene content is only measurable in the columella of mature-green fruit. The highly variable content of chlorophyll in mature-green fruit makes it an inaccurate measure of their maturity. A non-destructive evaluation of quality and maturity is needed to produce packs containing the uniform fruit

Fig. 10.4. A series of wide belts with holes of increasing diameter is used to sort fruit by size.

Fig. 10.5. Part of a packing line for mature-green tomatoes where fruit are inspected for external quality defects.

Fig. 10.6. Small packing house where breaker and turning fruit are inspected for external quality defects and packed with minimal handling.

demanded by food-service and retail consumers. There are many characteristics that could be measured non-destructively (Abbot *et al.*, 1997). Internal development of lycopene, evolution of ethylene and liquefaction of the locular contents have all been shown to be reliable indicators of fruit maturity. However, the instruments needed to rapidly measure these criteria in a commercial packing shed have not yet been developed.

Mature-green and pink fruit can be packed loose by weight or volume, or on trays in cardboard boxes. Since mature-green fruit are usually treated with ethylene and ripened while in these boxes, the boxes must be moisture resistant. They must also have a sufficient number of properly spaced holes to provide adequate ventilation. The enhanced production of heat, moisture and CO_2 by ripening fruit must be carried away by fresh air containing the low concentration of ethylene used to stimulate the natural ripening of the fruit. Air flow and exchange should keep CO_2 levels in the boxes below the 2% level that inhibits ripening. High levels of CO_2 inhibit the ripening action of ethylene (Saltveit, 1997). Forced air is used for cooling after packing and for maintaining the proper temperature and atmosphere during the application of ethylene (Sargent and Moretti, 2002).

The superior appearance and flavour of greenhouse tomatoes can be used to justify the higher selling price that is needed to offset the higher production costs. This may be difficult in the USA, where supermarket mark-up is higher than in European countries and where greenhouse fruit must compete with low-cost field-grown tomatoes for much of the year.

RIPENING

Tomatoes are climacteric, with the onset of a climacteric rise in respiration and ethylene production coinciding with the first appearance of red colour (Fig. 10.7). The manifold changes accompanying ripening are presented in detail in Chapter 5. This section will concentrate on the postharvest initiation of ripening with ethylene and how this step fits into the general postharvest handling of tomato fruit (Cantwell, 2000; Dorais *et al.*, 2001; Sargent and Moretti, 2002).

The period of growth and development from fertilization to maturity lasts about 45–55 days, depending on cultivar and climate. Ripening is the final stage of maturation, when fruit develop the characteristic colour, flavour, texture and aroma associated with optimum quality. Whether initiated naturally or induced by ethylene, ripening is accompanied by many changes, including an increased production of heat from respiration. Proper stacking of boxes on the pallet is important to ensure adequate air flow to all fruit (Fig. 10.8).

The ventilation and refrigeration capacity of the storage facility must match the increased production of heat during the climacteric rise in

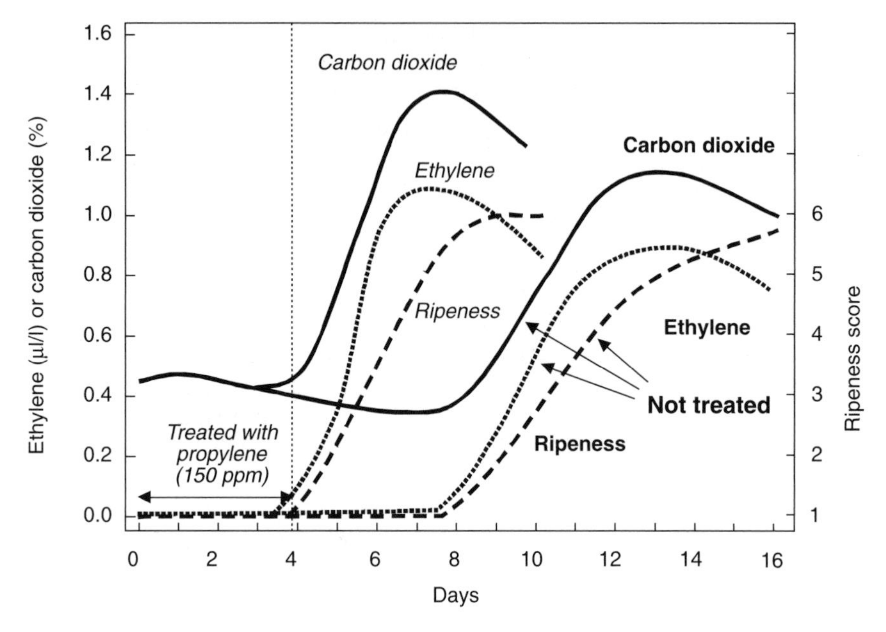

Fig. 10.7. Climacteric rise in respiration and ethylene production with the onset of ripening with and without treatment with 150µl/l propylene in air for 4 days at 20°C.

respiration. The increased rate of evaporation from ripening fruit due to this elevated respiratory production of heat must be minimized by decreasing the vapour pressure deficit between the tomatoes and their surrounding atmosphere. This is done by maintaining high relative humidities (e.g. 85–95%) in the ripening rooms.

Exposure to ppm (µl/l) levels of ethylene in air or to dips in ethylene-releasing compounds such as ethrel will promote the ripening of mature-green tomato fruit to acceptable levels of quality (Table 10.1). Ethylene promotes the destruction of chlorophyll and the synthesis of lycopene in ripening fruit (Chapter 5), but can also induce ripening (i.e. lycopene synthesis and softening) in fruit that are so immature that good quality is never attained. In one fresh-market cultivar, fruit that were about 90% of final size (i.e. more than 42 days after anthesis) were able to go through a normal climacteric and ripened to an acceptable level of quality after harvest. Less-mature fruit did turn red and soften if given enough time, but the ripened fruit were of inferior quality. Fruit that were older than 31 days coloured without added ethylene, while younger fruit required added ethylene to colour and soften. Exposure of immature fruit as young as 17 days after anthesis to 1000 ppm ethylene in air induced a respiratory and ethylene climacteric, colour change and softening.

Fig. 10.8. Properly stacked boxes of mature-green tomato fruit on a pallet being moved by a fork-lift to a controlled temperature room for ethylene treatment.

Table 10.1. Effect of ethylene treatment, either as 10 ppm (μl/l) ethylene in air or 2000 ppm (μg/g) ethrel dip, on the rate of ripening of mature-green tomato fruit.

Days of treatment	Days to breaker	Days from breaker to red-ripe	Days from red-ripe to discard	Total days
0	10.9	6.9	15.9	33.7
1	7.8	6.3	17.1	31.2
2	8.1	7.2	16.4	31.7
4	5.9	5.8	22.2	33.9
Continuous	4.4	6.3	15.2	25.9

Once ripening is initiated at the beginning of the breaker stage, internal levels of ethylene rise and continue to support the autocatalytic production of ethylene and ripening. Removal of ethylene from the external atmosphere does not significantly lower internal levels that are sustained by cellular synthesis or alter the pattern of ripening. Introducing inhibitors of ethylene action (e.g. carbon dioxide, 1-MCP) into the gaseous environment of the fruit can slow ripening, as can lowering the temperature to near the chilling threshold. A few hours of exposure to ethanol vapour can inhibit ripening of tomatoes harvested at various degrees of ripeness without affecting subsequent quality (Saltveit and Sharaf, 1992).

An ethylene concentration of 100–150 ppm is commonly used to initiate and hasten the ripening of harvested mature-green fruit at 18–22°C (Fig. 10.9). Advanced mature-green fruit treated in this way will reach the breaker stage in 24–36 h, depending on temperature. This atmosphere can be produced by a number of methods. The 'shot' method introduces a relatively large amount of ethylene into the ripening room by metering ethylene from compressed gas cylinders. Ethylene and air mixtures between 3.1% and 32% (31,000–320,000 ppm) are explosive. While these concentrations are more than 200-fold higher than the recommended levels, they have (rarely) been reached when metering equipment has malfunctioned. Use of compressed gas containing around 3.1% ethylene eliminates this problem. Catalytic converters are instruments that use a metal catalyst to convert certain alcohols into ethylene. They deliver a continuous flow of low-concentration ethylene gas into the storage room. Ethylene can also be applied in aqueous form from the decomposition of compounds such as ethrel. While stable at acidic pH, ethrel quickly breaks down to ethylene as the temperature and pH increase (Table 10.2).

Almost all fruit harvested at the mature-green stage are gassed with ethylene either at the packing warehouse or upon receipt at the terminal wholesale market or distribution centre. Tomatoes can be exposed to ethylene either before or after packing, but most are treated after packing. Treating before packing has some advantages. Fruit that develop decay while ripening

Table 10.2. Kinetics (half-time) for ethrel decomposition to ethylene at various pHs and temperatures.

	Half-time for ethrel decomposition to ethylene		
Temperature (°C)	pH 6	pH 7	pH 8
10	70 days	14 days	7 days
20	10 days	2 days	1 day
30	30 h	10 h	5 h
40	10 h	2 h	1 h

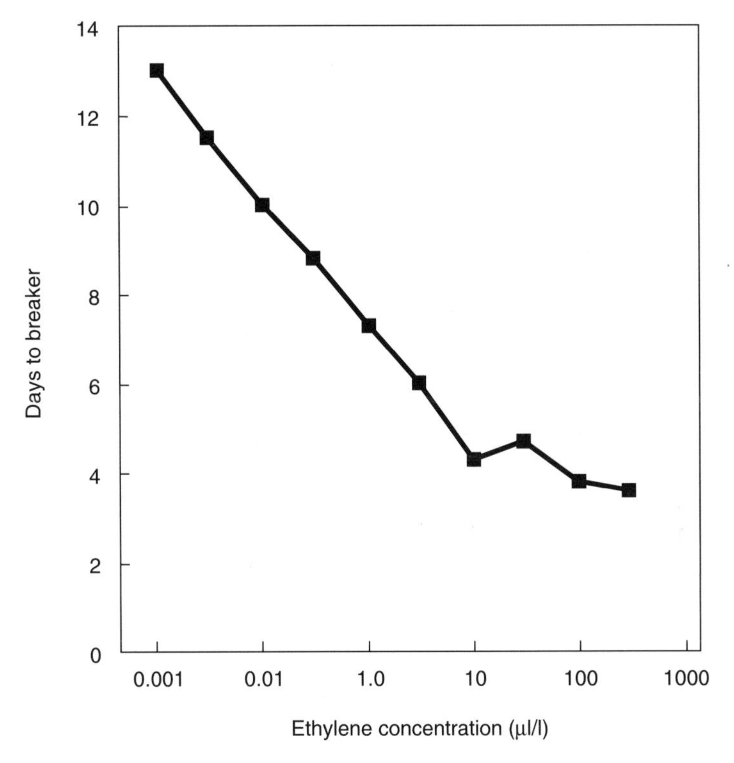

Fig. 10.9. Relationship between externally applied ethylene concentration and the days it takes mature-green tomato fruit to reach the breaker stage of ripeness.

in the warm, humid atmosphere can be eliminated before packing. After ripening, the uniformly ripe fruit can be selected from the less ripe to produce a more uniform pack-out. Although ripening is accelerated and made more uniform by gassing with ethylene, the inherent variability in maturity of mature-green tomatoes results in uneven ripening within a lot of fruit. Packed mature-green tomatoes may therefore produce a 'chequerboard' of fruit of different maturities after ripening and require repacking to produce the uniform pack-out demanded by many retailers (Fig. 10.10). Repacking is both expensive and damaging to the riper fruit.

Ethylene is used with both processing and fresh-market tomatoes to hasten and coordinate ripening. Sprays of aqueous solutions or application of viscous paste containing ethylene-releasing compounds (e.g. ethrel, ethephon, (2-chloroethyl) phosphonic acid) are used to hasten the ripening of fruit still attached to the plants. Ethrel is a compound that decomposes to innocuous products and ethylene. The rate of decomposition depends on pH, temperature, concentration, and penetration into the plant. Between 10°C and 20°C the

Fig. 10.10. Variable ripening of mature-green tomato fruit produces a 'chequerboard' pattern of ripe and green fruit within a box.

temperature coefficient Q_{10} is 6, while the Q_{10} is 5 between 30°C and 40°C. Decomposition follows first-order kinetics: in any given unit of time, the same fraction will decompose. The half-times for decomposition at different temperatures and pH values are given in Table 10.2. A spray or dip in a 2000 ppm aqueous ethrel solution produced the same effect as gassing the fruit with 10 ppm ethylene in air.

While ethrel can be used on processing tomatoes and fresh-market tomatoes before harvest, it is not approved for postharvest application in the USA. As multiple picks are usual for interdeterminate fresh-market tomatoes, pre-harvest ethrel application is not practical since the ethylene produced can accelerate flower abscission, leaf drop and general senescence of the plant. Localized application of pastes containing ethylene-releasing compounds to a single truss (e.g. on greenhouse-grown indeterminate varieties) eliminates this problem.

TEMPERATURE MANAGEMENT

Removing field heat before packing and storage and maintaining proper temperature during marketing are critical for proper ripening and quality retention. Since fruit warm as the day heats up, harvesting early in the day minimizes field-heat accumulation. Adequate foliage coverage can shield fruit from the heat of the sun, as can shading of harvested fruit. Forced-air cooling should be used quickly after harvest to remove field heat and cool the fruit to the desired temperature. Dumping fruit into water at the packing house may also cool them, but dumping warm fruit into cold water promotes the intrusion of water contaminated with microorganisms and/or chemicals into the fruit through the stem-scar and should be avoided (Bartz, 1988; Kader, 2002). Fluctuations in temperature promote both water loss, which causes shrivel, and condensation of water on the surface of the fruit, which promotes the growth of spoilage organisms. Maintaining the proper temperature in the cold chain from harvest to consumption maximizes quality retention and postharvest life.

The optimum storage temperature that slows ripening and also retains the quality of fruit ripened after storage varies with the stage of maturity. Mature-green fruit can be stored at 12–15°C, ripening fruit at 10–12°C and fully ripe fruit at 8–10°C. Ripening of mature-green fruit is best at 20–25°C. Improper temperature management is the primary cause for the development of many postharvest diseases and disorders (Snowdon, 1992; and see Chapter 5). Tomato fruit are chilling sensitive and suffer physiological damage when held at non-freezing temperatures below ~10°C (Fig. 10.11).

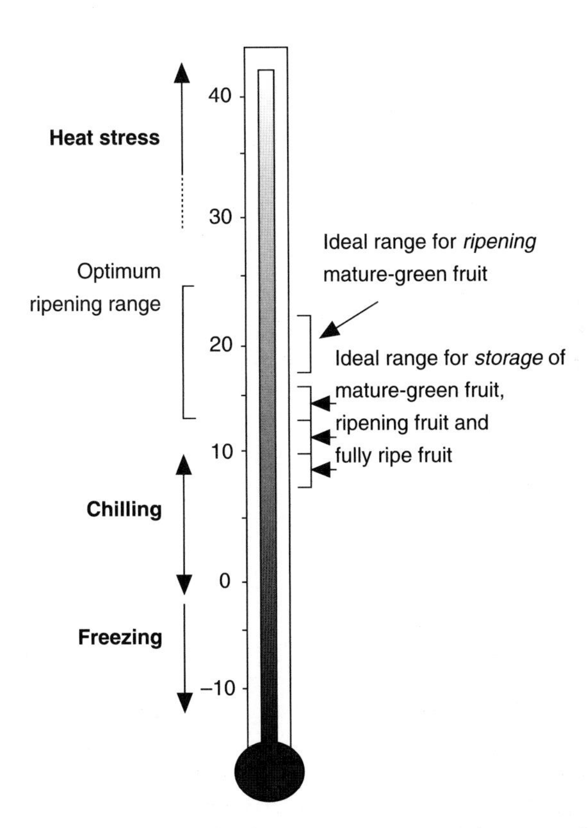

Fig. 10.11. Temperature limitations for the proper ripening and storage of fresh-market tomato fruit.

Chilling injury is characterized by slow and abnormal ripening, increased disease susceptibility (often to specific decay organisms), reduced acidity, accelerated water loss and surface pitting. As with other crops, the degree of chilling sensitivity is related to the seasonal temperature. Chilling can occur in the field before harvest and contribute to the sensitivity of the fruit to chilling after harvest. Tomatoes grown in a hot climate are more sensitive to chilling than are those grown in a cooler climate. Also, the temperature at harvest appears to influence the level of chilling sensitivity. Fruit harvested hot are more chilling sensitive than those harvested when cool. Preconditioning at low (non-chilling) and high (heat-shock) temperatures has been used to increase tolerance of mature-green fruit to subsequent chilling. Holding fruit at elevated temperatures may induce the heat-shock response that is protective against subsequent chilling injury. However, the shortened shelf-life resulting from holding the fruit at 35°C for 3 days must be offset by a greater extension of storage life at what would previously have been chilling temperatures.

After ethylene treatment, the optimum transport and storage temperature is between 13°C and 21°C. Within these limits, ripening will be faster at higher temperatures (Table 10.3). Ripening of mature-green fruit will be uneven and abnormal below 10°C while ripening may be too fast and decay more extensive above 21°C. Since chilling injury of mature-green fruit primarily affects subsequent ripening, riper pink fruit are almost by definition more chilling resistant than mature-green fruit and can be stored at lower, near-chilling temperatures. The shipping temperature for pink tomatoes will depend largely on the number of days in transit and the degree of ripeness that the person receiving the fruit may desire.

High temperatures (e.g. 30–35°C) inhibit ripening. Fruit stored at these high temperatures turn orange instead of red, because lycopene synthesis is inhibited above 30°C. Brief exposures of fruit on the plant to high temperatures (as during a hot day when shaded fruit may reach 30°C) are reversible, with normal ripening proceeding at cool temperatures. A brief exposure to higher temperatures (e.g. exposure to full sun) can cause permanent damage, cell death and sunscald. Exposure of harvested fruit to direct sunlight after harvest should be avoided by use of coverings and shading.

Table 10.3. Effect of temperature on number of days to table ripeness (stage 5) for mature-green tomato fruit (source: Kader, 1986).

	Temperature (°C)					
Ripeness stage	12.5	15.0	17.5	20.0	22.5	25.0
Mature-green	18 days	15 days	12 days	10 days	8 days	7 days
Breaker	16 days	13 days	10 days	8 days	6 days	5 days
Turning	13 days	10 days	8 days	6 days	4 days	3 days
Pink	10 days	8 days	6 days	4 days	3 days	2 days

CONTROLLED ATMOSPHERE

Controlled and modified atmospheres are slightly beneficial for both transit and short-term storage, but only as a supplement to proper temperature control. Tomatoes derive a slight benefit from storage under atmospheres containing 3–5% oxygen and 0–3% CO_2 at 12°C (Leshuk and Saltveit, 1990; Saltveit, 2001, 2003). The reduced oxygen level reduces the respiration rate, reduces production of ethylene and sensitivity to ethylene, and retards ripening. Injury occurs at oxygen levels < 2% and CO_2 concentrations > 5%. However, these values must be approached very cautiously. For example, elevated CO_2 atmospheres are problematic because atmospheres containing 2% or more CO_2 have caused increased softening and uneven ripening under some conditions. A controlled or modified atmosphere may be beneficial, but only when used in conjunction with established postharvest management practices used to create and maintain optimal temperature and humidity.

In many other fruits, gas diffusion takes places over the entire surface, but in a tomato gas diffusion occurs primarily through the stem scar. Any impediment to diffusion, such as excessive waxing or the development of unusual amounts of corking, can significantly alter the internal gas concentrations even in a constant ambient environment. The variable results reported for the storage of tomatoes under modified atmospheres may have resulted from the disparity between the imposed ambient atmosphere and the internal atmosphere that developed as the result of diffusion, respiration and metabolism (Saltveit, 2001).

Modified-atmosphere packaging can be used to produce atmospheres low in oxygen and high in CO_2 that slow ripening and extend the market life of the fruit. A semi-permeable plastic film is usually incorporated into the package. The combination of gas diffusion through the film and the respiration of the fruit produces and maintains a modified atmosphere. Changes in respiration resulting from ripening or changes in temperature can significantly alter the atmosphere in the package. Selecting fruit of the desired maturity and proper temperature management are crucial. The semi-permeable films also allow ethylene to accumulate and the relative humidity to reach saturation. Elevated levels of ethylene can stimulate ripening and nullify the inhibitory effects of the modified atmosphere on ripening. While the high relative humidity within the package lessens water loss from the fruit, it also fosters microbial growth. Packets of absorbents that reduce ethylene and water vapour in the package are useful, but they add cost to the package.

Although an atmosphere containing 5–10% carbon monoxide will reduce disease, its extreme toxicity has severely limited commercial application. Carbon monoxide has slight ethylene-like activity, and will promote ripening of mature-green fruit in air.

SUMMARY

There are many steps involved in the postharvest handling of mature-green and vine-ripe fresh-market tomatoes as they progress from the field or greenhouse to the consumer (Fig. 10.12). Each step must be maximized to retain quality, because the loss of quality is cumulative and accelerated by physical injuries, poor sanitation and improper temperature and humidity management.

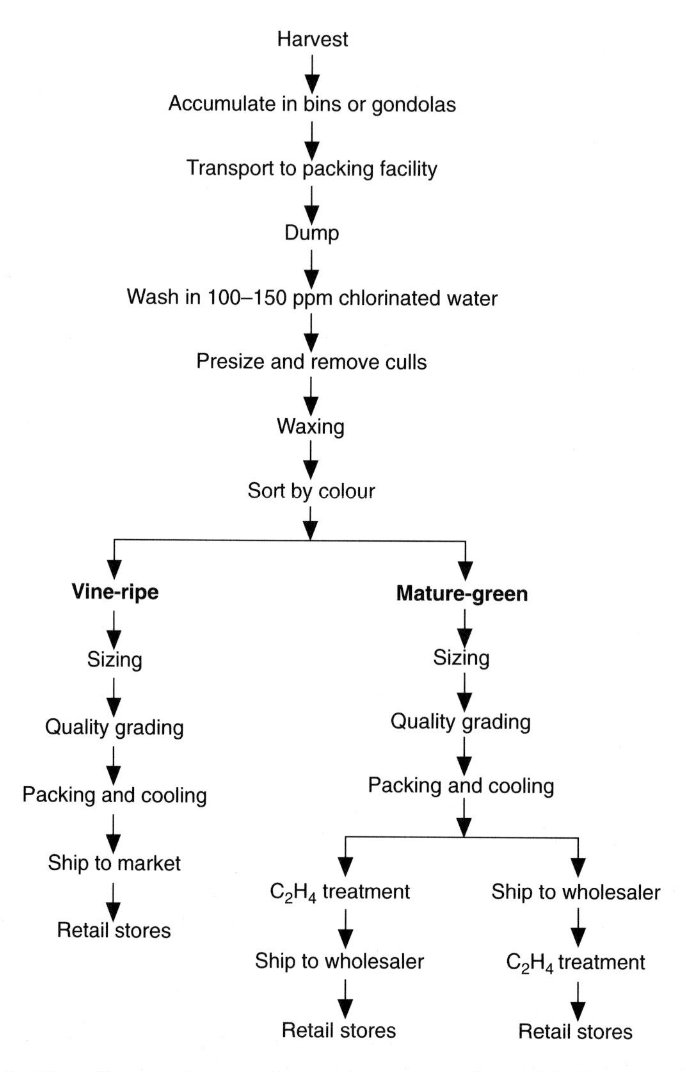

Fig. 10.12. Flow diagram showing the postharvest handling of fresh-market tomatoes harvested either mature-green or vine-ripe.

Harvesting tomatoes at the proper maturity and careful handling during harvest, transport, packing and marketing will ensure that high quality fruit reach the consumer. External colour changes are good indicators of the physiological maturity of the fruit of almost all tomato cultivars. Multiple picks by hand are common for interdeterminate fresh-market tomatoes, while determinate tomatoes are usually mechanically harvested only once. Application of ppm (μl/l) levels of ethylene can be used to hasten and coordinate the ripening of mature-green fruit. Once fruit have reached the breaker stage, internally synthesized ethylene promotes its own synthesis and sustains ripening even when external ethylene levels are reduced to zero. Temperature management is crucial, since temperatures below ~10°C cause chilling injury and temperatures above ~25°C inhibit lycopene synthesis. Controlled atmospheres have some benefit in maintaining quality, but they are usually minimal compared with the benefits of proper management of temperature, physical injury, relative humidity and sanitation.

REFERENCES AND FURTHER READING

Abbott, J.A., Lu, R.F., Upchurch, B.L. and Stroshine, R.L. (1997) Technologies for nondestructive quality evaluation of fruits and vegetables. *Horticultural Reviews* 20, 1–120.

Bartz, J.A. (1988) Potential for postharvest diseases in tomato fruit infiltrated with chlorinated water. *Plant Diseases* 72, 9–13.

Beaulieu, J.C. and Saltveit, M.E. (1997) Inhibition or promotion of tomato fruit ripening by acetaldehyde and ethanol is concentration-dependent and varies with initial fruit maturity. *Journal of the American Society for Horticultural Science* 122, 392–398.

Beerh, O.P. and Siddappa, G.S. (1959) A rapid spectrophotometric method for the detection and estimation of adulterants in tomato ketchup. *Food Technology* 13, 414–418.

Cantwell, M. (2000) Optimum procedures for ripening tomatoes. In: *Management of Fruit Ripening*. University of California Postharvest Horticultural Series No. 9, pp. 80–88.

Dorais, M., Papadopoulos, A.P. and Gosselin, A. (2001) Greenhouse tomato fruit quality. *Horticultural Reviews* 26, 236–319.

Fisher, R.L. and Bennett, A.B. (1991) Role of cell wall hydrolases in fruit ripening. *Annual Review of Plant Physiology and Plant Molecular Biology* 42, 675–703.

Hardenburg, R.E., Watada, A.E. and Wang, C.Y. (1986) *The Commercial Storage of Fruits, Vegetables and Florist and Nursery Stocks*. USDA, ARS, Washington, DC, Agriculture Handbook No. 66, 130 pp.

Kader, A.A. (1986) Effect of postharvest handling procedures on tomato quality. *Acta Horticulturae* 190, 209–221.

Kader, A.A. (ed.) (2002) *Postharvest Technology of Horticultural Crops*, 3rd edn. University of California, DANR Special Publication No. 3311 (Chapters 6, 22, 23 and 33).

Kader, A.A., Morris, L.L., Stevens, M.A. and Albright-Holton, M. (1978) Composition and flavor quality of fresh market tomatoes as influenced by some postharvest handling procedures. *Journal of the American Society for Horticultural Science* 103, 6–13.

Leshuk, J.A. and Saltveit, M.E. (1990) Controlled atmospheres and modified atmospheres for the preservation of vegetables. In: Calderon, M. (ed.) *Food Preservation by Modified Atmospheres.* CRC Press, Boca Raton, Florida, pp. 315–352.

Saltveit, M.E. (2001) A summary of CA requirements and recommendations for vegetables. In: *Optimal Controlled Atmospheres for Horticultural Perishables, Postharvest Horticulture Series* No. 22A. Postharvest Technology Center, University of California, Davis, California, pp. 71–94. Also available on CD at: http://postharvest.ucdavis.edu as *Postharvest Horticulture Series* No. 22.

Saltveit, M.E. (2003) A summary of CA requirements and recommendations for vegetables. *Acta Horticulturae* 600, 723–727.

Saltveit, M.E. and Sharaf, A.R. (1992) Ethanol inhibits ripening of tomato fruit harvested at various degrees of ripeness without affecting subsequent quality. *Journal of the American Society for Horticultural Science* 117, 793–798.

Sargent, S.A. and Moretti, C.L. (2002) Tomato. In: Gross, K.C., Wang, C.Y. and Saltveit, M.E. (eds) *Agricultural Handbook 66 – The Commercial Storage of Fruits, Vegetables, Florist and Nursery Crops.* Available at: http://www.ba.ars.usda.gov/hb66/contents.html

Sisler, E.C., Serek, M. and Dupille, E. (1996) Comparison of cyclopropene, 1-methylcyclopropene, and 3,3-dimethylcyclopropene as ethylene antagonists in plants. *Plant Growth Regulation* 18, 169–174.

Snowdon, A.L. (1992) *Color Atlas of Post-harvest Diseases and Disorders of Fruits and Vegetables,* Vol. 2. *Vegetables.* CRC Press, Boca Raton, Florida, 416 pp.

Stevens, M.A. (1994) Processing tomato breeding in the '90s: a union of traditional and molecular techniques. *Acta Horticulturae* 376, 23–34.

Zamir, D., Grandillo, S. and Tanksley, S.D. (1999) Genes from wild species for the improvement of yield and quality of processing tomatoes. *Acta Horticulturae* 487, 285–288.

INDEX

Note: page numbers in *italics* refer to figures and tables